Electrochemical Nanofabrication

Second Edition

Electrochemical Nanofabrication

Principles and Applications

Second Edition

edited by
Di Wei

Published by

Pan Stanford Publishing Pte. Ltd.
Penthouse Level, Suntec Tower 3
8 Temasek Boulevard
Singapore 038988

Email: editorial@panstanford.com
Web: www.panstanford.com

British Library Cataloguing-in-Publication Data
A catalogue record for this book is available from the British Library.

Electrochemical Nanofabrication: Principles and Applications (Second Edition)
Copyright © 2016 Pan Stanford Publishing Pte. Ltd.

All rights reserved. This book, or parts thereof, may not be reproduced in any form or by any means, electronic or mechanical, including photocopying, recording or any information storage and retrieval system now known or to be invented, without written permission from the publisher.

For photocopying of material in this volume, please pay a copying fee through the Copyright Clearance Center, Inc., 222 Rosewood Drive, Danvers, MA 01923, USA. In this case permission to photocopy is not required from the publisher.

ISBN 978-981-4613-86-6 (Hardcover)
ISBN 978-981-4613-87-3 (eBook)

Printed in the USA

Contents

Preface · xiii

1. Electrochemical Nanofabrications: A General Review · 1
Di Wei

  1.1 Electrochemical Atomic Layer Epitaxy · 2
  1.2 Electrochemical Synthesis of Quantum Dots and Semiconducting Nanocompounds · 4
  1.3 Electrochemical Deposition Methods for Metallic Nanostructures · 7
  1.4 Electrochemical Nanolithography · 9
  1.5 3D Electrochemical Nanoconstruction · 15
  1.6 Electrochemical Etching and LIGA Technique · 15
  1.7 Micro- and Nano-Machining by Ultrashort Voltage Pulsing Technique · 17
  1.8 Template-Free Methods for Conducting Polymer Nano-Architecture · 21
  1.9 Template Methods · 24
      1.9.1 Anodized Aluminum Oxide Membranes · 24
      1.9.2 Zinc Oxide · 27
      1.9.3 Titanium Dioxide · 30
      1.9.4 Electrochemical Fabrication of Soft Matters in Nanoscale Using Templates · 32
  1.10 Carbon Nanotube Templates · 35
  1.11 Colloidal Polystyrene Latex Templates · 36
  1.12 Electrochemistry and Self-Assembled Monolayers · 39
  1.13 Other Template Methods · 41
  1.14 Nanoscale Electrochemistry · 43
  1.15 Sonoelectrochemistry and Others · 44
  1.16 Conclusions · 45

2. Electrochemical Replication of Self-Assembled Block Copolymer Nanostructures   59

Edward Crossland, Henry Snaith, and Ullrich Steiner

2.1	Introduction	60
2.2	Principles of Block Copolymer Self-Assembly	62
2.3	Block Copolymer Thin Films	67
	2.3.1 Alignment of the Microphase	70
	2.3.1.1 Electric field alignment	71
2.4	Porous Block Copolymer Film Templates	75
	2.4.1 Nondegradative Routes to Porous Templates	77
	2.4.2 Accessible Pore Sizes	78
	2.4.3 Template Stability	79
	2.4.4 Nanowire Replication: Cylinder-Forming BCP Templates	80
	2.4.4.1 In-plane nanowires	80
	2.4.4.2 Standing nanowires arrays	81
	2.4.4.3 Polymeric nanowire replication	84
	2.4.5 Combination with Top-Down Lithography	85
	2.4.6 Bicontinuous Gyroid Copolymer Templates	86
	2.4.6.1 Viewing the porous gyroid morphology	88
	2.4.6.2 Replication of gyroid network arrays	89
2.5	Applications: The Bulk Heterojunction Solar Cell	96
	2.5.1 Block Copolymers in the Dye-Sensitized Solar Cell	97
2.6	Concluding Remarks	103

3. Synthesis of Organic Electroactive Materials in Ionic Liquids   113

Michal Wagner, Carita Kvarnström, and Ari Ivaska

3.1	Introduction	113
3.2	Structure and Properties of Ionic Liquids	114
3.3	Electropolymerization of Conducting Polymers in Ionic Liquids	117

	3.4	Synthesis of Polymer Composites and Carbon-Based Nanomaterials in Ionic Liquids	123
	3.5	Conclusions	128

4. Imidazolium-Based Ionic Liquid Functional Materials and Their Application to Electroanalytical Chemistry — 139

Yuanjian Zhang, Yanfei Shen, and Li Niu

	4.1	Introduction	139
	4.2	Electrosynthesis	140
	4.3	Functionalization of Ionic Liquids for Ease Immobilization	142
	4.4	Electrolyte-Free Electrochemistry	150
	4.5	ILs-Based Multifunctional Compounds for Electrocatalysis and Biosensors	152
	4.6	ILs-Protected Nanostructures as Electrocatalysts for Some Key Reactions	156
	4.7	Others	160
	4.8	Conclusions	163

5. Nanostructured TiO$_2$ Materials for New-Generation Li-Ion Batteries — 171

Gregorio F. Ortiz, Pedro Lavela, José L. Tirado, Ilie Hanzu, Thierry Djenizian, and Philippe Knauth

	5.1	Introduction	171
	5.2	Economic and Scientific Context of Battery Technology over the World	173
		5.2.1 Li-Ion Battery Technology: Past and Present	177
	5.3	Fabrication of Self-Assembled TiO$_2$ Nanotubes	182
		5.3.1 Anodization of Titanium: Experimental Aspects	182
		5.3.2 The Principle of Fabrication	185
	5.4	Characterization of Titania Nanotubes Layers	188
		5.4.1 Scanning Electron Micrograph and X-Ray Diffraction	188
		5.4.2 TiO$_2$ Polymorph: The Interest of Anatase and Amorphous Titania	192

5.5	Battery Applications	199
	5.5.1	Electrochemical Behavior of Samples in Lithium Cells	199
	5.5.2	SEM Study of Cycled Electrodes	211
5.6	Conclusions	212

6. Hierarchically Nanostructured Electrode Materials for Lithium-Ion Batteries — 223

Yu-Guo Guo, Sen Xin, and Li-Jun Wan

6.1	Brief Introduction to Lithium-Ion Batteries	224
6.2	Anode Materials	227
	6.2.1	Graphite	227
	6.2.2	Nongraphitized Carbon Materials	228
	6.2.3	Alloys	229
	6.2.4	Transition Metal Oxides	230
6.3	Cathode Materials	230
	6.3.1	Layered Structured Hexagonal Oxide	231
	6.3.2	Spinel Structured Oxide	231
	6.3.3	Olivine Structured Oxide	232
6.4	Nanostructured Electrode Materials	232
	6.4.1	Advantages of Nanomaterials	233
	6.4.2	Disadvantages of Nanomaterials	235
6.5	Hierarchically Nanostructured Electrode Materials	236
	6.5.1	Sphere-Like Nano/Micro Hierarchical Structures for Electrode Materials	237
	6.5.2	Flower-Like Nano/Micro Hierarchical Structures for Electrode Materials	240
	6.5.3	Hierarchical 3D Mixed Conducting Networks	242
6.6	Conclusions	244

7. Ionic Liquid-Assisted Fabrication of Graphene-Based Electroactive Composite Materials — 251

Pia Damlin, Bhushan Gadgil, and Carita Kvarnström

7.1	Introduction	251
7.2	Conducting Polymers	252

7.3	Ionic Liquids			254
	7.3.1	Ionic Liquids in Electrochemistry		255
7.4	Graphene			257
	7.4.1	Preparation of Graphene		258
	7.4.2	Production of Graphene Oxide		259
		7.4.2.1	Post treatment of GO	259
	7.4.3	Electrochemical Exfoliation of Graphite		261
7.5	Graphene/CP Composites			263
7.6	Applications of Composite Materials			269
	7.6.1	Supercapacitors		269
	7.6.2	Electrochromic Composite Materials		274

8. Chemically Converted Graphene: Functionalization, Nanocomposites, and Applications — 291

Li Niu, Yuanyuan Jiang, Yizhong Lu, Shiyu Gan, Fenghua Li, and Dongxue Han

8.1	Introduction		292
8.2	Graphene-Based Nanocomposite		293
	8.2.1	Graphene–Polymer Composites	293
	8.2.2	Graphene-Filled Polymer Composites	294
	8.2.3	Polymer-Functionalized Graphene	296
	8.2.4	Graphene–Nanoparticle Composites	297
	8.2.5	Composites with Metal Nanoparticles	298
	8.2.6	Composites with Metal Oxide Nanoparticles	305
	8.2.7	Graphene Quantum Dots Hybrids	307
	8.2.8	Composite with Other Nanoparticles	308
	8.2.9	Graphene Composite with Organic Molecules	309
8.3	Conclusions and Future Outlook		314

9. Development of Graphene-Based Nanostructures — 327

Huaqiang Cao

9.1	Introduction	328
9.2	Fluorescence Quenching of Hybrid Graphene Material Covalently Functional with Indolizine	329

9.3	Graphene-Based Materials Used as Electrodes in Ni-MH and Li-Ion Batteries	335
9.4	Removal of Dye from Water by Cu_2O@Graphene	349
9.5	Summary	354

10. Recent Advances in Multidimensional Electrode Nanoarchitecturing for Lithium-Ion and Sodium-Ion Batteries — 365

Gregorio Ortiz, Pedro Lavela, Ricardo Alcántara, and José L. Tirado

10.1	Introduction	365
10.2	Newly Developed Procedures for nt-TiO_2 Utilization	366
10.3	First-Row Transition Metal Oxide Nanocomposites with Unusual Performance	369
	10.3.1 Conversion Electrodes	369
	10.3.2 Composites with Carbon Materials	371
	10.3.3 Composites with Other Metal Oxides	373
	10.3.4 Composites with Metals	375
	10.3.5 Composites with Polymers	375
10.4	Metal Foams for 2D and 3D Battery Architectures	376
10.5	Graphene–Transition Metal Oxide Heterostructures for Battery Applications	380
	10.5.1 Synthetic Route	382
	10.5.2 Metal Oxides Involved in Energy Storage System	387
10.6	Surface Modification of Nanostructrures for Improved Battery Performance	392

11. Electrochemical Fabrication of Carbon Nanomaterial and Conducting Polymer Composites for Chemical Sensing — 417

Zhanna A. Boeva, Rose-Marie Latonen, Tom Lindfors, and Zekra Mousavi

11.1	Introduction	417
11.2	Composites of Carbon Nanotubes and Conducting Polymers	419

    11.2.1 Poly(3,4-Ethylenedioxythiophene)   420
    11.2.2 Polyaniline   427
    11.2.3 Polypyrrole   433
11.3 Composites of Graphene Derivatives and Conducting Polymers   439
    11.3.1 Poly(3,4-Ethylenedioxythiophene)   440
    11.3.2 Polyaniline   451
    11.3.3 Polypyrrole   452
    11.3.4 Other ECPs   457
11.4 Conclusions   459

Index   473

Preface

With the development of nanotechnology and nanomaterials, the arena of electrochemical nanofabrication has expanded significantly. The first version of the book was drafted in 2009. In 2010, the Nobel Prize in Physics was awarded to Prof. Konstantin Novoselov and Prof. Andre Geim from the University of Manchester for groundbreaking experiments regarding the two-dimensional material graphene. Three years later, the European Commission launched the European Union's biggest-ever research initiative, the Graphene Flagship, with a budget of 1 billion euro. This has boosted the research on graphene globally.

As Nokia's representative involved in the Graphene Flagship from the very beginning, I believe the two-dimensional wonder materials are defining a new scope of research in electrochemistry. This is also the motivation for us to collect new advances on applying graphene in different electrochemical devices such as electrochemical sensors and energy solutions. In this second edition, the book is enriched with the synthesis of graphene-based materials through electrochemical methods, the applications of graphene in lithium ion and sodium ion batteries, and using graphene composites for different sensing platforms. This will be of great interest to a broad audience in nanotechnology and electrochemistry.

Di Wei

Chapter 1

Electrochemical Nanofabrications: A General Review

Di Wei

Nokia Technologies c/o University of Cambridge,
21 JJ Thomson Avenue, CB3 0FA, Cambridge, UK

dw344@cam.ac.uk

Nanofabrications have been largely used in various applications such as photovoltaics, sensors, catalysts, integrated circuits, electronics, micro-optics, and countless others. The methodology of nanofabrication can be generally divided into two types of processes: top-down and bottom-up. Top-down process refers to approaching the nanoscale from larger dimensions, such as lithography, nanoimprinting, scanning probe, and E-beam technique. In contrast, the bottom-up fabrication process builds nanoscale artifacts from the molecular level up, through single molecules or collections of molecules that agglomerate or self-assemble. Using a bottom-up approach, such as self-assembly enables scientists to create larger and more complex systems from elementary subcomponents (e.g., atoms and molecules). Generally speaking, top-down processes that transfer minute patterns onto material are more matured than bottom-up processes. An

---

Electrochemical Nanofabrication: Principles and Applications (Second Edition)
Edited by Di Wei
Copyright © 2016 Pan Stanford Publishing Pte. Ltd.
ISBN 978-981-4613-86-6 (Hardcover), 978-981-4613-87-3 (eBook)
www.panstanford.com

exception is epitaxial processes that create layers through layer-by-layer growth with registry at the atomic level.

Electrodeposition has traditionally been used to form high-quality, mostly metallic, thin films for decades. By carefully controlling the electrons transferred, the weight of material formed can be determined according to Faraday's law of electrolysis. It was shown that high-quality copper interconnects for ultra-large-scale integration chips can be formed electrochemically on Si wafer [1, 2]. Electrodeposition has thus been shown compatible with state-of-the-art semiconductor manufacturing technology. The largest semiconductor companies, for example, IBM, Intel, AMD, and Motorola, are installing wafer-electroplating machines on their fabrication lines [1]. The electrodeposition of Cu with the line width 250 nm was used in the mass production of the microprocessor Pentium III in 1998. In 2003, the line width of the CPU was reduced to 130 nm in Pentium IV. Electrochemistry was largely used in chip fabrication [3] and the packaging of microelectronics [4]. However, compared with other nanofabrication techniques, electrochemical nanofabrication is still a maiden area, which needs further development and fulfillment.

1.1 Electrochemical Atomic Layer Epitaxy (EC-ALE)

Electrochemical atomic layer epitaxy (EC-ALE) is the combination of underpotential deposition (UPD) and ALE. UPD is the formation of an atomic layer of one element on a second element at a potential under, or prior to, that needed to deposit the element on itself [5, 6]. The shift in potential results from the free energy of the surface compound formation. Early UPD studies were carried out mostly on polycrystalline electrode surfaces [7]. This was due, at least in part, to the difficulty of preparing and maintaining single-crystal electrodes under well-defined conditions of surface structure and cleanliness [8]. The definition of epitaxy is variable but focuses on the formation of single crystal films on single crystal substrates. This is different from other thin film deposition methods where polycrystalline or amorphous film deposits are formed even on single crystal substrates. Homoepitaxy is the formation of a compound on itself. Heteroepitaxy is the formation of a compound on a different compound or element and is much more prevalent. The principle of ALE is to grow the deposit one atomic layer at a time [9]. EC-ALE uses UPD for the surface-limited reactions in an ALE cycle [10]. Surface-limited reactions are developed for the deposition of each component element in a cycle to directly form a compound via layer-by-layer growth, avoiding 3D nucleations. With cycle, a compound monolayer is formed, and the deposit thickness is controlled by the number of cycles. In an EC-ALE process, each reactant has its own solution and deposition potential, and there are generally rinse solutions as well. In synthesis of CdTe, for example, atomic layers of tellurium and cadmium are alternatively electrodeposited to build up a thin layer of CdTe [11, 12]. The necessary atomic level control over the electrodeposition of these two elements is obtained by depositing both elements using UPD. The thickness of the CdTe layer prepared by EC-ALE can be specified by controlling the number of Cd and Te layers that are deposited.

Figure 1.1 Transmission electron micrograph (TEM) of a CdTe deposit formed using 200 cycle of CdTe via EC-ALE [13]. Reproduced by kind permission from the publisher.

Figure 1.1 shows the TEM figure of the CdTe using EC-ALE with 200 cycles [13]. The regular layered structure, parallel with the substrate gold (Au) lattice planes, suggests the epitaxial nature of the deposit. There are a number of ways to introduce dopants into an EC-ALE deposit and they can be introduced homogeneously throughout the deposit. Such doping studies of ZnS were initially run with the idea of forming phosphor screens for flat panel display applications [14]. Control of growth at the nanoscale is a major frontier of materials science. The manipulation of a compound's dimensions and unit cell, at the nanoscale, can result in materials with unique properties. By forming nanocrystalline materials or constructing superlattices, nanowires, and nanoclusters, the electronic structure (bandgap) of a semiconductor can be engineered. EC-ALE has been developed as an electrochemical methodology to grow semiconducting compounds with atomic layer control.

1.2 Electrochemical Synthesis of Quantum Dots and Semiconducting Nanocompounds

Electrodeposition normally leads to small particle size, largely because it is a low-temperature technique, thereby minimizing grain growth. However, it possesses the additional feature of a very high degree of control over the amount of deposited material through Faraday's law, which relates the amount of material deposited to the deposition charge. This feature is particularly desirable when isolated nanocrystals are to be deposited on a substrate.

Quantum dots are semiconductor particles having diameters that are smaller than about 10 nm. Such semiconductor nanoparticles exhibit a bandgap that depends on the particle diameter: the smaller the nanoparticle, the larger the bandgap. Because quantum dots possess a "size-tunable" bandgap, these diminutive particles have potential applications in detectors, light-emitting diodes, electroluminescent devices, and lasers. Electrochemical methods can synthesize size-monodisperse quantum dots on graphite surfaces, which provide an electrical connection to the graphite *in situ*. The essential features of these methods can be depicted as in Fig. 1.2.

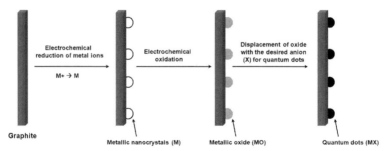

Figure 1.2 The electrochemical and chemical method to synthesize the quantum dots and other semiconductor nanocrystals on graphite.

The first step involves the electrodeposition of metal nanoparticles onto a graphite surface from a solution containing the corresponding metal ions. The metal nanoparticles are electrochemically oxidized to yield a metal oxide (MO), in which the oxidation state of the metal matches the oxidation state in the final product. Finally metal oxide nanoparticles are converted into nanoparticles of a semiconducting salt (MX) via a displacement reaction in which oxide or hydroxide is replaced by the desired anions (X). Examples of using these methods can be shown in the synthesis of CuI [15], CdS [16], and ZnO [17] quantum dots. Ultrathin films of quantum dots with deposits of nonconnected nanocrystals and thick films of more than 10 nm in average thickness can be made by electrochemical methods [18–20]. Besides quantum dots, several methods and variations have been developed to electrodeposit other semiconducting nanocompounds. Oxides are probably the largest group of electrodeposited compounds (e.g., aluminum anodization). The electrodeposition of II-VI compounds has been extensively studied in a number of articles [21–23]. Various semiconductor electrodeposition methods were also reviewed [24, 25]. In summary, the most prominent electrodeposition methods for semiconducting compounds include: codeposition, precipitation, and various two-stage techniques.

Semiconductor film can be electrodeposited either by EC-ALE or by codeposition [26]. The most successful methodology to form II-VI compounds has been codeposition [27–31], where both elements are deposited at the same time from the same solution. Stoichiometry is maintained by having the more inactive element as the limiting reagent, and poising the potential where the less noble element will underpotentially deposit only on the more noble element. A classic example is CdTe formation [27], where the solution contains Cd^{2+} and $HTeO^{2+}$, usually at pH 2. The potential is set to reduce $HTeO^{2+}$ to Te on the surface at a limiting rate, while Cd^{2+} is reduced on the Te at an underpotential, a potential where no bulk Cd is formed. Cd^{2+} ions are present in a large excess, to deposit quantitatively on Te as it is formed, resulting in stoichiometric CdTe. Although the structure and morphology of codeposited compounds are variable, some having been described as "cauliflower" like, high-quality deposits have been formed [32]. There are a number of papers in the literature concerning the formation of compound semiconductor diodes by electrodeposition, the most popular structure being a CdS-CdTe-based photovoltaic. CdS was generally deposited first on an ITO/glass substrate, followed by a layer of CdTe, usually by codeposition [33–39]. Sailor and Martin et al. grew an array of CdSe-CdTe nano-diodes in 200 nm pore alumite [40–42] using a compound electrodeposition methodology called sequential monolayer electrodeposition [43]. A commercial process is being developed by BP Solar to form CdTe-based photovoltaics using codeposition. Relatively rapid deposition rate has been achieved by codeposition and it is presently the most practical compound semiconductor electrodepostion methodology. Codeposition holds great promise if greater control can be achieved. At present, the main parameters of control are solution composition and deposition potential. There have been a number of attempts to improve the process by using variations in reactant concentration, pH [44, 45] and the potential program [43, 46–50]. In most cases, the deposits are improved by annealing. In the application of photovoltaic applications, annealing is used to convert CdTe from the as-deposited n-type material to the desired p-type [39, 51]. Ideally lattice-matched semiconductor substrates could be used to form deposits. For instance, InSb is lattice-matched with CdTe and could be used as a substrate. Good-quality deposits of CdSe have been formed on InP and GaAs substrates using codeposition [52, 53]. This clearly shows the applicability of high-quality commercial compound semiconductors wafers as substrates for compounds electrodeposition.

The precipitation method involves electrochemical generation of a precursor to one of the constituent elements, in a solution containing precursors to the other elements [54–57]. The reaction is essentially homogeneous, but as one reactant is formed at the electrode surface, most of the product precipitates on the surface. This method resembles passive oxide film formation on reactive metals, where metal ions react with the solvent, oxygen or hydroxide. The film thickness is controlled by the amount of electrogenerated precursor. However, as the method resembles precipitation, the quality of the resulting deposit is questionable, and the process is difficult to control. Film thickness is necessarily limited by the need for precursor transport through the deposit. A classic example is the formation of CdS by oxidizing a Cd electrode in a sulphide solution [54, 58–62]. Two stage methods are where thin films of the compound element, or an alloy, are first deposited, at least one by electrodeposition [63]. A second stage, annealing, then results in interdiffusion and reaction of the elements to form the compound. The deposits are annealed in air, an inert gas, or a gaseous precursor to one of the compound's component elements. For instance, electrodeposited CuIn alloys have been annealed in H_2S to form $CuInS_2$ [64]. Given the need for annealing, this methodology has limitations for the formation of more involved device structures. In general, annealing has been used to either form or improve the structures of compound films formed by the electrodeposition methods described above. The primary tool for understanding compound electrodeposition and for improving control over the process has been the methodology of EC-ALE [65, 66].

1.3 Electrochemical Deposition Methods for Metallic Nanostructures

In organic nanoparticles can be fabricated by many different techniques. Electrochemical and wet-chemical methods are demonstrated to be effective approaches to make metal nanostructures under control without addition of a reducing agent or protecting agent. An *in situ* electrochemical reduction method for fabricating metal nanoparticles on carbon substrates simultaneously assembling into ordered functional nanostructures was developed [67]. Silver cation (Ag^+) was adsorbed on a highly oriented pyrolytic graphite surface modified by 4-aminophenyl monolayer with coordination interaction, and then homogeneously dispersed Ag nanoparticles could be obtained through pulsed potentiostatic reduction. Multilayered metal nanostructures on glassy carbon electrodes have been obtained by extending this method [68, 69]. Room temperature ionic liquids are molten salts at room temperature. The larger electrochemical window of ionic liquids in comparison with aqueous electrolytes enables the broader investigation of electrodeposition of metal and semiconductor elements and compounds in nanoscale. Nanoscale electrocrystallization of metals such as nickel (Ni) and cobalt (Co) and the electrodeposition of semiconductors (Ge) on Au (111) and Si (111):H have been studied in the underpotential and overpotential range from ionic liquids [70]. For example, 3D growth in Co electrodeposition on Au (111) from ionic liquids based on imidazolium cations starts at potentials below −0.17 V vs. Co/Co(II). 3D and 2D structures of Co and Ni deposition in the nanoscale were also illustrated. In addition, nanocrystalline aluminum can be obtained by electrodeposition from ionic liquids containing imidazolium cations without additives [71]. The crystal refinement is due to a cathodic decomposition of the imidazolium ions to a certain extent giving rise to nanocrystalline aluminum.

Metal nanowires can be obtained using solution phase reduction [72], template synthesis [73–76] and physical vapor deposition (PVD) [77] onto carbon nanotubes. Metal wires with widths down to 20 nm and lengths of millimeters can be prepared on silicon surface using electron beam lithography [78] or by PVD [79]. However, none of these methods is useful to prepare free-standing metal nanowires that are longer than 20 μm. Penner et al. [80] have used the step edge defects on single crystal surfaces as templates to form metal nanowires by electrochemical step edge decoration. Metallic molybdenum (Mo) wires with diameters ranging from 15 nm to 1 μm and lengths of up to 500 μm were prepared in a two-step procedure on freshly cleaved graphite surfaces [81]. Such nanowires can be obtained size selectively because the mean wire diameter was directly proportional to the square root of the electrolysis time. Parallel arrays of long (>500 μm), dimensionally uniform nanowires composed of Mo, copper (Cu), Ni, Au, and palladium (Pd) can also be electrodeposited by the same strategy [80]. They were first prepared by electrodepositing nanowires of a conductive metal oxide such as NiO, Cu_2O, and MoO_2. Nanowires of the parent metal were then obtained by reducing the metal oxide nanowires in hydrogen at an elevated temperature. Nanowires with diameters in the range 15 nm to 750 nm were obtained by electrodeposition onto the step edges present on the surface of highly oriented pyrolytic graphite electrode. After embedding the nanowires in a polymer film, arrays of nanowires could be lifted off the graphite surface, thereby facilitating the incorporation of these arrays in devices such as sensors. Vertical arrays of metal nanowire hold promise for making chemical and biological sensors in addition to electron emitters in field-emission displays. The difficulty of growing well-defined arrays has kept these technologies at bay. However, electrochemical nanofabrication using crystalline protein masks solved such problem [82]. A simple and robust method was developed to fabricate nanoarrays of metals and metal oxides over macroscopic substrates using the crystalline surface layer (S-layer) protein of *deinococcus radiodurans* as an electrodeposition mask. Substrates are coated by adsorption of the S-layer from a detergent-stabilized aqueous protein extract, producing insulating masks with 2–3 nm diameter solvent-accessible openings to the deposition substrate. The coating process can be controlled to achieve complete or fractional surface coverage. The general applicability of the technique was demonstrated by forming arrays of Cu_2O, Ni, Pt, Pd, and Co exhibiting long-range order with the 18 nm hexagonal periodicity of the protein openings. This protein-based approach to electrochemical nanofabrication should permit the creation of a wide variety of two-dimensional inorganic structures.

1.4 Electrochemical Nanolithography

In addition to its well-known capabilities in imaging and spectroscopy, scanning probe microscopy (SPM) has shown great potentials for patterning of material structures in nanoscales with precise control of the structure and location. Electrochemical nanolithography using SPM, which includes scanning tunneling microscopy (STM) and atomic force microscope (AFM), has been used to fabricate patterned metal structures [83–86], semiconductors [86, 87] and soft matter such as conducting polymers [88, 89]. The electrochemical processing of material surfaces at the nanoscale both laterally and vertically can be conducted by scanning probe anodization/cathodization, which uses the tip-sample junction of a scanning probe microscope connected with an adsorbed water column as a minute electrochemical cell. Reviews on SPM nanolithography examined its various applications in modification, deposition, removal and manipulation of materials for nanoscale fabrication [90].

STM has tremendous potential in metal deposition studies. The initial stages of metal deposition and the Ag adlayer on Au(111) have been studied by Kolb et al. [91]. The inherent nature of the deposition process which is strongly influenced by the defect structure of the substrate, providing nucleation centers, requires imaging in real space for a detailed picture of the initial stage. With STM, the atomic resolution helps understand these processes on a truly atomistic level. Underpotentially deposited Ag revealed a series of ordered adlayer structures with an increasing density of adatoms when the applied potentials were decreased. Nanostructuring can be also achieved by involving electrochemical processes into the overall procedures. For example, Li et al. [92, 93] deposited Ag and Pt clusters on a graphite surface by applying positive voltage pulses to the STM tip in a solution containing the respective metal cations. This effect was attributed to nucleation within holes that were created on the graphite substrate surface by the voltage pulses. Kolb et al. [94], on the other hand, were able to detach Cu clusters from a tip, where they had been previously deposited electrochemically, onto a Au substrate by mechanical contact. However, all of these techniques suffer from restrictions, which could be largely avoided if controlled nanostructuring could be achieved by a direct local electrochemical reaction on the substrate, with the geometry being determined by the location of the tip that acts as a local counterelectrode. By applying ultrashort voltage pulses (≤ 100 ns), holes of about 5 nm in diameter and 0.3 to 1 nm depth on an Au substrate can be created by local anodic dissolution, while cathodic polarization leads to the deposition of small Cu clusters [95]. The development of allowing the generation of small metal clusters, with the help of an STM tip, and placing them at will onto single crystal electrode surfaces was reported [94, 96]. This so-called tip-induced metal deposition involves conventional electrolytic deposition of a metal onto the tip of an STM, followed by a controlled tip approach, during which metal is transferred from the tip to the surface [97]. Small copper clusters, typically two to four atomic layers in height, were precisely positioned on a Au(111) electrode by a process in which copper was first deposited onto the tip of the STM, which then acted as a reservoir from which copper could be transferred to the surface during an appropriate approach of the tip to the surface [98]. Tip approach and position were controlled externally by a microprocessor unit, allowing the fabrication of various patterns, cluster arrays, and "conducting wires" in a very flexible and convenient manner [94]. The formation of such clusters with the tip of a STM is simulated by atom dynamics, and subsequently the stability of these clusters is investigated by Monte Carlo simulation in a grand canonical ensemble. It leads to the conclusion that optimal systems for nanostructuring are those where the participating metals have similar cohesive energies and negative heats of alloy formation. In this respect, the system Cu-Pd(111) is predicted as a good candidate for the formation of stable clusters [99]. In addition to producing the metal nano-clusters, STMs can also be applied to form nanoscale pits in thin conducting films of thallium (III) oxide [100] as well as to write stable features on an atomically flat Au(111) surface [101]. Pit formation was only observed when the process was performed in humid ambient conditions. The mechanism involved in pit formation was attributed to localized electrochemical etching reactions beneath the STM tip. By applying voltage pulses (close to 3 V) across the tunneling junction in controlled atmosphere with the presence of water or ethanol vapor, a nano-hole can be produced. The smallest hole formed is 3 nm in diameter and 0.24 nm in depth. This nano-hole represents the loss of about 100 Au atoms in the top atomic layer of gold surface, and no atomic perturbation is seen inside and outside the nano-hole. Different nanostructures (lattice of dots, legends, map, etc.) can also be fabricated. The threshold voltage for the formation of a nano-hole depends on the relative humidity; however, the relationship between threshold voltage and relative humidity is basically independent of the tip material.

The application of conducting AFM probe anodization to nanolithography was also used in the fabrication and patterning of materials in a similar manner as STM probes. AFM-tip-induced and current-induced local oxidation of silicon and metals was reported, and this novel local oxidation process can be used to generate thin oxide tunneling barriers of 10–50 nm [87]. The direct modification of silicon and other semiconductor and metal surfaces by the process of anodization using the electric field from an SPM is one promising method of accomplishing direct-writing lithography for the electronic device fabrication [85]. This technique involves the application of an electrical bias to both the conducting probe and the sample substrate to locally oxidize selected regions of a sample surface. Nanostructures down to a few tens of nanometers in size have been also fabricated with Langmuir–Blodgett (LB) films and self-assembled monolayers (SAMs) using AFM lithography [102]. AFM anodization was successfully carried out with SAMs on the silicon substrate.

Applied voltage between the AFM tip and sample, the scanning speed, surface group, and the relative humidity in air are very important factors for nanometer-scale lithography of the ultrathin films. The high structural order and perfect thickness of ultrathin organic molecular assemblies are the major advantages of nanoscale lithography. Soft matter such as conducting polymers can also be patterned in the nanoscale by such lithographic method using AFM. Cai et al. described the first observation of localized electropolymerization of pyrrole and aniline on highly oriented pyrolytic graphite (HOPG) substrates under AFM tip-sample interactions [88]. A scanning or oscillating AFM tip, providing the horizontal scratching force and the vertical tapping force, is essential as the driving force for the surface modification with the conducting polymer. It was shown that under the AFM tip interaction, electropolymerization can be blocked on the bare HOPG substrate or enhanced on the as-polymerized film. The localized electropolymerization in selected surface areas enables the nanomodification of lines, square platforms or hollows of polypyrrole and polyaniline on the substrates. Nano-writing of intrinsically conducting polymer can be also achieved via a novel electrochemical nanolithographic technique using electrochemical AFM. Conducting polymer (polythiophene derivatives) nanolines as small as 58 nm in width were obtained and the line width was controlled as a function of the writing speed and writing potential [89]. The electrochemical nanolithography process is shown in Fig. 1.3. Conductive AFM probes and gold-coated silicon nitride (SiN_4) are used as working electrode. Silver wire and platinum wire were used as reference electrode and counter electrode, respectively. Higher writing potential and slower writing speed produce wider conducting polymer nanolines due to enhanced propagation. The great benefit of this method lies in no specific restriction in the choice of substrates and the ease of controlling feature size, which is expected to facilitate to fabrication of all plastic nanoelectronic devices.

Application of an AFM tip as a "nib" to directly deliver organic molecules onto suitable substrate surfaces, such as Au, is referred to as "dippen" nanolithography (DPN) method [103–105]. When AFM is used in air to image a surface, the narrow gap between the tip and surface behaves as a tiny capillary that condenses water from the air. This tiny water meniscus is actually an important factor that has limited the resolution of AFM in air. "Dip-pen" AFM lithography uses the water meniscus to transport organic molecules from tip to surface. By using this technique, organic monolayers can be directly written on the surface with no additional steps, and multiple inks can be used to write different molecules on the same surface. By coupling electrochemical techniques, the DPN are not limited to deliver organic molecules to the surface. Electrochemical "dip-pen" nanolithography (EC-DPN) technique can be used to directly fabricate metal and semiconductor nanostructures on surfaces.

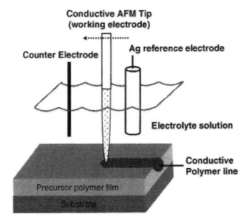

Figure 1.3 Electrochemical nanolithography [89]. Reproduced by kind permission from the publisher.

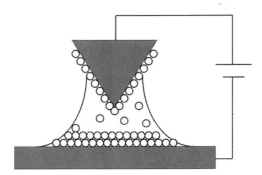

Figure 1.4 Schematic sketch of the EC-DPN experimental setup.

The tiny water meniscus on the AFM tip was used as a nano-sized electrochemical cell, in which metal salts can be dissolved, reduced into metals electrochemically and deposit on the surface as shown in Fig. 1.4. In a typical experiment, an ultrasharp silicon cantilever coated with H_2PtCl_6 is scanned on a cleaned p-type Si(100) surface with a positive DC bias applied on the tip. During this lithographic process, H_2PtCl_6 dissolved in the water meniscus is electrochemically reduced from Pt(IV) to Pt(0) metal at the cathodic silicon surface and deposits as Pt nano-features as shown in Fig. 1.5 [106].

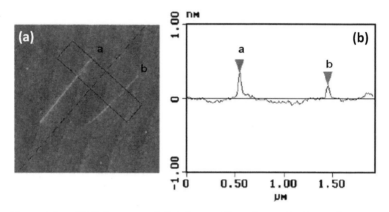

Figure 1.5 AFM image and height profile of two Pt lines drawn at different scan speed. (a) Line at 10 nm/s and (b) line at 20 nm/s. The voltage applied at the tip is 3 V for both lines and the relative humidity is 43% [106]. Reproduced by kind permission from the publisher.

Electrochemical AFM "dip-pen" nanolithography has significantly expanded the scope where DPN nanofabrication can be applied. It combines the versatility of electrochemistry with the simplicity and power of the DPN method to produce nanostructures with high resolution. Electrochemical STM-based methods require that the substrates be metallic, but substrates used in EC-DPN do not have to be metallic since the control feedback of the AFM does not rely on the current between the tip and surface. Si wafers coated with native oxide provides enough conductivity for the reduction of the precursor ions. This development significantly expands the scope of DPN lithography, making it a more general nanofabrication technique that not only can be used to deliver organic molecules to surfaces but is also capable of fabricating metallic and semiconducting structures with precise control over location and geometry. Local electrochemical deposition of freestanding vertically grown platinum nanowires was demonstrated with a similar approach, electrochemical fountain pen nanofabrication (EC-FPN) [107]. The EC-FPN exploits the meniscus formed between an electrolyte-filled nanopipette ("the fountain pen") and a conductive substrate to serve as a confined electrochemical cell for reducing and depositing metal ions. Freestanding Pt nanowires were continuously grown off the substrate by moving the nanopipette away from the substrate while maintaining a stable meniscus between the nanopipette and the nanowire growth front. High-quality and high-aspect-ratio polycrystalline Pt nanowires with diameter of 150 nm and length over 30 μm were locally grown with EC-FPN. The EC-FPN technique is shown to be an efficient and clean technique for localized fabrication of a variety of vertically grown metal nanowires and can potentially be used for fabricating freeform 3D nanostructures.

1.5   3D Electrochemical Nanoconstruction

In combination with lithographic patterns, electrochemistry has taken a key position in products and manufacturing processes of micro and nanotechnology, which has established a multibillion dollar market with applications in information, entertainment, medical, automotive, telecom, and many other technologies such as lab on a chip [108]. Different strategies and techniques such as electrochemical etching, LIGA (German acronym for Lithographie, Galvanoformung, Abformung [Lithography, Electrodeposition, and Molding]), and ultrashort voltage pulsing, which have been used in microtechnology, were also applied to construct 3D structures in the nanoscale.

1.6   Electrochemical Etching and LIGA Technique

Electrochemical etching with ultrashort voltage pulses allows to dissolve electrochemically active materials within an extremely narrow volume and to manufacture three dimensional (3D) microstructures. Micro- and nanoporous silicon can be generated by anodization of silicon wafers in hydrofluoric acid. Since etching proceeds preferably in the (100) direction of the single-crystalline silicon wafer, the pore shape is nearly straight and the depth is equal for all pores. The pores start to grow on a polished wafer in a random pattern, and their arrangement is usually defined by transferring a suitable lithographic pattern and generating a corresponding pattern of pits by alkaline etching as shown in Fig. 1.6. The pattern of the deep pores generated subsequently by the electrochemical etching process corresponds to the pattern of the shallow starter pits. The cross section of the pores is usually square with rounded corners and their size can be varied by changing the etching current. More complex cross sections can be generated by overlapping of pores using additional etching steps and corresponding pattern definition (e.g. by means of lithography).

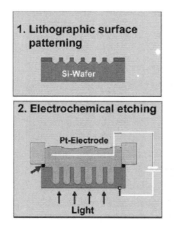
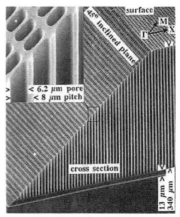

Figure 1.6 Macroporous silicon structure generated by means of electrochemical etching of single-crystalline silicon (*Source*: V. Lehmann, Siemens AG) [108]. Reproduced by kind permission from the publisher.

Such confined etchant layer technique has been applied to achieve effective three-dimensional (3D) micromachining on n-GaAs and p-Si. It operates via an indirect electrochemical process and is a maskless, low-cost technique for microfabrication of arbitrary 3D structures in a single step [109]. It has also been presented that free-standing Si quantum wire arrays can be fabricated without the use of epitaxial deposition or lithography by electrochemical and chemical dissolution of wafers [110]. This novel approach uses electrochemical and chemical dissolution steps to define networks of isolated wires out of bulk wafers.

Electrochemical methods, either alone or in combination with other techniques, have been developed for shaping materials. 3D microstructures with extremely high precision and aspects ratio can be manufactured by means of LIGA technology, which combines deep lithography, electrodeposition, and molding process steps. LIGA offers high potential regarding miniaturization, freedom of design, and mass production. Micro-gear system produced from a nickel iron alloy by means of LIGA has been shown in Fig. 1.7.

Figure 1.7 Micro-gear system produced from a nickel iron alloy by means of LIGA technology (*Source*: Micromotion, IMM) [108]. Reproduced by kind permission from the publisher.

However, the ultraprecise microstructures with extreme aspect ratio could only be generated by deep X-ray lithography. Difficulties, for example, in the access to synchrotron radiation facilities have limited the commercialization of LIGA technique in mass fabrication.

1.7 Micro- and Nano-Machining by Ultrashort Voltage Pulsing Technique

Use of electrochemistry in micro-machining has been a matured area [111]. In contrast to the conventional processes of electrochemical micromachining, where the gap between the electrodes is usually 0.1 mm and direct current is applied, a novel electrochemical microfabrication method using a gap in "μm" range and very short voltage pulses of some tens of nanoseconds was developed. The short pulse confines electrochemical processes correspondingly, removal of material to a very narrow volume to enable a precise nanomachining. Ultrashort pulses can be employed to machine conducting materials with lithographic precision [112]. Resolution can be improved significantly through the use of ultrashort voltage pulses, compared with the use of conventional direct current anodization. Three-dimensional complex nanostructures, lines, curved features, and arrays can be machined in substrates in single-step processing.

The method is based on the application of ultrashort voltage pulses of nanosecond duration, which leads to the spatial confinement of electrochemical reactions, e.g., dissolution of material. The electrochemical dissolution rate of the material has to be intentionally varied over the workpiece surface by applying inhomogeneous current density distribution in the electrolyte and at the workpiece surface. This can be achieved by the geometric shape of the tool, locally very small tool-workpiece distances, partial insulation of the tool or workpiece, and high overall current densities, etc. This situation is illustrated in Fig. 1.8. The workpiece is preferentially etched within the gap region between the front face of the tool and the workpiece surface. This approach for local confinement of electrochemical reactions is based on the local charging of the electrochemical double layer (DL) and the resulting direct influence on the electrochemical reaction rates.

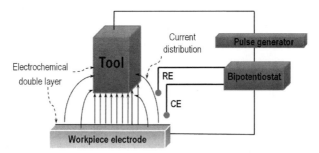

Figure 1.8 Sketch of the experimental setup and principle of electrochemical micromachining with ultrashort pulses. RE and CE are abbreviations for reference and counter electrode, respectively.

The potentials of the workpiece and tool are controlled by the low-frequency bipotentiostat. The voltage pulses are supplied by the high-frequency pulse generator. An ultrashort voltage pulse limits the charging of double layer capacitance to the vicinity of the tool. The current distribution between the DL is also illustrated in Fig. 1.10. The pulsing time constant is given by the DL capacitance multiplied by the resistance of the electrolyte along the current path. The latter factor is locally varying, depending on the local separation of the electrode surfaces [112]. Therefore, on proper choice of the pulse duration, DL areas where the tool and workpiece electrodes are in close proximity are strongly charged by the voltage pulses, whereas at further distances, the charging becomes progressively weaker. The pulse duration provides a direct control for the setting the machining accuracy. Machining precisions below 100 nm were achieved by the application of 500 ps voltage pulses [113].

The application of ultrashort voltage pulses between a tool electrode and a workpiece in an electrochemical environment allows the 3D machining of conducting materials with nanoscale precision. The principle is based on the finite time constant for DL charging, which varies linearly with the local separation between the electrodes. During nanosecond pulses, the electrochemical reactions are confined to electrode regions in close proximity. The performance of electrochemical micromachining with ultrashort voltage pulses was demonstrated in a number of experiments where microstructures were manufactured from various materials like copper, silicon and stainless steels [112]. 3D structures with high aspect ratios can be achieved by using suitable microelectrodes and piezo-driven micropositioning stages. Small tools can be used to make very small features. Tools (Fig. 1.9a) were first fabricated by focused ion beam (FIB) milling and then used in the machining. High aspect ratio nanometer accurate features were machined in nickel using ultrashort voltage pulse electrochemical machining [114]. The potentials of the shaped tungsten tool and nickel substrate electrodes were controlled with a bipotentiostat that kept the potentials of the tool and substrate constant versus an Ag/AgCl reference electrode by applying a potential to Pt counter electrode. Experiments were conducted in an aqueous HCl solution with various concentrations. Supplementary circuitry was present to allow the additional of ultrashort (order of 1 ns) pulses to the potential of the tool electrode. The separation of the tool and the substrate workpiece electrodes was controlled with piezoactuators and the tool was fed into the workpiece with a constant feeding speed, avoiding mechanical contact between the electrodes by monitoring the DC current between tool and workpiece. Examples of patterns made with the 2 × 2 array of cubes tool are shown in Fig. 1.9b. This indicates that the feature resolution improves with decrease in pulse duration.

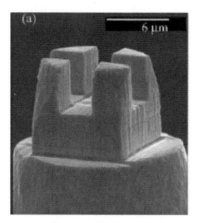

Figure 1.9 Scanning electron micrographs. (a) Tungsten tool; (b) machined Ni substrate. Experimental conditions: U_{sub} = −1.7 V, U_{tool} = −1.0 V, pulse duration indicated, 2 V amplitude, 1:10 pulse to pause ratio, and 0.2 M HCl electrolyte. Feature resolution and edge sharpness increased as pulse duration decreased [114]. Reproduced by kind permission from the publisher.

3D machining of electrochemically active materials, including the construction of unconventional island patterns on a surface with nanoscale resolution, was also realized by this method [95, 115–117]. Thus, electrochemical machining can be applied to microelectromechanical systems (MEMS) [118] and even in the nanoelectromechanical systems (NEMS). Electrochemical methods can realize the nanofabrication in a selective place and make the complicated 3D nanostructures. Conducting polymers can also be fabricated in this way. Similar to the electrochemical machining, by application of short voltage pulses to the tool electrode in the vicinity of the workpiece electrode, the electropolymerization of pyrrole can be locally confined with micrometer precision [119]. As the produced nuclei of conducting polymers will grow preferentially vertically to the surface, fiber-like morphologies were found in the local polypyrrole electrosynthesis with short voltage pulses. A polypyrrole ring with needle-like feature can be selectively nanofabricated, which was grown with 1 µs pulses in 0.2 M H_2SO_4/0.2M Pyrrole. To obtain the ring structure, a 50 µm diameter flattened cylindrical Pt wire tool was brought to a distance of 1 µm from the workpiece substrate before applying the pulses. The small distance between tool and the substrate strongly hinders the supply of monomer at the reaction site. The consequent relatively low supersaturation leads to a low density of nuclei, which grow and form isolated polypyrrole fibers. With the increase of the distance, a more compact ring-like structure can be obtained.

1.8 Template-Free Methods for Conducting Polymer Nano-Architecture

The assembly of large arrays of oriented nanowires containing molecularly aligned conducting polymers without using a porous membrane template to support the polymer was reported [120]. The uniform oriented nanowires were prepared through controlled nucleation and growth during a stepwise electrochemical galvanostatic deposition process, in which a large number of nuclei were first deposited on the substrate using a large current density. After the initial nucleation, the current density was reduced stepwise in order to grow the oriented nanowires from the nucleation sites created in the first step. The usefulness of these new polymer structures is demonstrated with a chemical sensor device for H_2O_2, the detection of which is widely investigated for biosensors. It offers a general approach to control nucleation and has potential for growing oriented nanostructures of other materials.

Figure 1.10 demonstrates the steps to electrochemically fabricate arrays of oriented conducting polyaniline (PANI) nanowires by well controlled nucleation and growth without templates. After initiating the nucleation of the conducting polymer at high current densities the current density is reduced to avoid formation of further nuclei. The existing nuclei grew preferentially vertically to the surface. A typical procedure involves electrochemical deposition in an aniline-containing electrolyte solution, by using the substrate as the working electrode. This process involves 0.08 mAcm^{-2} for 0.5 h, followed by 0.04 mAcm^{-2} for 3 h, which was then followed by another 3 h at 0.02 mAcm^{-2}. The stepwise growth produced uniform, oriented nanowires on a variety of flat and rough surfaces [120].

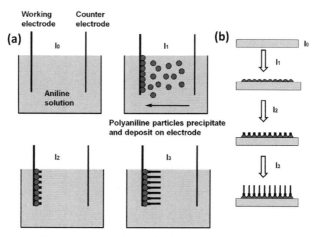

Figure 1.10 Schematic drawing of the steps for growing oriented conducting polymer nanowires. (a) Schematics of the reactions in the electrochemical cell. (b) Schematics of the nucleation and growth.

The PANI films prepared by the above method appear to be fairly uniform with the diameters of the tips ranging from 50 nm to 70 nm. The direct electrochemical synthesis of large arrays of uniform and oriented nanowires of conducting polymers with a diameter much smaller than 100 nm, on a variety of substrates (Pt, Si, Au, carbon, silica), without using a supporting template was studied [121]. Ordered PANI nanowires tailored by such stepwise electrochemical depositions showed remarkably enhanced capacitance [122]. The superior capacitive behaviors of PANI nanowires show great potential in the application of supercapacitors and rechargeable batteries.

The same galvanostatic template free, site-specific electrochemical method was developed to precisely fabricate individual and addressable conducting polymer nanowires on electrode junctions in a parallel-oriented array [123].

Electrochemical polymerization at low and constant current levels was used to fabricate 10 nanoframework-electrode junctions simultaneously with uniform diameter (ca. 40 to 80 nm) PANI nanowires intertwined to nanoframeworks. Electropolymerization was carried in an aqueous solution containing 0.5 M aniline and 1.0 M HCl. First, a constant current (50 nA) was applied for ca. 30 min to introduce the PANI nuclei onto the Pt working junction electrodes. The current was then reduced to 25 nA for 180 min. The PANI nanoframeworks begin to propagate from the working junction electrodes to the other set of the junction electrodes. In the last step, the current was decreased to 12 nA for 180 min, and the 10 polyaniline nanoframework-electrode junctions were obtained simultaneously with each PANI nanoframework positioned precisely within the 2 μm gap between its electrodes. All these nanoframework electrode junctions can be covered by such conducting polymer wires simultaneously and site-specifically in a parallel fashion. This device can be used as miniaturized resistive sensors for real-time detection of NH_3 and HCl gases. Such two-terminal devices can be easily converted to three-terminal transistors by simply immersing them into an electrolyte solution along with a gate electrode. Electrolyte-gated transistors based on conducting polymer nanowires junction arrays was developed and the field-induced modulation can be applied for signal amplification to enhance the device performance [124]. Conducting polymer nanowires, including PANI (ca. 50–80 nm), polypyrrole (PPy) (ca. 60–120 nm), and poly(3,4-ethylenedioxythiophene) (PEDOT) (ca. 80–150 nm), were introduced to the 10 paralleled 2-μm-wide gaps by the template free electropolymerization method. The preparation of electrolyte gated transistors can be completed simply immersing the conducting polymer nanowire-based two-terminal resistive device along with a gate electrode (a Pt wire or Ag/AgCl electrode) into a buffered electrolyte solutions containing NaCl. P-channel transistor characteristics at pH 7 and n-type behavior in basic media were observed for both PANI and PPy nanowires, whereas PEDOT nanowire-based device only exhibit depletion mode behaviors in neutral solutions. These open new opportunities to fabricate sensor arrays with conducting polymer nanowires to realize the ultrasensitive, real-time, and parallel detection of analytes in solutions.

1.9 Template Methods

The growth of thin films displaying special features like aligned pores perpendicularly to the substrate surface and nanoporous structures has attracted the attention of many research groups. A lot of bottom-up techniques, particularly those in which, self-assembly processes play a relevant role in the growth mechanisms of that nanostructures have been reported. Among them, electrochemical techniques constitute one of the most used to fabricate highly ordered nanostructures for replicating other nanostructured materials and for growing functionalized material arrays. Such electrochemical nanofabrication is mainly based on ordered and nanoporous anodic aluminum oxide membranes (AAO), anodic titania membranes, colloidal polystyrene (PS) latex spheres, self-assembled monolayers (SAMs) and self-assembled block copolymers and even carbon nanotubes. Template methods offer very important strategies for complicated nanostructures.

1.9.1 Anodized Aluminum Oxide (AAO) Membranes

A simple and completely nonlithographic preparation technique for free-standing nanostructured films with a close-packed hexagonal array of nanoembossments has been developed by porous anodic aluminum oxide (AAO) membrane templates with different pore diameters. They have been largely used to construct different freestanding inorganic and organic nanowires [125–127], nanotubes [128] and ordered arrays of nanoparticles [129] since invented.

Aluminum anodization provides a simple and inexpensive way to obtain nanoporous templates with uniform and controllable pore diameters and periods over a wide range. The usual electrochemical method for producing the AAO film is the anodization of high-purity Al plates at constant voltage (e.g., anodized at 22 V from an Al foil and detached by the reverse-bias method) [130]. Membranes with several different pore sizes can be made, for example, in the following electrolytes: Aqueous solutions of H_2SO_4 at 10–20 V for pores ~10–25 nm, $H_2C_2O_4$ at 40–80 V for pores ~40–100 nm, and H_3PO_4 at 100–140 V for pores ~100–170 nm. The pore diameter is linearly related to the anodizing voltage (1.2 nm/V). A voltage reduction was done to the thin barrier layer that inhibits anodic current during electrodeposition [131]. Other attempts have been made to create nanoporous symmetries other than hexagonal packing [132]. Recently, a novel AAO membrane with a six-membered ring symmetry coexisting with the usual hexagonal structure has been fabricated by constant current anodization [133]. The pore sizes of this structure can be tailored by changing the processing conditions. Ordered arrays of nanodots with novel structure have been fabricated by such AAO template. In the final stage, the porous alumina substrate can be removed by etching in KOH. Moreover, one of the interesting possibilities afforded by the anodization process is that the anodization can take place on arbitrary surfaces, such as curved surfaces. Unique features including cessation, bending, and branching of pore channels are observed when fabricating AAO templates on curved surfaces [134]. The new structures may open new opportunities in optical, electronic, and electrochemical applications.

Many strategies have been ingeniously implemented to fabricate complicated nanostructures based on the AAO templates. For example, hexagonally ordered Ni nanocones have been fabricated using an a porous AAO template where the pores are of a cone shape [135]. The conical AAO film was found to exhibit hexagonal order with a period of 100 nm. The Ni nanocones and the surface morphology of the nanoconical film exhibit the same periodic structure of the template as shown in Fig. 1.11.

The hexagonally ordered Ni nanocones and nanoconical film were produced using anodization and metal plating techniques. The conical AAO template was produced using a process of repeated applications of the anodization and pore-widening steps, applying the two steps alternately. The Ni nanocones were produced by electroless Ni deposition onto the conical AAO template, with the pores filled with Ni particles. The resulting Ni nanocones exhibit the same ordered structure. The Ni nanoconical film was produced by detaching the deposited Ni layer, with the surface morphology also a hexagonally ordered array with a period of 100 nm. These nanostructures were produced using the wet-processes of anodization, a pore widening treatment, the pulsed deposition of Pd particles and electrochemical Ni deposition. Clearly, this complete process can produce good-quality results.

Figure 1.11 (a) Surface normal, (b) surface angled, and (c) cross-sectional view of the hexagonally ordered conical Ni film [135]. Reproduced by kind permission from the publisher.

A double-templating approach using simple electrochemical methods to create aligned arrays of nanotubes on substrates was also developed [136]. The method used to fabricate nanotube arrays is shown schematically in Fig. 1.12. Initially, nanorod arrays are fabricated by electrodeposition into AAO templates. By varying the anodization conditions, the pore diameter, spacing, and height can be tuned, and the pore ordering can also be controlled. Nickel nanorods are then electrodeposited into the pores of the alumina. After deposition, the exposed ends of the nanorods are modified by anodization in a dilute KOH solution. Because the anodization is performed when the alumina is still in place, only the top ends of the nanorods are anodized. Then the alumina template is removed by a selective chemical etch, leaving an array of nickel nanorods with anodized tips. This nanorod array is then used for electrodeposition of gold nanotube arrays. The nanotube material deposits uniformly across the entire surface of the nanorod arrays, except at the anodized tips of the nanorods. Finally, the nickel nanorod array template is selectively removed, resulting in an array of open-ended nanotubes on the substrate. This approach allows both fabricate and organize nanostructures over large areas on substrates in the same process. This method is demonstrated to prepare arrays of electrodeposited, open-ended nanotubes aligned vertically on substrates. The nanotube inner diameter, spacing, and height are determined by the nanorod dimensions, which in turn correspond to the alumina film characteristics. The nanotube morphology and wall thickness can be controlled by the electrodeposition parameters.

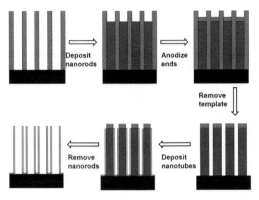

Figure 1.12 Schematic of method to fabricate nanotube arrays on substrates. Nanorods are electrodeposited into nanoporous AAO template films, the alumina is removed, and then the nanorods are used as secondary templates for nanotube electrodeposition.

Utilizing electrochemically prepared textured aluminum sheets as a replication master in conjunction with electrochemical deposition of metals revealed a highly facile and economical way for the production of periodic metallic nanostructures in a large area with high fidelity in pattern transfer as well as with a good degree of flexibility in materials [137]. The growth of metal nanowires using AAO membranes as hard templates has been reviewed [138]. These kinds of metal nanowires can be applied to superconductivity, optical spectroscopy, sensing, and catalytic conversion, and energy harvesting. The AAO template method provides access to arrays of single-crystal metal nanowires and to quasi-one-dimensional metal nanostructures with controlled compositional variation along their length. Semiconductors such as ZnO and TiO_2 with nanostructures can also be manufactured with such template or by related anodization techniques.

1.9.2 Zinc Oxide (ZnO)

ZnO exhibits many unusual properties including uniaxial piezoelectric response and n-type semiconductor characteristics.

Such properties can be used in applications such as field-emission materials [139], light-emitting diodes (LEDs) [140], solar cells [141], and gas sensors [142]. Electro deposition of ZnO films has been reported by several groups and used in fabrication of oriented nanowire and nanorod arrays [143–148]. ZnO films can be electrodeposited cathodically in aqueous chloride solutions using dissolved oxygen as a precursor. The deposition reaction in the electrolyte is: $Zn^{2+} + \frac{1}{2}O_2 + 2e^- \rightarrow ZnO$. The deposition mechanism was analyzed in termhemical induced surface precipitation due to an increase of local pH resulting from the oxygen reduction reaction. A dramatic effect of temperature on the formation of ZnO was observed. When the temperature increases ($T_{transition} \approx 50°C$), a transition between amorphous insulating zinc oxyhydroxide ($ZnO_x(OH)_y$) to well-crystallized and conducting ZnO happens [149]. The dimensions and the deposition rate of electrodeposited ZnO nanowire arrays can be controlled by changing the chemical nature of the anions in solution. A significant variation of the diameter (65 to 110 nm) and length (1.0 to 3.4 μm) of the nanowires can be obtained by changing only the nature of the anions in solution [150]. Recently, large-scale, single-crystalline ZnO nanotube arrays were directly fabricated onto F-doped SnO_2 (FTO) glass substrate via an electrochemical deposition method from an aqueous solution [151]. The tubes had a preferential orientation along the [0001] direction and hexagon-shaped cross sections. The novel nanostructure could be easily fabricated without a prepared layer of seeds on the substrate. The surface condition of substrate material and the experimental conditions played a key role in the nanotube formation. The growth of ZnO arrays and the deposition reaction take place by applying a cathodic potential (–0.7 V vs. SCE) to the FTO substrate at 80°C. The dissolved oxygen serves as the oxygen source for the growth of ZnO arrays in the electrolyte. SEM image of the result ZnO nanotubes is shown in Fig. 1.13.

The electrochemically deposited single crystalline ZnO nanowires can be applied in LEDs [147, 148]. ZnO nanowire films can be embedded in an insulating spin-coated polystyrene layer. The spin-coating parameters are carefully fine-tuned to completely fill out the space between the ZnO nanowires and produce only a very thin coverage of the nanowire tips. The polystyrene layer thickness at the tips can be further reduced by the plasma etching treatment to make the n-type ZnO tip junctions outside. A top contact consisting of a thin p-type poly(3,4-ethylene dioxythiophene): poly(styrenesulfonate) (PEDOT:PSS) layer and an evaporated Au film are provided to serve as the hole injection anode in the LED. A flexible LED can be realized when electrochemical depositing the ZnO nanowire on flexible transparent polymer substrates (e.g. polyethylene terephthalate, PET) coated with indium-tin oxide (ITO) [140]. The infrastructure of such flexible LED is illustrated in Fig. 1.14. This device exhibits electroluminescence over most of the visible spectrum at moderate forward bias.

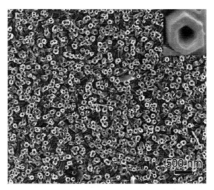

Figure 1.13 SEM image of the ZnO nanotube film [151]. Reproduced by kind permission from the publisher.

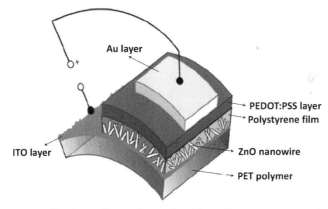

Figure 1.14 Design scheme for a flexible LED structure consisting of vertically oriented single crystalline nanowires grown electrochemically on a polymeric ITO-coated substrate. The top contact consists of p-type polymer (PEDOT:PSS) and an evaporated Au layer. Light is emitted through the transparent polymer [140]. Reproduced by kind permission from the publisher.

1.9.3 Titanium Dioxide (TiO$_2$)

Another semiconductor that has similar bandgap with ZnO, titanium dioxide (TiO$_2$) thin films have been widely exploited in many applications such as microelectronics [152], highly efficient catalysts [153], microorganism photolysis [154], antifogging and self-cleaning coatings [155], biosensors [156], gate oxides in metal-oxide-semiconductor field effect transistors (MOSFET) [157], and more recently in dye-sensitized solar cells (DSSCs) [158].

The geometry of anodic TiO$_2$ nanotubes can be controlled over a wide range by the applied potential in H$_3$PO$_4$/HF aqueous electrolytes in contrast to other electrolytes. It was found that for potentials between 1 and 25 V, tubes could be grown with any desired diameter ranging from 15 to 120 nm combined with tube length from 20 nm to 1 µm. The diameter and the length depend linearly on the voltage [159]. Aqueous HCl electrolyte can be used as an alternative to fluoride containing electrolytes to obtain the vertically oriented TiO$_2$ nanotube arrays by anodization of titanium thin films [160]. Nanotube arrays up to 300 nm in length, 15 nm inner pore diameter, and 10 nm wall thickness can be made using 3 M HCl aqueous electrolyte for anodization potentials between 10 and 13 V.

A highly ordered TiO$_2$ nanotube array with a unique surface morphology can be fabricated by electrochemical anodization of titanium in an organic electrolyte (e.g., 1:1 mixture of DMSO and ethanol) containing 4% HF. The TiO$_2$ nanotube arrays with improved photochemical response can be obtained using electrochemical anodization of titanium in fluorinated organic electrolytes by optimizing etching time, applied potential, solvents and the HF concentration [161]. Combining the electrochemical parameters in an optimal way, ordered TiO$_2$ nanotube layers with a length of over 250 µm have been obtained at 120 V with 0.2 M HF [162]. The tubes can grow as a hexagonal close-packed pore array. Although the TiO$_2$ nanotubes fabricated in organic solution have the longest length and the largest surface area, its conductivity may be lower than the one synthesized in aqueous solutions [156]. Crucial parameters that decide on the dimensions are the fluoride ion concentration, the voltage and the anodization time. The different length of TiO$_2$ nanotubes synthesized by electrochemical anodization can be controlled, and is shown in Fig. 1.15.

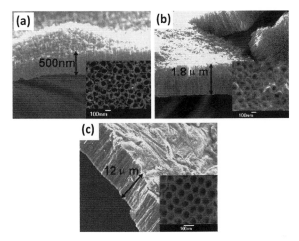

Figure 1.15 SEM images of TiO_2 nanotubes prepared in (a) 0.1 M HF acid solution at 20 V (b) 1.0 M $NaHSO_4$ containing 0.1 M KF at 20 V and (c) ethylene glycol containing 0.25% NH_4F at 60 V for 1 h [156]. Reproduced by kind permission from the publisher.

Highly ordered transparent TiO_2 nanotube arrays produced by electrochemical anodization have been used in DSSCs. It suggests superior electron transport in nanotubular TiO_2 based DSSCs [163]. Remarkable photoconversion efficiencies were expected to be obtained with increase in the length of the nanotube arrays. Carboxylated polythiophene derivatives can be self-assembled onto the TiO_2 arrays produced by anodizing titanium foils in ethylene-glycol-based electrolytes [164]. Such self-assembled hybrid polymer-TiO_2 nanotube array heterojunction solar cells can yield power efficiency of 2.1% under AM 1.5 without dyes. It was found that the formation mechanism of TiO_2 nanotubes is similar to the porous alumina case under high electrical field. TiO_2 nanotube arrays can be fabricated by anodic oxidation of titanium foil in different electrolytes. The produced TiO_2 nanotube arrays possess large surface area and good uniformity and are ready for enzyme immobilization [165–167], which can be used as biosensors. Furthermore, different length of TiO_2 nanotube arrays fabricated by anodic oxidation in different electrolytes were studied for their sensitivities to hydrogen peroxide after co-immobilized horseradish peroxidase (HRP) and thionine chloride. The nanotube arrays fabricated in potassium fluoride solution has the best sensitivity to H_2O_2 with a detection range from 10^{-5} to 3×10^{-3} M [156].

With the use of the template of an AAO, TiO_2 nanowires can be also obtained by cathodic electrodeposition [168, 169] where the metallic ions are attracted to the AAO cathode electrode and reduced to metallic form. For example, in a typical process, the electrodeposition is carried out in 0.2 M $TiCl_3$ solution with pH = 2 with a pulsed electrodeposition approach, and titanium and/or its compound are deposited into the pores of the AAO. By heating the above deposited template at 500°C for 4 h and removing the template, pure anatase TiO_2 nanowires can be obtained. In addition, highly ordered TiO_2 single crystalline (pure anatase) nanowire arrays can be fabricated within the pores of anodic aluminum oxide (AAO) template by a cathodically induced sol–gel method [170]. During this electrochemically induced sol–gel process, both the formation of sol particles and the gelation process take place in the AAO pores. Therefore, TiO_2 nanowires with very small diameters (less than 20 nm or even smaller) can be obtained by this technique. In addition, the length of the nanowires can be well controlled by varying the deposition time and potential of the working electrode.

1.9.4 Electrochemical Fabrication of Soft Matters in Nanoscale Using Templates

Conducting polymer nanowires are promising one-dimensional nanostructured materials for application in nanoelectronic devices and sensors [171, 172] due to their light weights, large surface areas, chemical specificities, easy processing with low costs, and adjustable transport properties. Nanofiber, nanospheres, and other nanoscales of soft matters such as conducting polymers have been fabricated in traditional chemistry way and/or via self-assembly [173–182]. Basically, 1D conducting polymer nanostructures can also be synthesized chemically or electrochemically by using "hard" and "soft" template methods. Obviously, the hard-template method (e.g., AAO) is an effective and straightforward route for fabricating conducting polymer nanostructures with diameters determined by the diameter of the pores in the template. Controlled conducting PANI nanotubes and nanofibers have been fabricated in the AAO templates and find promising applications in lithium/

PANI rechargeable batteries [183]. Nanotubes and nanowires of conducting polymers, including PANI, polypyrrole (PPy), and poly(3,4-ethylenedioxythiophene) (PEDOT), can be synthesized by electrochemical methods using the AAO templates [184]. By changing the doping level, dopant, and template-dissolving solvents, the electrical and optical properties of the nanotubes and nanowires can be controlled. The diameters of the conducting polymer nanotubes and nanowires are in the range 100–200 nm, depending on the diameter of the nanoporous template used. It was found that the polymerization was initiated from the wall-side of the AAO template. The synthesized nanotubes have an open end at the top with the filled end at the bottom. As polymerization time increases, the nanotubes will be filled and nanowires will be formed with the length increased. For example, PPy nanotube can be synthesized by applying current of 2–3 mA for 1 min. When the time is increased to 15–40 min, PPy nanowires will be produced. Conducting polymer nanotube and nanowires prepared by this electrochemical method using AAO templates can be applied in field emitting applications [185, 186]. Figure 1.16 shows a uniform layer of polyaniline nanowires produced at constant potential at 1.0 V for 10 min through the AAO template [187].

Figure 1.16 SEM of electrochemically synthesized (a) PANI nanowires and (b) composite nanowires in AAO membranes (etched with NaOH) [187]. Reproduced by kind permission from the publisher.

PPy nanotubes and nanowires can be also electrochemically synthesized through nanoporous AAO template in ionic liquids

[188]. The electrolyte consisted of pyrrole monomer, solvent, and room temperature ionic liquid dopant such as 1-butyl-3-methyl imidazolium tetrafluoroborate ([BMIM][BF$_4$]) and 1-butyl-3-methyl imidazolium hexafluorophosphate ([BMIM][PF$_6$]), which is stable in air and moisture. The length and diameter of PPy nanotubes and nanowires were determined by the synthetic conditions such as polymerization time, current, and dopant. The formation of nanotube and nanowire of PPy sample was confirmed by using field emission scanning electron microscope and transmission electron microscope. Formation of PANI nanotubules in room temperature ionic liquids by means of electrochemical polymerization without any template has also been reported [189]. PANI nanotubules were synthesized electrochemically on a modified ITO glass in [BMIM][PF$_6$] containing 1M trifluoroacetic acid. Tubular structures of PANI with the diameter of ca. 120 nm were shown by scanning electron microscopy.

Variable inorganic nanoparticles of different sizes can be combined with conducting polymers, giving rise to a novel composite material with interesting physicochemical properties and possibilities for important applications. Electrochemical methods have proved to be effective in incorporating metal nanoparticles in either predeposited polymers or in growing polymer films. Metals can be electrodeposited at conducting polymer electrodes [190]. Nanocables consisting of Ag nanowires sheathed by polyaniline were fabricated in porous AAO template [187]. Silver/polyaniline (Ag/PANI) nanowires were prepared by simultaneous oxidative electropolymerization of aniline and reduction of Ag$^+$ in porous AAO from an acidic electrolyte containing Ag$^+$ and aniline. One-step electrochemical fabrication of devices based on such inorganic-organic hybrid materials offers a new strategy to make electronic devices. Electrochemical fabrication of nonvolatile memory device based on polyaniline and gold particles was reported [191]. Au nanoparticles are synthesized and embedded into the PANI polymer matrix simultaneously during the electropolymerization. The impedance states of the PANI:Au composite films are nonvolatile in nature and can be read and switched several times with minimal degradation in air. Such electrochemical fabrication method can produce multistable nonvolatile memory device in one step, which simplifies fabrication of the memory device significantly.

1.10 Carbon Nanotube Templates

Carbon nanotubes (CNTs) can be used as templates for conducting polymers. Composite films of CNTs with PANI, PPy, or PEDOT were prepared via electrochemical codeposition from solutions containing acid treated CNTs and the corresponding monomer [192]. Electrochemical synthesis of PPy/CNT nanoscale composites using well-aligned carbon nanotube arrays was reported [193]. The thickness of the PPy film can be easily controlled by the value of the film-formation charge. Aligned coaxial nanowires of carbon nanotubes can be sheathed with conducting polymers shown in Fig. 1.17 [194]. In addition, the aligned MWNT electrode arrays can be given additional robustness by electrodepositing conducting polymer around the tubes and then employing them as enzyme sensors. Nanostructuring electrodes with carbon nanotubes for its sensing applications have been reviewed [195]. Glucose oxidase, the most popular oxidase enzyme, can be immobilized onto CNTs by electrodeposition within conducting polymers [196].

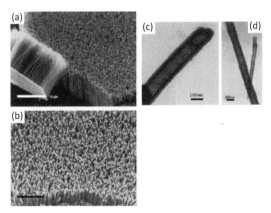

Figure 1.17 SEM images of (a) aligned nanotubes after transfer onto a gold foil (a small piece of the as-synthesized aligned nanotube film is included at the bottom-left corner to show the amorphous carbon layer as well) and (b) the conducting polymer-CNT coaxial nanowires produced by cyclic voltammetry (25 mV/s) on the aligned carbon nanotube electrode in an aqueous solution of 0.1 M $NaClO_4$ containing 0.1 M pyrrole. TEM images of the CP-CNT coaxal nanowire formed from the cyclic voltammetry method. The images (c) are in the tip region and (d) on the wall [194]. Reproduced by kind permission from the publisher.

Electrochemical functionalization of single-walled carbon nanotubes (SWNTs) with PANI has also been done in ionic liquids [197]. SWNTs are covalently functionalized during the electropolymerization of aniline in ionic liquids. This methodology provides a novel way by which large amount of SWNTs (15 mg/ml) can be modified by aniline electrochemically.

1.11 Colloidal Polystyrene (PS) Latex Templates

In nanoelectrodeposition, the aim is to place only a single layer or more of coverage on a surface in a very controlled way. Colloidal crystal such as polystyrene (PS) latex templates have been widely used to synthesize highly ordered macroporous ceramics [198], metals [199], and polymers [200, 201] where a particular interest has been in the optical, magnetic and photonic bandgap properties of the resultant structures. PS latex nanospheres can be self-assembled on hydrophobic surfaces such as unoxidized silicon or gold and used as templates [202, 203]. Two- or three-dimensional highly ordered macroporous cobalt, iron, nickel, gold, platinum, and palladium films containing close packed arrays of spherical holes of uniform size (an inverse opal structure) can be prepared by such simple and versatile template technique [202–204]. The films were prepared by electrochemical reduction of metal cations (e.g., gold ions) dissolved in aqueous solution within the interstitial spaces of preassembled colloidal templates assembled on gold surfaces. Following the electrochemical deposition of the metal films, the polystyrene templates were removed by dissolution in toluene. The resulting films will show well-formed two- or three-dimensional porous structures consisting of interconnected hexagonal close packed arrays of spherical voids. The diameter of the spherical voids within the structures can be varied by changing the diameter of the PS latex spheres used to form the template. The thickness of the films is controlled by varying the charge passed in their deposition.

Electrochemical deposition is ideal for the production of thin supported layers for applications such as photonic mirrors, because the surface of the electrochemically deposited film can be very uniform. Electrochemical deposition occurs from the electrode surface out through the overlying template, the first layer of templated material, deposited out to a thickness comparable with the diameter of the template spheres used, has a different structure from subsequent layers. The subsequent growth of the film by electrodeposition out through the template leads to a modulation of the surface topography of the film in a regular manner that will depend on the precise choice of deposition bath and deposition conditions. This is clearly shown in the following figure if more than one layer of PS is used in templates.

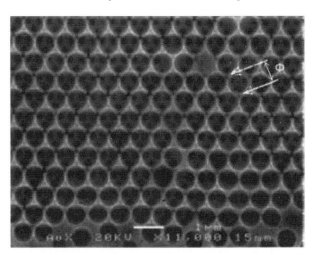

Figure 1.18 SEM images of regions of macroporous gold films grown with a thickness gradient by electrochemical deposition through templates assembled from either 750-nm-diameter polystyrene spheres [204]. The electrodeposition was performed at a potential of −0.90 V_{SCE}. Φ is the diameter of the sphere on top layer. The three dotted circle lines represent the spheres beneath. Reproduced by kind permission from the publisher.

The SEM image shows that the spherical voids left in the gold films after the removal of the PS spheres are arranged in well-ordered, single domain, close-packed structures. Figure 1.18 gives an image for a macroporous gold film prepared through a template of 750-nm-diameter spheres in a region where more than one layer of PS is self-assembled. Within each hemispherical void in the top layer there are again three smaller dark circles (diameter ca. 100 nm). These correspond to the interconnections to the three spherical voids in the layer below (marked as dotted circles in Fig. 1.18) that are left around the regions where the original polystyrene spheres in the two layers were in contact. Semiconductors such as PbO_2 can also be electropolymerized by such template methods [205]. Highly ordered magnetic nanoscale dot arrays of Ni can be fabricated from double-templated electrodeposition [206, 207]. Patterns of ordered arrays of spheres with controlled spacing can be electrochemical deposited by two steps. The double templates were firstly prepared by self-assembly of PS latex spheres on a gold coated glass substrate. This primary template was used for the electrodeposition of the conducting polymer resulting in a macroporous polymer template. After the deposition of PPy, the PS spheres were dissolved in toluene leaving a secondary polymer template. The PPy was then converted into an insulator either by over oxidation or by undoping at a sufficiently negative potential. This insulating structure was used as the template for electrochemical deposition of magnetic material such as Ni. Electrodeposition of magnetic materials gradually fills the spherical cavities of the PPy template. Ordered arrays of Ni dots with quasi-spherical geometry can be fabricated by this way and the diameter of the dots can be set from 20 nm [207].

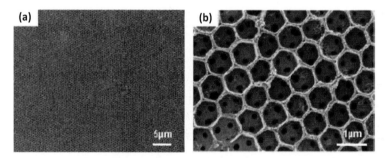

Figure 1.19 SEM images of PANI inverse opals prepared via cyclic voltammetry, at low (a) and higher (b) magnification. Scan rate 20 mV/s, 10 scan cycles [208]. Reproduced by kind permission from the publisher.

Ordered 3D arrays of polyaniline (PANI) inverse opals can also fabricated via electrochemical methods by using colloidal crystals of polystyrene beads as sacrificial templates as shown in Fig. 1.19 [208]. Compared with films obtained by chemical synthesis, the inverse opaline samples obtained by electrochemistry had a much higher structural quality. Such PANI inverse opals were prepared by a galvanostatic method at a current density 0.05 mA/cm^2. By adjusting the polymerization time and applied current, this method allows for the exact control over the structure formation and film thickness of the obtained PANI inverse opline films. To explore potential biosensing applications, PANI composite inverse opals were fabricated by modifying the structure with different dopants, such as poly(acrylic acid) (PAA) and poly(styrene sulfonate) (PSS). It was found that these dopants had a significant effect on the structure and the mechanical stabilities of the obtained opaline films. With selection of suitable dopants, PANI composite inverse opals could be fabricated with very high quality. The obtained films remained electroactive in buffer solutions of neutral pH. Together with their huge surface area, they would be ideal candidates for biosensing applications. Such macroporous structures were used as electrocatalysts for the oxidation of reduced β-nicotinamide adenine dinucleotide (NADH). This showed that the electrocatalytic efficiency of the inverse opline film was much higher than that of an unpatterned film.

Using PS templates is a clear example to show the significant advantages of electrochemical deposition methods. It produces a high-density deposited material and no shrinkage of the material takes place when the template is removed. Also it can be used to prepare a wide range of materials and allows fine control over the thickness of the resulting macroporous film through control of the total charge passed to deposit the film.

1.12 Electrochemistry and Self-Assembled Monolayers (SAMs)

Electron transfer cannot occur in blocking films. However, for very thin self-assembled monolayers (SAMs) of alkane thiols or oxide films, electrons can tunnel through the film and cause Faradaic reactions. Monolayer of alkane thiols can form spontaneously ordered adlayers on substrate like Au(111) due to a strong interaction between the sulphur of the thiol and the gold substrate. These SAMs have received tremendous attention in recent years [209, 210].

Electrochemical deposition onto self-assembled monolayers gives new insights into nanofabrication [211]. Pattern transfer with high resolution is a frontier topic in the emerging field of nanotechnologies. Electrochemical molding is a possible route for nanopatterning metal, alloys and oxide surfaces with high resolution in a simple and inexpensive way. This method involves electrodeposition onto a conducting master covered by a alkanethiolate SAMs. This molecular film enables direct surface-relief pattern transfer from the conducting master to the inner face of the electrodeposit, and also allows an easy release of the electrodeposited film due to their excellent anti-adherent properties. Replicas of the original conductive master can be also obtained by a simple two-step procedure. SAM quality and stability under electrodeposition conditions combined with the formation of smooth electrodeposits are crucial to obtain high-quality pattern transfer with sub-50 nm resolution. Figure 1.20 demonstrates the steps involved in metal electrodeposition on SAMs covered substrates. Further in-depth investigations are required for improving SAM quality reducing the defect size and density, and accordingly increasing the lateral resolution of the method.

Electrochemical work comprised mainly copper deposition onto alkanethiol SAMs. Under UPD, Cu would go down on the bare substrate only [212]. The chain length of the alkane thiol on gold has an influence on the deposition over potential [213]. The long-chain alkane thio (C18) on Au (111) is demonstrated to have a high blocking power [214]. However, it appears to remain one of the challenges for the near future to find the experimental conditions under which metal can be deposited on top of a SAM, preferably as a 2D overlayer. On the other hand, electrochemistry can induce self-assembly of surface-templated (organo) silica thin films on various conducting supports, with mesopore channels oriented perpendicular to the solid surface over wide areas [215]. This method is intrinsically simple and fast and does not require any pretreatment of the support. It consists of combining the electrochemically driven self-assembly of surfactants at solid-liquid interfaces and an electroassisted generation method to produce sol–gel films. The method of electrochemically assisted self-assembly of mesoporous silica thin films involves the application of a suitable cathodic potential to an electrode immersed in a surfactant-containing hydrolyzed solution to generate the hydroxyl ions that are necessary to catalyse polycondensation of the precursors and self-assembly of hexagonally packed one-dimensional channels that grow perpendicularly to the electrode surface. The method opens the way to electrochemically driven nanolithography for designing complex patterns of widely accessible meso-structured materials.

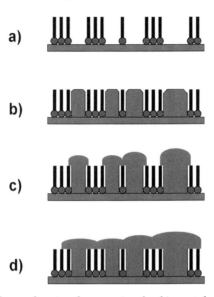

Figure 1.20 Scheme showing the steps involved in metal electrodeposition on SAMs covered substrates. (a) Defective sites at SAMs (the molecules are indicated in black); (b) nucleation and growth of the metal (red) within the SAM; (c) three-dimensional growth outside the SAM; (d) formation of a continuous metallic film on the SAM.

1.13 Other Template Methods

Molecular templates such as modified cyclodextrin have been used for electrochemical nanofabrication. Conducting polymer nanowires and nanorings can be electrochemically synthesized using the molecular templates (thiolated cyclodextrins) on gold [216]. The strategy is to apply electrochemical growth on gold electrodes modified with SAMs of well-separated thiolated cyclodextrins in an alkanethiol forest. Thiolated aniline monomer is anchored to the surface within the cyclodextrin cavity and forms an initiation point for polymer wire growth. Nanosized conducting polymer wires several tens of nanometer thick and a few micrometers long can be synthesized electrochemically by this template. Even though the polymer wires appear to be made of numerous single strands, it was the first time that a single nanowire thread of a conducting polymer has been isolated.

Block copolymers are a class of materials that self-assemble on macromolecular dimensions and have enormous potential in nanoscale patterning and nanofabrication as templates [217, 218]. Some complicated nanoporous structures can be replicated by electrochemical deposition with the help of self-assembly of block-copolymers, for example, poly(4-fluorostyrene)-b-poly(D,L-lactide) [PFSPLA] [219]. Free-standing nanowires were obtained after removal of the block copolymer template by either dissolution or by UV irradiation. Such mild etch method is generally useful to the nanostructures that are sensitive to more aggressive template removal processes. Furthermore, the electrochemical deposition method guarantees a conformable and stable interface between the electrode and the electrodeposited materials. PPy nanotubules have been chemically and electrochemically synthesized inside the pores of nanoporous polycarbonate (PC) particle track-etched membranes (nano-PTM) [220]. Other templates to produce the nano-structured conducting polymers such as DNA [221] and living neural tissue [222] have been reported for its potential applications in biosensors. Nanofabrication technologies are useful for developing highly sensitive, reproducible nanobiosensors. A nanometric system that is composed of well-oriented nanowell arrays can be used for highly sensitive electrochemical DNA detection, it obtained a two-orders-of-magnitude enhancement in sensitivity [223].

The step edge defects on single crystal surfaces can be also used as templates to form conducting polymer nanostructures. Polypyrrole nanostructures with diameters less than or equal to 10 nm have been electropolymerized using step and pit defects on highly ordered pyrolytic graphite (HOPG) as templates for electropolymerization [224]. Step defects were naturally occurring, and pits were formed via oxidation of freshly cleaved surfaces of an HOPG wafer by heating at similar to 640°C. Underpotential deposition of approximately similar to 80 mV caused polypyrrole to form only on the step and pit edges of HOPG at and not on the basal plane. The size of these nanostructures could be controlled by limiting the pyrrole polymerization time at anodic potentials.

In addition, traditional top-down nanofabrication methods such as focused ion beam (FIB), can be used to fabricate nanopore array electrodes [225]. FIB milling thus represents a simple and convenient method for fabrication of prototype nanopore electrode arrays. These electrode nano-arrays can be used in electrochemical nanofabrication for applications in sensing and fundamental electrochemical studies.

1.14 Nanoscale Electrochemistry

Individual carbon nanotubes have been modified selectively on one end with metal using a bipolar electrochemical technique [226]. A stable suspension of nanotube was introduced in a capillary containing 10 mM $HAuCl_4$ aqueous electrolyte, and a high electric field is applied to orientate and polarize the individual tubes. During their transport through the capillary under sufficient polarization (30 kV), each nanotube is the site of water oxidation on one end and the site of metal ion reduction on the other end with the size of the formed metal cluster being proportional to the potential drop along the nanotube.

Bipolar electrochemistry occurs when an external electrical field polarizes an object that is not physically connected to the electrodes and thus generates an anodic and a cathodic area on the same object. The substrate can be any kind of material, but its conductivity must be higher than that of the surrounding medium. The induced potential difference between the two extremities of the object, and therefore the kinetics of the associated redox reactions, is directly proportional to the particle's effective length [227]. When the capillary is exposed to a high electric field (10–30 kV), an electroosmotic flow is generated inside the capillary, transporting the CNT/$AuCl_4$ suspension from the anodic capillary inlet toward the cathodic compartment. A 1 mm long carbon fiber was inserted into a glass capillary connected to two reservoirs filled with $HAuCl_4$ electrolyte. The fiber was observed under the microscope during the application of different potential values. It was found that potentials higher than 40 V are needed to generate a visible metal deposit on the cathodic side of the fiber. After less than 5 min, a visible gold deposit was clearly formed on the negativel polarized end of the fiber. In the course of 1 h, the metal continued growing and its morphology, as revealed by SEM, was dominated by an agglomeration of small crystallites. The Au particle formed after 45 min was illustrated in Fig. 1.21.

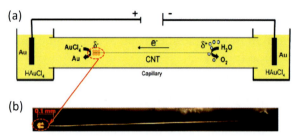

Figure 1.21 (a) Capillary filled with an aqueous CNT/AuCl$_4$ suspension dipping in the two reservoirs of a capillary electrophoresis setup. A high electric field is applied, leading to the polarization of the individual nanotubes, thus triggering different electrochemical reactions on either end. Optical micrographs of a carbon fiber inside a glass capillary during dissymmetric gold deposition by bipolar electrochemistry. (b) The applied voltage was 70 V for a capillary with a length of 10 cm, filled with 10 mM HAuCl$_4$ aqueous electrolyte [226]. Reproduced by kind permission from the publisher.

Looking to the future, this capillary-assisted bipolar electrodeposition can be generalized to other types of nano-objects and also deposits of a very different nature such as other metals, semiconductors, or polymers. The approach, therefore, opens up the way to a whole new family of experiments leading to complex nano-objects with an increasingly sophisticated design allowing original applications.

1.15 Sonoelectrochemistry and Others

Single crystalline CdSe nanotubes have been successfully synthesized by a sonoelectrochemical route in aqueous solution at room temperature [228]. The sonoelectrochemical method is accomplished by applying an electric current pulse to nucleate and perform the electrodeposit, followed by a burst of ultrasonic energy that removes the products from the sonic probe cathode [229, 230]. The growth progress suggested that the CdSe nanotubes were fabricated by sonication-induced rolling-up of CdSe nanosheets and the resulting CdSe products are in tubular structure. This method is simple, convenient, and environmentally benign. Such sonoelectrochemical way is plausible to be extended to synthesis of other nanomaterials.

1.16 Conclusions

Electrochemical nanofabrication itself includes both top-down nanofabrication (e.g., electrochemical lithography) and bottom-up process such as electrochemical atomic layer epitaxy (EC-ALE). Nano-templates, including anodized aluminum oxide (AAO) membranes, colloidal polystyrene (PS) latex spheres, single/aligned carbon nanotubes, graphene, self-assembled monolayers (SAMs), blocked copolymers, and cyclodextrin molecules, can be used for the preparation of various types of nanowires, nanotubes, ordered arrays of nanoparticles and nanodots electrochemically. 3D nanomaterials can be synthesized using LIGA and ultrashort voltage pulsing method. Combining electrochemistry with other nanofabrication techniques such as FIB provides many novel strategies in the fabrication of nanomaterials with specific design. Selective areas in the nanoscale can be modified by electrochemical nanostructuring with metals, metal oxides, semiconducting (e.g., ZnO, TiO_2) nanotubes, nanorods, quantum dots, and conducting polymers using a bipolar electrochemical technique. In the construction of soft matters such as conducting polymers, traditional spin casting cannot guarantee nanostructures due to the fast speed of solvent evaporation. The traditional lithography and pattern technique is costly. Electrochemical technique provides an innovative and economic way of nanofabrication. It especially offers better alternative to construct the soft matter nanostructures in a controllable manner. Various organic and inorganic materials with nanoscale architectures can be fabricated by electrochemical nanofabrication. It is a versatile method for fabricating nanostructures with its simplicity, low-temperature processing, cost-effectiveness, and precise control of the deposit thickness through control of the total charge passed, which are the essential advantages than other nanofabrication techniques till date. It is an exciting era to witness the emerging of nanotechnology. Electrochemistry will definitely contribute to its development independently and interdisciplinarily with other nanofabrication methods.

References

1. Andricacos, P. C., Uzoh, C., Dukovic, J. O., Horkans, J., and Deligianni, H. (1998). *IBMJ. Rev. Dev.*, **42**, pp 567.
2. Andricacos, P. C. (1999). *Interface*, **8**, pp 32.
3. Datta, M. (2003). *Electrochim. Acta*, **48**, pp 2975.
4. Sun, J. J., Kondo, K., Okamura, T., Oh, S., Tomisaka, M., Yonemura, H., Hoshino, M., and Takahashi, K. (2003). *J. Electrochem. Soc.*, **150**, pp G355.
5. Herrero, E., Buller, L. J., and Abruna, H. D. (2001). *Chem. Rev.*, **101**, pp 1897.
6. Gewirth, A. A. and Niece, B. K. (1997). *Chem. Rev.*, **97**, pp 1129.
7. Kolb, D. M., in Gerisher, H., and Tobias, C. W. (Eds.), 1978. *Advances in Electrochemistry and Electrochemical Engineering*, Vol. 11, Wiley Interscience, New York.
8. Herrero, E., Buller, L. J., and Abruna, H. D. (2001). *Chem. Rev.*, **101**, pp 1897.
9. Bedair, S. (1993). *Atomic Layer Epitaxy*, Elsevier, Amsterdam.
10. Huang, B. M., Lister, T. E., and Stickney, J. L. (1997). *Surf. Sci.*, **392**, pp 27.
11. Suggs, D. W., and Stickney, J. L. (1993). *Surf. Sci.*, **290**, pp 362.
12. Suggs, D. W., Villegas, I., Gregory, B. W., and Stickney, J. L. (1992). *J. Vac. Sci. Technol. A*, **10**, pp 886.
13. Villegas, I., and Napolitano, P. (1992). *J. Electrochem. Soc.*, **146**, pp 117.
14. Colletti, L. P., Slaughter, R., and Stickney, J. L. (1997). *J. Soc. Info. Display*, **5**, pp 87.
15. Hsiao, G. S., Andersen, M. G., Gorer, S., Harris, D., and Penner, R. M. (1997). *J. Am. Chem. Soc.*, **119**, pp 1439.
16. Andersen, M. G., Gorer, S., and Penner, R. M. (1997). *J. Phys. Chem.*, **101**, pp 5859.
17. Nyffenegger, R. M., Craft, B., Shaaban, M., Gorer, S., and Penner, R. M. (1998). *Chem. Mater.*, **10**, pp 1120.
18. Baranski, A. S., Fawcett, W. R., and McDonald, A. C. (1984). *J. Electroanal. Chem.*, **160**, pp 271.
19. Baranski, A. S., and Fawcett, W. R. (1984). *J. Electrochem. Soc.*, **131**, pp 2509.
20. Gal, D., Hodes, G., Hariskos, D., Braunger, D., and Schock, H. W. (1998). *Appl. Phys. Lett.*, **73**, pp 3135.

21. Hodes, G. (1994). *Sol. Energy Mater.*, **32**, pp 323.
22. Rajeshwar, K. (1992). *Adv. Mater.*, **4**, pp 23.
23. Fulop, G. F., and Taylor, R. M. (1985). *Ann. Rev. Mater. Sci.*, **15**, pp 197.
24. DeMattei, R. C., and Feigelson, R. S., in McHardy, J., and Ludwig, F. (Eds.), 1992. *Electrochemistry of Semiconductors and Electrodes*, Noyes, Park Ridge, NJ.
25. Hodes, G., in Rubinstein, I. (Eds.), 1995. *Electrochemistry: Principles, Methods and Applications*, Marcel Dekker, New York.
26. Stickney, J. L., in Alkire, R. C., and Kolb (Eds.), D. M. 2001. *Advances in Electrochemical Science and Engineering*, Vol. 7, Wiley-VCH, New York.
27. Panicker, M. P. R., Knaster, M., and Kroger, F. A. (1978). *J. Electrochem. Soc.*, **125**, pp 566.
28. Danaher, W. J., and Lypns, L. E. (1978). *Nature*, **271**, pp 139.
29. Hodes, G., Manassen, J., and Cahen, D. (1976). *Nature*, **261**, pp 403.
30. Cuomo, J. J., and Gambino, R. J. (1968). *J. Electrochem. Soc.*, **115**, pp 755.
31. Gobrecht, H., Liess, H. D., and Tausend, A. (1963). *Ber. Bunsenges. Phys. Chem.*, **67**, pp 930.
32. Tomkiewicz, M., Ling, I., and Parsons, W. S. (1982). *J. Electrochem. Soc.*, **129**, pp 2016.
33. Dennison, S. (1994). *J. Mater. Chem.*, **4**, pp 41.
34. Das, S. K., and Morris, G. C. (1993). *J. Appl. Phys.*, **73**, pp 782.
35. Das, S. K. (1993). *Thin Solid Films*, **226**, pp 259.
36. Yoshida, T. (1992). *J. Electrochem. Soc.*, **139**, pp 2353.
37. Morris, G. C., Das, S. K., and Tanner, P. G. (1992). *J. Cryst. Growth*, **117**, pp 929.
38. Das, S. K., and Morris, G. C. (1992). *J. Appl. Phys.*, **72**, pp 4940.
39. Bhattacharya, R. N., and Rajeshwar, K. (1985). *J. Appl. Phys.*, **58**, pp 3590.
40. Metzger, R. M., Konovalov, V. V., Sun, M., Xu, T., Zangari, G., Xu, B., Benakli, M., and Doyle, W. D. (2000). *IEEE Trans. Magnetics*, **36**, pp 30.
41. Zangari, G., and Lambeth, D. N. (1997). *IEEE Trans. Magnetics*, **33**, pp 3010.
42. Diggle, J. W., Downie, T. C., and Goulding, C. W. (1969). *Chem. Rev.*, **69**, pp 365.

43. Klein, J. D., Herrick, R. D., Palmer, D., Sailor, M. J., Brumlik, C. J., and Martin, C. R. (1993). *Chem. Mater.*, **5**, pp 902.
44. Murase, K., Watanabe, H., Mori, S., Hirato, T., and Awakura, Y. (1999). *J. Electrochem. Soc.*, **146**, pp 4477.
45. Murase, K., Uchida, H., Hirato, T., and Awakura, Y. (1999), *J. Electrochem. Soc.*, **149**, pp 531.
46. Kressin, A. M., Doan, V. V., Klein, J. D., and Sailor, M. J. (1991). *Chem. Mater.*, **3**, pp 1015.
47. Das, S. K., and Morris, G. C. (1993). *Sol. Energy Mater.*, **30**, pp 107.
48. Verbrugge, M. W., and Tobias, C. W. (1987). *Am. Inst. Chem. Eng. J.*, **33**, pp 628.
49. Lokhande, C. D. (1987). *J. Electrochem. Soc.*, **134**, pp 1728.
50. Verbrugge, M. W., and Tobias, C. W. (1985). *J. Electrochem. Soc.*, **132**, pp 1298.
51. Basol, B. M. (1985). *J. Appl. Phys.*, **58**, pp 3809.
52. Beaunier, L., Cachet, H., Froment, M., and Maurin, G. (2000). *J. Electrochem. Soc.*, **147**, pp 1835.
53. Cachet, H., Cortes, R., Froment, M., and Maurin, G. (1999). *Philos. Mag. Lett.*, **79**, pp 837.
54. Miller, B., Menezes, S., and Heller, A. (1978). *J. Electroanal. Chem.*, **94**, pp 85.
55. Vedel, J., Soubeyrand, M., and Castel, E. (1977). *J. Electrochem. Soc.*, **124**, pp 177.
56. Miles, M. H., and McEwan, W. S. (1972). *J. Electrochem. Soc.*, **119**, pp 1188.
57. Panson, A. J. (1964). *Inorg. Chem.*, **3**, pp 940.
58. Krebs, M., and Heusler, K. E. (1992). *Electrochim. Acta*, **37**, pp 1371.
59. Briss, V. I., and Kee, L. E. (1986). *J. Electrochem. Soc.*, **133**, pp 2097.
60. Peter, L. M. (1978). *Electrochim. Acta*, **23**, pp 165.
61. Peter, L. M. (1978). *Electrochim. Acta*, **23**, pp 1073.
62. Miller, B., and Heller, A. (1976). *Nature*, **262**, pp 680.
63. Kapur, V. K., Choudary, V., and Chu, A. K. P. (1986). *Patent*, pp 4581108.
64. Hodes, G., Engelhard, T., Herrington, C. R., Kazmerski, L. L., and Cahen, D. (1985). *Progr. Cryst. Growth Charact.*, **10**, pp 345.
65. Gregory, B. W., and Stickney, J. L. (1991). *J. Electroanal. Chem.*, **300**, pp 543.

66. Gregory, B. W., Suggs, D. W., and Stickney, J. L. (1991). *J. Electrochem. Soc.*, **138**, pp 1279.
67. Tang, Z., Liu, S., Dong, S., and Wang, E. (2001). *J. Electroanal. Chem.*, **502**, pp 146.
68. Shen, Y., Liu, J., Wu, A., Jiang, J., Bi, L., Liu, B., Li, Z., and Dong, S. (2003). *Langmuir*, **19**, pp 5397.
69. Liu, J., Cheng, L., Song, Y., Liu, B., and Dong, S. (2001). *Langmuir*, **17**, pp 6747.
70. Freyland, W., Zell, C. A., Abedin, S. Zein El, and Endres, F. (2003). *Electrochim. Acta*, **48**, pp 3053.
71. Liu, Q. X., Abedin, S., Zein El, and Endres, F. (2008). *J. Electrochem. Soc.*, **155**, pp D357.
72. Jana, N. R., Gearheart, L., and Murphy, C. J. (2001). *Adv. Mater.*, **13**, pp 1389.
73. Loiseau, A., and Pascard, H. (1996). *Chem. Phys. Lett.*, **256**, pp 246.
74. Martin, C. R. (1991). *Adv. Mater.*, **3**, pp 457.
75. Sun, L., Searson, P. C., and Chien, C. L. (1999). *Appl. Phys. Lett.*, **74**, pp 2803.
76. Brumlik, C. J., and Martin, C. R. (1991). *J. Am. Chem. Soc.*, **113**, pp 3174.
77. Zhang, Y., Franklin, N. W., Chen, R. J., and Dai, H. (2000). *Chem. Phys. Lett.*, **331**, pp 35.
78. Peckerar, M., Bass, R., and Rhee, K. W. (2000). *J. Vacuum Sci. Technol. B*, **18**, pp 3143.
79. Himpsel, F. J., Jung, T., and Ortega, J. E. (1997). *Surf. Rev. Lett.*, **4**, pp 371.
80. Walter, E. C., Zach, M. P., Favier, F., Murray, B., K. Inazu, J. C. Hemminger, and R. M. Penner, in Zhang, J. Z., and Wang, Z. L. (Eds.), 2002. *Physical Chemistry of Interfaces and Nanomaterials*, Proc. SPIE, 2002.
81. Zach, M. P., Ng, K. H., and Penner, R. M. (2000). *Science*, **290**, pp 2120.
82. Allred, D. B., Sarikaya, M., Baneyx, F., and Schwartz, D. T. (2005). *Nano Lett.*, **5**, pp 609–613.
83. Kolb, D. M. (2001). *Angew. Chem. Int. Ed.*, **40**, pp 1162.
84. Kolb, D. M., Ullmann, R., and Ziegler, J. C. (1998). *Electrochim. Acta*, **43**, pp 2751.
85. Sugimura, H., and Nakagiri, N. (1996). *Thin Solid Films*, **282**, pp 572–575.
86. Mardare, A. I., Wieck, A. D., and Hassel, A. W. (2007). *Electrochim. Acta*, **52**, pp 7865.

87. Avouris, P., Martel, R., Hertel, T., and Sandstrom, R. (1998). *Appl. Phys. A-Mater. Sci. Process.*, **66**, pp S659–S667.
88. Cai, X. W., Gao, J. S., Xie, Z. X., Xie, Y., Tian, Z. Q., and Mao, B. W. (1998). *Langmuir*, **14**, pp 2508–2514.
89. Jang, S. Y., Marquez, M., and Sotzing, G. A. (2005). *Synthetic Metals*, **152**, pp 345.
90. Tseng, A. A., Notargiacomo, A., and Chen, T. P. (2005). *J. Vacuum Sci. Technol. B*, **23**, pp 877–894.
91. Esplandiu, M. J., Schneeweiss, M. A., and Kolb, D. M. (1999). *Phys. Chem. Chem. Phys.*, **1**, pp 4847.
92. Li, W., Virtanen, J. A., and Penner, R. M. (1992). *Appl. Phys. Lett.*, **60**, pp 1181.
93. Li, W., Hsiao, G. S., Harris, D., Nyffenegger, R. M., Virtanen, J. A., and Penner, R. M. (1996). *J. Phys. Chem.*, **100**, pp 20103.
94. Kolb, D. M., Ullmann, R., and Will, T. (1997). *Science*, **275**, pp 1097.
95. Schuster, R., Kirchner, V., Xia, X. H., Bittner, A. M., and Ertl, G. (2008). *Phys. Rev. Lett.*, **80**, pp 5599.
96. Ullmann, R., Will, T., and Kolb, D. M. (1993). *Chem. Phys. Lett.*, **209**, pp 238.
97. Landman, U., Luedtke, W. D., Burnkam, N. A., and Colton, R. J. (1990). *Science*, **248**, pp 454.
98. Kolb, D. M., Ullmann, R., and Will, T. (1997). *Science*, **275**, pp 1097–1099.
99. Popolo, M. G. Del, Leiva, E. P. M., Mariscal, M., and Schmickler, W. (2005). *Surface Science*, **597**, pp 133–155.
100. Hung, C. J., Gui, J. N., and Switzer, J. A. (1997). *Appl. Phys. Lett.*, **71**, pp 1637–1639.
101. Lebreton, C., and Wang, Z. Z. (1996). *Microelectronic Eng.*, **30**, pp 391–394.
102. Lee, H., Jang, Y. K., Bae, E. J., Lee, W., Kim, S. M., and Lee, S. H. (2002). *Curr. Appl. Phys.*, **2**, pp 85.
103. Hong, S. H. and Mirkin, C. A. (2000). *Science*, **288**, pp 1808.
104. Piner, R. D., Zhu, J., Xu, F., Hong, S. H., and Mirkin, C. A. (1999). *Science*, **283**, pp 661.
105. Hong, S. H., Zhu, J., and Mirkin, C. A. (1999). *Science*, **286**, pp 523.
106. Li, Y., Maynor, B. W., and Liu, J. (2001). *J. Am. Chem. Soc.*, **123**, pp 2105–2106.

107. Suryavanshi, A. P. and Yu, M. F. (2007). *Nanotechnology*, **18**.
108. Ehrfeld, W. (2003). *Electrochim. Acta*, **48**, pp 2857.
109. Zhang, L., Ma, X. Z., Lin, M. X., Lin, Y., Cao, G. H., Tang, J., and Tian, Z. W. (2006). *J. Phys. Chem. B*, **110**, pp 18432–18439.
110. Canham, L. T. (1990). *Appl. Phys. Lett.*, **57**, pp 1046.
111. Schulze, J. W., and Osaka, T. 2002. *Electrochemical Microsystem Technologies*, Taylor & Francis, London.
112. Schuster, R., Kirchner, V., Allongue, P., and Ertl, G. (2000). *Science*, **289**, pp 98–101.
113. Kock, M., Kirchner, V., and Schuster, R. (2003). *Electrochim. Acta*, **48**, pp 3213.
114. Trimmer, A. L., Hudson, J. L., Kock, M., and Schuster, R. (2003). *Appl. Phys. Lett.*, **82**, pp 3327.
115. Kirchner, V., Xia, X., and Schuster, R. (2001). *Acc. Chem. Res.*, **34**, pp 371.
116. Schuster, R. (2001). *Chem. Phys. Chem.*, **2**, pp 411.
117. Schuster, R. (2007). *Chem. Phys. Chem.*, **8**, pp 34.
118. Cohen, A., in Gad el Hak, M. (Ed.), 2001. *The MEMS Handbook*, CRC Press, Boca Raton, Florida.
119. Giacomini, M. T., and Schuster, R. (2005). *Phys. Chem. Chem. Phys.*, **7**, pp 518.
120. Liu, J., Lin, Y., Liang, L., Voigt, J. A., Huber, D. A., Tian, Z. R., Coker, E., Mckenzie, B., and Mcdermott, M. J. (2003). *Chemistry — A European Journal*, **9**, pp 605.
121. Liang, L., Liu, J., Windisch Jr, C. F., Exarhos, G. J., and Lin, Y. (2002). *Angew. Chem. Int. Ed.*, **41**, pp 3665.
122. Zou, X., Zhang, S., Shi, M., and Kong, J. (2007). *J. Solid State Electrochem.*, **11**, pp 317.
123. Wang, J., Chan, S., Carlson, R. R., Luo, Y., Ge, G., Ries, R. S., Heath, J. R., and Tseng, H. R. (2004). *Nano Lett.*, **4**, pp 1693.
124. Alam, M. M., Wang, J., Guo, Y., Lee, S. P., and Tseng, H. R. (2005). *J. Phys. Chem. B*, **109**, pp 12777.
125. Qin, J., Nogues, J., Mikhaylova, M., Roig, A., Munoz, J. S., and Muhammed, M. (2005). *Chem. Mater.*, **17**, pp 1829.
126. Xu, D. S., Xu, Y. J., Chen, D. P., Guo, G. L., Gui, L. L., and Tang, Y. Q. (2000). *Chem. Phys. Lett.*, **325**, pp 340.
127. Huang, Y., Duan, X. F., Cui, Y., Lauhon, L. J., Kim, K. H., and Lieber, C. M. (2001). *Science*, **294**, pp 1313.

128. Lahav, M., Weiss, E. A., Xu, Q., and Whitesides, G. M. (2006). *Nano Lett.*, **6**, pp 2166.
129. Masuda, H., Yasui, K., and Nishio, K. (2000). *Adv. Mater.*, **12**, pp 1031.
130. Nielsch, K., Choi, J., Schwirn, K., Wehrspohn, R. B., and Gosele, U. (2002). *Nano Lett.*, **2**, pp 677.
131. Furneaux, R. C., Rigby, W. R., and Davidson, A. P. (1989). *Nature*, **337**, pp 147.
132. Sun, Z., and Kim, H. K. (2002). *Appl. Phys. Lett.*, **81**, pp 3458.
133. Zhao, S., Chan, K., Yelon. A., and Veres, T. (2007). *Adv. Mater.*, **19**, pp 3004.
134. Yin, A. J., Guico, R. S., and Xu, J. (2007). *Nanotechnology*, **18**.
135. Nagaura, T., Takeuchi, F., Yamauchi, Y., Wada, K., and Inoue, S. (2008). *Electrochem. Commun.*, **10**, pp 681.
136. Sander, M. S., and Gao, H. (2005). *J. Am. Chem. Soc.*, **127**, pp 12158.
137. Lee, W., Jin, M. K., Yoo, W. C., Jang, E. S., Choy, J. H., Kim, J. H., Char, K., and Lee, J. K. (2004). *Langmuir*, **20**, pp 287–290.
138. Kline, T. R., Tian, M., Wang, J., Sen, A., Chan, M. W. H., and Mallouk, T. E. (2006). *Inorg. Chem.*, **45**, pp 7555.
139. Lee, C. J., Lee, T. J., Lyu, S. C., Zhang, Y., Ruh, H., and Lee, H. J. (2002). *Appl. Phys. Lett.*, **81**, pp 3648.
140. Nadarajah, A., Word, R. C., Meiss, J., and Koenenkamp, R. (2008). *Nano Lett.*, **8**, pp 534.
141. Law, M., Greene, L. E., Johnson, J. C., Saykally, R., and Yang, P. D. (2005). *Nature Mater.*, **4**, pp 455.
142. Hu, Y., Zhou, X., Han, Q., Cao, Q., and Huang, Y. (2003). *Mater. Sci. Eng. B*, **99**, pp 41.
143. Xu, L. F., Guo, Y., Liao, Q., Zhang, J. P., and Xu, D. S. (2005). *J. Phys. Chem. B*, **109**, pp 13519.
144. Cui, J. B., and Gibson, U. J. (2005). *Appl. Phys. Lett.*, **87**, pp 133108.
145. Koenenkamp, R., Boedecher, K., Lux-Steiner, M. C., Poschenrieder, M., and Wagner, S. (2000). *Appl. Phys. Lett.*, **77**, pp 2575.
146. Chen, Z. G., Tang, Y. W., Zhang, L. S., and Luo, L. J. (2006). *Electrochim. Acta*, **51**, pp 5870.
147. Koenenkamp, R., Word, R. C., and Schlegel, C. (2004). *Appl. Phys. Lett.*, **85**, pp 6004.
148. Koenenkamp, R., Word, R. C., and Godinez, M. (2005). *Nano Lett.*, **5**, pp 2005.

149. Peulon, S., and Lincol, D. (1998). *J. Electrochem. Soc.*, **145**, pp 864.
150. Elias, J., Tena-Zaera, R., and Levy-Clement, C. (2008). *J. Phys. Chem. C*, **112**, pp 5736.
151. Tang, Y., Luo, L., Chen, Z., Jiang, Y., Li, B., Jia, Z., and Xu, L. (2007). *Electrochem. Commun.*, **9**, pp 289.
152. Burns, G. P. (1989). *J. Appl. Phys.*, **65**, pp 2095.
153. Carlson, T., and Giffin, G. L. (1986). *J. Phys. Chem.*, **90**, pp 5896.
154. Matsunaga, T., Tomoda, R., Nakajima, T., Nakamura, N., and Komine, T. (1988). *Appl. Environ. Microbiol.*, **54**, pp 1330.
155. Wang, R., Hashimoto, K., and Fujishima, A. (1997). *Nature*, **388**, pp 431.
156. Xiao, P., Garcia, B. B., Guo, Q., Liu, D., and Cao, G. (2007). *Electrochem. Commun.*, **9**, pp 2441.
157. Peercy, P. S. (2000). *Nature*, **406**, pp 1023.
158. Gratzel, M. (2003). *Nature*, **421**, pp 586.
159. Bauer, S., Kleber, S., and Schmuki, P. (2006). *Electrochem. Commun.*, **8**, pp 1321.
160. Allam, N. K., and Grimes, C. A. (2007). *J. Phys. Chem. C*, **111**, pp 13028.
161. Ruan, C., Paulose, M., Varghese, O. K., Mor, G. K., and Grimes, C. A. (2005). *J. Phys. Chem. B*, **109**, pp 15754.
162. Albu, S. P., Ghicov, A., Macak, J. M., and Schmuki, P. (2007). *Phys. Stat. Sol. (RRL)*, **1**, pp R65.
163. Mor, G. K., Shankar K., Paulose, M., Varghese, O. K., and Grimes, C. A. (2006). *Nano Lett.*, **6**, pp 215.
164. Shankar, K., Mor, G. K., Prakasam, H. E., Varghese, O. K., and Grimes, C. A. (2007). *Langmuir*, **23**, pp 12445.
165. Tsuchiya, H., Macak, J. M., Muller, L., Kunze, J., Muller, F., Greil, P., Virtanen, S., and Schmuki, P. (2006). *J. Biomed. Mater. Res.*, **77A**, pp 534.
166. Liu, S., and Chen, A. (2005). *Langmuir*, **21**, pp 8409.
167. Bavykin, D. V., Milson, E. V., Marken, F., Kim, D. H., Marsh, D. H., Riley, D. J., Walsh, F. C., El-Abiary, K. H., and Lapin, A. A. (2005). *Electrochem. Commun.*, **7**, pp 1050.
168. Lei, Y., Zhang, L. D., and Fan, J. C. (2001). *Chem. Phys. Lett.*, **338**, pp 231.
169. Liu, S., and Huang, K. (2004). *Sol. Energy Mater. Sol. Cells*, **85**, pp 125.

170. Miao, Z., Xu, D., Ouyang, J., Guo, G., Zhao, X., and Tang, Y. (2002). *Nano. Lett.*, **2**, pp 717.

171. Liu, H., Kameoka, J., Czaplewski, D. A., and Craighead, H. G. (2004) *Nano Lett.*, **4**, pp 671.

172. Yun, M., Myung, N. V., Vasquez, R. P., Lee, C., Menke, E., and Penner, R. M. (2004). *Nano Lett.*, **4**, pp 419.

173. Zhang, Z., Deng, J., Shen, J., Wan, M., and Chen, Z. (2007). *Macromol. Rapid Commun.*, **28**, pp 585.

174. Li, X., Shen, J., Wan, M., Chen, Z., and Wei, Y. (2007). *Synthetic Metals*, **157**, pp 575.

175. Zhu, Y., Hu, D., Wan, M., Jiang, L., and Wei, Y. (2007). *Adv. Mater.*, **19**, pp 2092.

176. Long, Y., Chen, Z., Ma, Y., Zhang, Z., Jin, A., Gu, C., Zhang, L., Wei, Z., and Wan, M. (2004). *Appl. Phys. Lett.*, **84**, pp 2205.

177. Zhang, L., and Wan, M. (2005). *Thin Solid Films*, **477**, pp 24.

178. Qiu, H., Wan, M., Matthews, B., and Dai, L. (2001). *Macromolecules*, **34**, pp 675.

179. Zhang, Z., Wan, M., and Wei, Y. (2006). *Adv. Funct. Mater.*, **16**, pp 1100.

180. Zhang, L., and Wan, M. (2002). *Nanotechnology*, **13**, pp 750.

181. Zhang, L., Long, Y., Chen, Z., and Wan, M. (2004). *Adv. Funct. Mater.*, **14**, pp 693.

182. Ding, H., Wan, M., and Wei, Y. (2007). *Adv. Mater.*, **19**, pp 465.

183. Cheng, F., Tang, W., Li, C., Chen, J., Liu, H., Shen, P., and Dou, S. (2006). *Chemistry — A European Journal*, **12**, pp 3082.

184. Kim, B. H., Park, D. H., Joo, J., Yu, S. G., and Lee, S. H. (2005). *Synthetic Metals*, **150**, pp 279.

185. Joo, J., Kim, B. H., Park, D. H., Kim, H. S., Seo, D. S., Shim, J. H., Lee, S. J., Ryu, K. S., Kim, K., Jin, J. I., Lee, T. J., and Lee, C. J. (2005). *Synthetic Metals*, **153**, pp 313.

186. Kim, K., and Jin, J. I. (2001). *Nano Lett.*, **11**, pp 631.

187. Drury, A., Chaure, S., Kroell, M., Nicolosi, V., Chaure, N., and Blau, W. J. (2007). *Chem. Mater.*, **19**, pp 4252.

188. Koo, Y. K., Kim, B. H., Park, D. H., and Joo, J. (2004). *Molecular Crystals and Liquid Crystals*, **425**, pp 55.

189. Wei, D., Kvarnstroem, C., Lindfors, T., and Ivaska, A. (2006). *Electrochem. Commun.*, **8**, pp 1563.

190. Atta, N. F., Galal, A., and Khalifa, F. (2007). *Appl. Surf. Sci.*, **253**, pp 4273.
191. Wei, D., Baral, J. K., Oesterbacka, R., and Ivaska, A. (2008). *J. Mater. Chem.*, **18**, pp 1853.
192. Peng, C., Jin, J., and Chen, G. Z. (2007). *Electrochimica Acta*, **53**, pp 525.
193. Chen, J. H., Huang, Z. P., Wang, D. Z., Yang, S. X., Wen, J. G., and Ren, Z. F. (2001). *Appl. Phys. A*, **73**, pp 129.
194. Gao, M., Huang, S., Dai, L., Wallace, Gao, G., R., and Wang, Z. (2000). *Angew. Chem. Int. Ed.*, **39**, pp 3664.
195. Gooding, J. J. (2005). *Electrochim. Acta*, **50**, pp 3049.
196. Gao, M., Dai, L. M., and Wallace, G. G. (2003). *Electroanalysis*, **15**, pp 1089.
197. Wei, D., Kvarnstroem, C., Lindfors, T., and Ivaska, A. (2007). *Electrochem. Commun.*, **9**, pp 206.
198. Holland, B. T., Blanford, C. F., and Stein, A. (1998). *Science*, **281**, pp 538.
199. Velev, O. D., Tessier, P. M., Lenhoff, A. M., and Kaler, E. W. (1999). *Nature*, **401**, pp 548.
200. Park, S. H., and Xia, Y. (1998). *Adv. Mater.*, **10**, pp 1045.
201. Johnson, S. A., Ollivier, P. J., and Mallouk, T. E. (1999). *Science*, **283**, pp 963.
202. Bartlett, P. N., Birkin, P. R., and Ghanem, M. A. (2000). *Chem. Commun.*, **17**, 1671.
203. Bartlett, P. N., Ghanem, M. A., Hallag, I. S. El, de Groot, P., and Zhukov, A. (2003). *J. Mater. Chem.*, **13**, pp 2596.
204. Bartlett, P. N., Baumberg, J. J., Birkin, P. R., Ghanem, M. A., and Netti, M. C. (2002). *Chem. Mater.*, **14**, pp 2199.
205. Bartlett, P. N., Dunford, T., and Ghanem, M. A. (2002). *J. Mater. Chem.*, **12**, pp 3130.
206. Zhukov, A., Ghanem, M. A., Goncharov, A. V., Bartlett, P. N., and de Groot, P. (2004). *J. Magn. Magn. Mater.*, **272–276**, pp e1369.
207. Goncharov, A. V., Zhukov, A., Bartlett, P. N., Ghanem, M. A., Boardman, R., Fangohr, H., and de Groot, P. (2005). *J. Magn. Magn. Mater.*, **286**, pp 1.
208. Tian, S., Wang, J., Jonas, U., and Knoll, W. (2005). *Chem. Mater.*, **17**, pp 5726.
209. Poirier, G. E. (1997). *Chem. Rev.*, **97**, pp 1117.

210. Ulman, A. (1996). *Chem. Rev.*, **96**, pp 1533.
211. Schilardi, P. L., Dip, P., Claro, P. C. D., Benitez, G. A., Fonticelli, M. H., Azzaroni, O., and Salvarezza, R. C. (2005). *Chemistry—A European Journal*, **12**, pp 38–49.
212. Sun, L., and Crooks, R. M. (1991). *J. Electrochem. Soc.*, **138**, pp L23.
213. Sondag-Huethorst, J. A. M., and Fokkink, L. G. J. (1995). *Langmuir*, **11**, pp 2237.
214. Schneeweiss, M. A., and Kolb, D. M. (1999). *Phys. Stat. Sol. (A)*, **173**, pp 51.
215. Walcarius, A., Sibottier, E., Etienne, M., and Ghanbaja, J. (2007). *Nature Mater.*, **6**, pp 602.
216. Choi, S. J., and Park, S. M. (2000). *Adv. Mater.*, **12**, pp 1547.
217. Grosso, D., Boissiere, C., Smarsly, B., Brezesinski, T., Pinna, N., Albouy, P. A., Amenitsch, H., Antonietti, M., and Sanchez, C. (2004). *Nature Mater.*, **3**, pp 787.
218. Park, M., Harrison, C., Chaikin, P. M., Register, R. A., and Adamson, D. H. (1997). *Science*, **276**, pp 1401.
219. Crossland, E. J. W., Ludwigs, S., Hillmyer, M. A., and Steiner, U. (2007). *Soft Matter*, **3**, pp 94.
220. Demoustier-Champagne, S., Duchet, J., and Legras, R. (1999). *Synthetic Metals*, **101**, pp 20.
221. Lin, X., Kang, G., and Lu, L. (2007). *Bioelectrochemistry*, **70**, pp 235.
222. Richardson-Burns, S. M., Hendricks, J. L., and Martin, D. C. (2007). *J. Neural Eng.*, **4**, pp L6.
223. Lee, H. Y., Park, J. W., and Kim, J. M. (2006). *Appl. Phys. Lett.*, **89**, pp 113901.
224. Noll, J. D., Nicholson, M. A., Patten, P. G. Van, Chung, C. W., and Myrick, M. L. (1998). *J. Electrochem. Soc.*, **145**, pp 3320–3328.
225. Lanyon, Y. H., De Marzi, G., Watson, Y. E., Quinn, A. J., Gleeson, J. P., Redmond, G., and Arrigan, D. W. M. (2007). *Anal. Chem.*, **79**, pp 3048–3055.
226. Warakulwit, C., Nguyen, T., Majimel, J., Delville, M. H., Lapeyre, V., Garrigue, P., Ravaine, V., Limtrakul, J., and Kuhn, A. (2008). *Nano Lett.*, **8**, pp 500.
227. Bradley, J. C., Babu, S., and Ndungu, P. (2005). *Fullerenes Nanotubes Carbon Nanostruct.*, **13**, pp 227.

228. Shen, Q., Jiang L., Miao, J., Hou, W., and Zhu, J. (2008). *Chem. Commun.*, **14**,1683.
229. Qiu, X. F., Lou, Y. B., Samia, A. C. S., Devadoss, A., Burgess, J. D., Dayal, S., and Burda, C. (2005). *Angew. Chem. Int. Ed.*, **44**, pp 5855.
230. Qiu, X. F., Burda, C., Fu, R. L., Pu, L., Chen, H. Y., and Zhu, J. J. (2004). *J. Am. Chem. Soc.*, **126**, pp 16276.

Chapter 2

Electrochemical Replication of Self-Assembled Block Copolymer Nanostructures

Edward Crossland, Henry Snaith, and Ullrich Steiner

Department of Physics, Cavendish Laboratories and Nanoscience Centre, University of Cambridge, CB3 0HE, UK

us222@hermes.cam.ac.uk, e.crossland1@physics.ox.ac.uk

Block copolymers have attracted great attention as a patterning tool in nanoscience because of their ability to self-assemble into highly ordered patterns on the 10 nm length scale. Many potential applications for these fascinating structures have been concieved, although surprisingly few have so far been realized. Development of polymer synthesis has widened the choice and availability of copolymer combinations and made possible the transferal of block copolymer morphologies from model systems into functional materials for application. Electrochemical synthesis inside porous templates made from selectively sacrificial block copolymer films is well suited to the replication of such small-scale architectures in an enormous range of potential materials. Here we examine the principles and main challenges posed by this relatively new technique and highlight successful reports of its application.

Electrochemical Nanofabrication: Principles and Applications (Second Edition)
Edited by Di Wei
Copyright © 2016 Pan Stanford Publishing Pte. Ltd.
ISBN 978-981-4613-86-6 (Hardcover), 978-981-4613-87-3 (eBook)
www.panstanford.com

Particular emphasis is given to the patterning of quasi 1D nanowire and bicontinuous gyroid arrays, both of which are of particular interest for device applications. The first study of these highly ordered electrochemically patterned semiconductor arrays used in a real device application is described in a dye-sensitized bulk heterojunction solar cell.

2.1 Introduction

The engineering of material architectures and composite structures on the nanometer length scale presents an extremely versatile route to creating enhanced and novel functionalities. Spontaneous self-assembly of constituent parts offers a rather attractive way around the ever-increasing demands of patterning from the top down on ever-smaller length scales. Block copolymers (BCPs), consisting of block sequences of two or more distinct monomer units, are a class of macromolecule that exhibit such self-assembly on the 10 nm scale. The assembly is readily tunable; depending on molecular characteristics (chemical composition and molecular weights) and processing parameters (temperature or application of external fields), the copolymer molecules organize themselves into a rich spectrum of highly ordered, chemically heterogeneous phases on the scale of a single polymer chain. The topology of these microphases ranges from close-packed spheres, hexagonally packed cylinders and bicontinuous gyroids to sheet-like lamellae and perforated lamellae.

Many proposed applications exploiting BCP patterns, for example high-density memory, photovoltaics, battery electrodes, and sensor technologies, are surface-supported, that is, in a film geometry. Low-cost, large-area solution processing of films is no problem for many polymers. We must, however, contend with the additional complication of interactions with external interfaces, which strongly influence the behavior of the molecular assembly. The grand challenge to the application of BCP patterning in novel film devices lies in simultaneous control of the microphase, coupled with a way of transferring such model structures into active materials and composites with desirable intrinsic properties. This is rather more difficult than it sounds; despite over three decades of development, there remain surprisingly few examples of functioning device applications that truly exploit ordered BCP assemblies. Broadly speaking, there are three approaches to overcome this challenge:

(1) *Direct synthesis*: a BCP made by chemically tethering two or more electronically functional polymers.
(2) *Sacrificial templating*: a BCP in which one component can be selectively removed after phase separation to leave a mesoporous matrix of the second. This material acts as a template to be subsequently filled with a chosen functional material. The template can be removed later to leave a freestanding replicate of the BCP morphology.
(3) *Structure directing agent*: a blend of the BCP with chemical precursors of functional materials, relying on microphase separation as an in situ structure director.

While route I is perhaps the most intuitive and direct approach, these copolymers are generally rather difficult to synthesize, which limits the available choice of composite materials to synthentically compatible copolymers. Route II, on the other hand, is more within the grasp of today's materials and allows us to concentrate efforts to control self-assembly on relatively well-understood model copolymers, while significantly expanding the selection of functional materials that can be synthesized in the template. Electrochemical deposition, already well developed in the replication of other nano and microscale porous systems such as anodized alumina and track-etched membranes, is ideally suited to reproducing the extremely small and high-aspect-ratio pores typical of BCP structures. Figure 2.1 outlines the concept of sacrificial copolymer templating.

Electrochemical replication of porous copolymer templates was first demonstrated by Thurn-Albrecht et al. in 2000. They produced a hexagonally ordered array of 10 nm diameter cobalt nanowires, standing vertically on a substrate some 25 nm apart with a density in excess of 1.9×10^{11} cm^{-2}. This work not only demonstrated the great potential of porous BCP templates but also highlighted the main practical difficulties of this method and how they can be overcome. First, the BCP must be selectively degradable, that is, one block can be completely removed in an etching process, which leaves the second component mechanically sound, acting as a self-supporting mesoporous matrix. Second, the pore structure left by etching must span the full thickness of the template from surface to substrate, such that any electrolyte flowing into the pores is able to achieve electronic contact with an underlying working electrode. That relatively few further examples of high-aspect-ratio electrochemical BCP replication exist owes much to the difficulties in satisfying this condition. The following discussion is, therefore, largely dedicated to describing important aspects of porous template formation in diblock copolymers—the characteristics of thin film self-assembly, microphase alignment techniques, selective degradation, and suitability to large area electrochemical replication. Progress in the patterning of standing and lying nanowire arrays and recent replication of spontaneously bicontinuous networks using this technique are reviewed. Finally, the first application of such highly ordered nanostructures in functioning photovoltaic devices is described in nanostructured solar cells based on electrochemically synthesized titanium dioxide arrays.

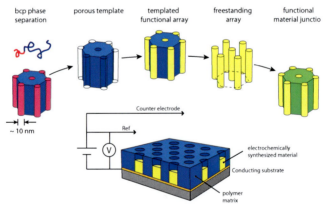

Figure 2.1 Schematic process of templating porous materials, highlighting the film-based electrochemical replication of a porous template in a standard electrochemical cell.

2.2 Principles of Block Copolymer Self-Assembly

The term block copolymer refers to the copolymers in which the two different monomer units are arranged in linear blocks. The simplest example of a block copolymer is an AB diblock copolymer (Fig. 2.2) [1].

| linear homopolymer | diblock copolymer | triblock copolymer | starblock copolymer |

Figure 2.2 Simple homopolymer and block copolymer architectures.

Very few pairs of homopolymers are miscible in the melt; indeed, the tendency of a blend of polymers to macrophase separate is much greater than that of an equivalent blend of the unconnected monomers. The thermodynamics of polymer melts is governed by the competing influence of energetic and entropic terms in the free energy of mixing. The energy of mixing (ΔU_{mix}) is proportional to the number of monomers present, while the entropic contribution (ΔS_{mix}), which favors homogenous mixing, is proportional to the number of polymer chains. The entropy of mixing per monomer is, therefore, decreased by a factor of the degree of polymerization (N) for a polymer blend compared with a blend of the unconnected monomers. The Flory-Huggins free energy of mixing per monomer (ΔF_{mix}) for a blend of polymers A and B at temperature T, with degree of polymerization N and volume fractions f_A and f_B, respectively, is given by

$$\Delta F_{mix} = \Delta U_{mix} - T \Delta S_{mix} \tag{2.1}$$

$$\frac{\Delta F_{mix}}{k_B T} = \frac{f_A}{N} \ln f_A + \frac{f_B}{N} \ln f_B + \chi f_A f_B, \tag{2.2}$$

where χ is the dimensionless Flory interaction parameter describing the energetic cost per monomer of contacts between A and B monomers

$$\chi = \chi_{AB} = \frac{Z}{k_B T}\left(\varepsilon_{AB} - \frac{1}{2}(\varepsilon_{AA} + \varepsilon_{BB})\right), \tag{2.3}$$

where Z is the number of nearest-neighbor contacts in a lattice model of the polymer, and ε_{AB} is the interaction energy per monomer between A and B monomers. Polymer mixing is, therefore, controlled by the product χN rather than χ as is the case for a simple monomer blend. A critical value of $\chi N = 2$

separates the case in which mixtures of all compositions are stable ($\chi N < 2$) from the case in which mixtures at some compositions will phase separate ($\chi N \geq 2$). Since N can be extremely large, only a very small positive value of χ (i.e., very weak repulsive monomer interactions) is required for the free energy of mixing to favor phase separation over a single-phase mixture of polymer chains.

In block copolymer melts, repulsive monomer interactions are again a strong driving force for phase separation into A and B rich domains. However, since each block is covalently tethered to its reluctant partner, the scale of phase separation is limited to macromolecular (mesoscopic) dimensions. The result is spontaneous formation of a spectrum of periodic, ordered structures on the 5–100 nm length scale, known as *microphase separation*. At sufficiently elevated temperatures, as with homopolymer blends, entropic terms overwhelm the energetic interaction terms in the free energy and result in a single disordered phase. The transition from a homogeneous melt phase of copolymer chains to chemically heterogeneous microdomains occurs on cooling at a critical value of χN known as the order–disorder transition (ODT). The morphology of the microphase and the temperature of the ODT depend on the composition of the copolymer, i.e., the volume fraction of one block f_A (with $f_B = 1 - f_A$). The microphase morphology adopted must balance minimizing unfavorable A–B interactions with the entropic penalty of stretching the two blocks away from each other as the chain adopts an extended configuration. For example, a symmetric diblock ($f_A = f_B$) will microphase separate into an alternating lamellar morphology with a flat interface, as illustrated in Fig. 2.3a. A simplified but illuminating treatment of this situation was given by Bates and Fredrickson [2]. At low temperatures (large χ), the lamellar microdomains are nearly pure in A or B components, separated by interfaces much narrower than the lamellar period λ—so-called "strong segregation." Assuming that the chains are uniformly stretched, an approximate expression for the free energy per chain for the lamellar phase (F_{LAM}) is given by:

$$\frac{F_{LAM}}{k_B T} = \frac{3(\lambda/2)^2}{2Na^2} + \frac{\gamma A}{k_B T} \qquad (2.4)$$

The first term in Eq. 2.4 is the entropic stretching penalty for a Gaussian chain of N monomers, each of size a, extended to a distance equal to half the lamellar domain period (λ). The second term represents the repulsive energetic interactions confined to the (sharp) A–B interface, as a product of the contact area per chain (A) and the interfacial tension (γ). A simple treatment of the polymer–polymer interface for highly immiscible polymers[1] gives us an estimate for the interfacial tension $\gamma = (k_B T/a_2)\sqrt{\chi}$. The volume-filling condition (assuming incompressibility) constrains the contact area per chain, $A\lambda/2 = Na_3$. Now, minimizing Eq. 2.4 with respect to λ, we obtain an expression for the equilibrium lamellar period (λ_0):

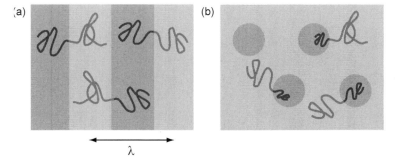

Figure 2.3 Microphase separation in diblock copolymers. The shaded regions represent domains rich in A or B monomers. (a) $f_A = f_B = 1/2$, lamellar phase separation with flat domain interfaces. (b) $f_A > 1/2$, spontaneous curvature of domains toward the minority phase in gyroidal, cylindrical, or spherical morphologies. The characteristic domain spacing λ is typically in the range 10–100 nm.

$$\lambda_0/2 \sim a\chi^{1/6} N^{2/3} \qquad (2.5)$$

The corresponding free energy of the lamellar phase can be used to estimate the order–disorder boundary. The free energy per chain in a homogeneous disordered phase (F_{DIS}) can be approximated simply by the A-B contact energy:

$$\frac{F_{DIS}}{k_B T} \approx \chi f_A f_B N = \chi N/4 \qquad (2.6)$$

---

[1] See, for example, reference [3].

Setting $F_{DIS} = F_{LAM}(\lambda_0)$ predicts a critical value of $\chi N \sim 10.5$ as the location of the order–disorder transition. From this simple argument (in good agreement with more rigorous mean field calculations), we find that symmetric diblocks of high molecular weight or with strong incompatibility ($\chi N > 10.5$) are predicted to microphase separate into ordered lamellae with a period that scales as the two-thirds power of the molecular weight. Smaller diblocks, on the other hand, with more compatible blocks ($\chi N < 10.5$), are expected to form a single homogeneous phase.

More sophisticated mean-field and analytical theories have been developed to describe microphase separation [4, 5]. Three regimes are generally defined according to the extent of segregation (or "degree of incompatibility") of the blocks: weak ($\chi N \sim 10$), intermediate ($\chi N \sim 10\text{–}100$), and strong ($\chi N \geq 100$). The above discussion applies, strictly speaking, to the strong segregation limit. These theories have been used to map "phase diagrams" for copolymers, such as that shown in Fig. 2.4, which include the microphase behavior below the ODT for nonsymmetric copolymers. As the copolymer is made steadily asymmetric by increasing f_A, there comes a point when the free energy is reduced by a curving of the interface toward the shorter block. The morphology changes from lamellar (L) to hexagonally packed cylinders (C) of the B block surrounded by a matrix

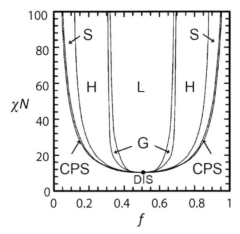

Figure 2.4 Diblock copolymer phase diagram parameterized by f and χN, calculated by Cochran et al. Reproduced with permission from Ref. [6].

of the majority A block (Fig. 2.3b). With further asymmetry, a body-centered cubic spherical phase (S) forms, followed by a narrow region of close-packed spheres (CPS) before a single disordered phase at the compositional extremes. A more complex, bicontinuous gyroid (G) phase is stable in a small region of phase space between the L and C phases, bringing the total to five ordered microphases with regions of thermodynamic stability (Fig. 2.5)[2].

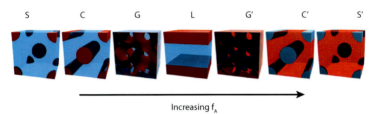

Figure 2.5 Equilibrium phase morphologies of AB diblock copolymers as a function of component A volume fraction f_A: (S) Spheres (body-centered cubic lattice), (C) Cylinders (hexagonal lattice), (G) Gyroid, (L) Lamellar. Reproduced with permission from Ref. [7].

2.3 Block Copolymer Thin Films

The term "thin film" often refers to film thicknesses comparable to the characteristic microphase domain spacing (10–50 nm). Here, in contrast to bulk materials, at least a significant minority of chains lie close to or at an interface [8]. On the other hand, when high-aspect-ratio, film-spanning structures are desirable, for example in fabricating high-surface-area arrays, the copolymer film may be many times thicker than the domain spacing. The effects of the substrate typically penetrate several domain spacings into the film—perhaps only a small percentage of its thickness. Even here, though, since a completely continuous internal pore structure connected everywhere to the substrate is required for electrochemical replication, these details of microphase morphology and its orientation are extremely important, and an understanding of behavior in thin films and at interfaces is crucial.

In the region of an external interface, differences in the surface-free energy or affinity of each block toward the substrate

---

[2]Mathematical Sciences Research Institute (MSRI) http://www.msri.org.

($\Delta\gamma$) have a strong influence on copolymer morphology. When the asymmetry is sufficiently large, the free energy of the system is reduced by preferential segregation of one or other component to the interface. Russell and coworkers first studied these effects experimentally in lamellae-forming poly(styrene)-b-(methylmethacrylate) (PS-b-PMMA) copolymers using neutron reflectivity [9] and dynamic secondary ion mass spectroscopy (SIMS) [10]. After phase separation, the lamellae were found to orient parallel to both gold and silicon substrates (a so-called $L_\parallel$ morphology). On silicon substrates, the total film thickness is $(n + 1/2)\lambda$, where λ is the lamellar period and n is the number of periods, while on gold, the thickness is $n\lambda$. The differences are explained by the symmetry or asymmetry of affinities for each block at the free surface and substrate surface, illustrated in Fig. 2.6. If the initial film is not commensurate with these quantized thicknesses, then on thermal annealing, islands and holes will form on the free surface to avoid the elastic penalty of stretching or compressing chains required to alter the domain spacing.

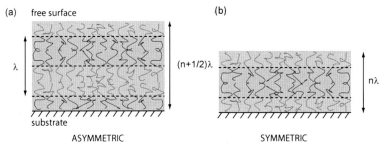

Figure 2.6 Quantization of film thickness in lamellae-forming copolymer thin films. (a) *Asymmetric* affinity of one block to the substrate and another to the free-surface results in film thicknesses quantized in units of $(n + 1/2)\lambda$. (b) *Symmetric* affinity of one block to both the substrate and free surface results in quantization in units of $n\lambda$.

Parallel domain alignment at an interface will propagate into the bulk of a copolymer film to a degree dependent on the magnitude of the substrate affinity asymmetry or "surface field" $\Delta\gamma$. TEM studies on a PS-b-PMMA lamellar system found a critical thickness (linearly dependent on $\Delta\gamma$) below which surface fields are enough to cause the entire film to adopt an $L_\parallel$ alignment [11]. For example, the critical thickness was ~10 lamellar periods when

$\Delta\gamma \sim 50$ nJ cm^{-2}, corresponding to a film roughly 400 nm thick. In films thicker than this, the coherence of the parallel alignment was lost some 3 lamellar periods from the interface. Substrate interactions can be tuned in an extremely well-controlled manner to favor either copolymer block, by surface grafting random copolymer brushes with varying fractions of each component [12]. A "neutral substrate," with $\Delta\gamma \sim 0$ has no energetic preference for either copolymer block. Approximately neutral surfaces can also be made using self-assembled monolayers SAMs such as alkane thiol layers on gold [13]. On neutral surfaces, the lamellar microdomains are found to spontaneously orient at 90° to the surface [12, 14], which removes commensurability constraints between the film thickness and λ, and the surface of the film remains smooth.

The behavior of cylinder-forming copolymers in the vicinity of a planar interface is perhaps less intuitive than lamellae. In this system, the bulk symmetry is broken regardless of orientation so that the microdomains are forced to adjust. Here the balance of film thickness and surface fields leads to various alignments of the phase, or even a complete change from the bulk morphology known as "surface reconstruction." Knoll et al. studied this effect in great detail using poly(styrene)-b-(butadiene)-b-(styrene) triblock copolymer thin films in conjunction with molecular dynamic simulations [15]. In addition to parallel alignment of the cylinders ($C_{||}$) similar to the lamellar copolymer, they found wetting layers, perforated lamellae, and lamellar reconstructions with increasing strength of the surface field. The averaged mean curvature of the domain interfaces decreases in order to adapt to the planar symmetry of the surface. In light of the modulating confinement effects of film thickness, cylinder-forming copolymers can be induced to form spontaneous standing morphologies ($C_\perp$) spanning the entire film with judicial choice of film thickness and weak surface fields [16, 17]. While simulations and theory predict that $C_\perp$ could be formed in much thicker films [15], in practice, the highest aspect ratio of a film-spanning, equilibrium $C_\perp$ microphase in a pure copolymer on a neutral planar surface is around two. This can be increased by incorporating a small amount of additional minority component homopolymer, which segregates in the blend to the center of the cylinders. Spanning vertical cylinders with aspect ratios of up to $\sim$10 have been achieved in this way using films of PS-*b*-PMMA blended with PMMA homopolymer [18].

2.3.1 Alignment of the Microphase

As described in Section 2.3, a film-spanning pore structure in films of the order of one domain spacing thick can be achieved using neutral substrate coatings and choice of film thickness. This is often all that is required when making lithographic etching masks or patterning arrays of dots or line arrays. However, when the microphase must be aligned throughout a much thicker film, additional external alignment fields are required. Some applications, such as addressable memory structures, also demand long-range in-plane order and/or alignment. A great deal of research effort has been dedicated to developing generalized approaches to microphase alignment, and a number of comprehensive reviews exist [7, 8].

Alignment and optimization of long-range *in-plane* order comprises a fascinating field of research in itself. Topographically (graphioepitaxy) [19, 20] and chemically (heteroepitaxy) [21, 22] prepatterned substrates exploit surface fields and in-plane confinement to guide and order standing cylinders, spheres, or lying nanowire structures. Similar confinement-alignment effects can be achieved by patterning the copolymer film itself using, for example, nanoimprint lithography (NIL) [17]. Long-range order and alignment can be encouraged using zone casting [23] or temperature gradients [24], which direct microdomains via a sweeping propagation of an ordering front.

Russell and coworkers reported an interesting method of inducing standing alignment in strongly immiscible cylinder-forming copolymer films by exposure to a controlled solvent atmosphere ("solvent annealing") followed by solvent extraction [25, 26]. It was proposed that an ordering front propagates down through the layer on removal of the solvent, as opposed to the horizontal sweeping gradient present during zone-casting. After solvent annealing, the vertical alignment is coupled to extremely high degrees of in-plane hexagonal ordering.

Shear flow fields are a remarkably effective means of orienting copolymer microdomains. Large amplitude oscillatory shear (LAOS) and extrusion are common methods used to align bulk copolymer melts and solutions [27]. Thomas and coworkers developed the use of roll casting to expose relatively thick films to a shear flow alignment field [28, 29]. Recently, a step-like deformation of precast films confined under a rubber stamp was shown to align a single-layer film of molten cylinders [30]. Shear flow fields are, however, not well suited to producing film-spanning vertical alignment in films many times thicker than the domain spacing.

Extremely precise local control of domain alignment can be achieved in a wide range of sample geometries using electric fields. While electric fields apply a weaker alignment force than shear flow, they make up for this in the ability to apply spatially specific fields either in-plane or out-of-plane by control of electrode geometry.

2.3.1.1 Electric field alignment

Directing the orientation of an anisotropic dielectric object in an electric field relies on the coupling of the electric field to induced polarization charges, whose distribution depends on the orientation of the object. When an external electric field is applied to a dielectric material by means of a constant potential difference across external electrodes, the electrostatic contribution to the free energy of the system (F) is given by

$$F = F_0 - \frac{1}{2}\int \varepsilon(\mathbf{r})|\mathbf{E}(\mathbf{r})|^2 d^3\mathbf{r}, \tag{2.7}$$

where F_0 is the free energy in the absence of the field, $\varepsilon(\mathbf{r})$ is the local dielectric constant, and $\mathbf{E}(\mathbf{r})$ the electric field [31]. We assume a linear dielectric response.

Thurn-Albrecht et al. gave an informative evaluation of Eq. 2.7 for the simplified picture of an infinitely long, isolated cylinder enclosed in an infinite matrix, with dielectric constants ε_1 and ε_2, respectively [32]. Solving Maxwell's equation ($\nabla \cdot [\varepsilon(\mathbf{r})\mathbf{E}(\mathbf{r})] = \rho_{\text{free}}$) to find the field distribution and substituting into Eq. 2.7, the difference in free energy per unit volume between states with the long axis parallel ($F_{\parallel}$) or perpendicular ($F_{\perp}$) to the field vector is

$$\Delta F = F_{\parallel} - F_{\perp} = -\frac{1}{2}|\mathbf{E}_0|^2 \varepsilon_0 \frac{(\varepsilon_1 - \varepsilon_2)^2}{\varepsilon_1 + \varepsilon_2}. \tag{2.8}$$

Since ΔF is always negative, the orientation of the cylinder axis parallel to the applied field is always the lower free energy state, whether $\varepsilon_1 > \varepsilon_2$ or $\varepsilon_1 < \varepsilon_2$. In this orientation, all spatial gradients of dielectric constant (sharp interfaces in this picture) are perpendicular to the field vector; that is, *dielectric interfaces tend to align parallel to the electric field*. The material heterogeneity of a microphase-separated copolymer has an associated spatial variation in the dielectric constant. As a consequence, anisotropic copolymer phases such as L and C will favor an orientation with their long axes parallel to the electric field vector. Admundson et al. calculated the free-energy contribution for a more realistic description of arrays of block copolymer microdomains by expanding the dielectric constant in powers of the local composition amplitude (the volume fraction of one component minus its mean value) [33, 34]. The essential physics of electric field alignment for domains with sharp interfaces is, however, captured by Eq. 2.8. Most important, the driving force is proportional to both the square of the dielectric contrast ($\Delta\varepsilon$), and the square of the electric field strength. In practice, Thurn-Albrecht found that a cylinder-forming PS-*b*-PMMA copolymer film (with dielectric constants at elevated temperatures of $\varepsilon_{PMMA} = 6$ and $\varepsilon_{PS} = 2.45$) could be completely aligned on a gold substrate in a 40 V/μm field.

There are numerous theoretical [33–35] and experimental [36–39] studies of the electric field alignment of block copolymers using both in-plane and vertical fields. Wang et al. found that ion complexation to one block can cause a significant increase in dielectric contrast and a corresponding increase in the electric field alignment response [40, 41]. Tsori et al. have also proposed that mobile ions contained in the copolymer contribute a further free energy reduction for domain alignment [42]. Electric fields can also induce complete phase morphology changes such as the spheres-to-cylinder transition observed by Xu et al. [43].

In a thin film, a vertical electric field such as that applied in a parallel-plate capacitor geometry drives a standing alignment of $L_\perp$ or $C_\perp$ microdomains. There is, therefore, a competition between the electric field and the surface fields that favor parallel domain alignment ($L_{||}$ or $C_{||}$). The free energy of Eq. 2.7 is modified in a simple model to include distinct contributions from the surface interactions and the electric field response:

$$F = F_0 + F_{el} + F_{surf} = F_0 - \frac{1}{2}\int_{V_{el}} \varepsilon(\mathbf{r})\,|\mathbf{E}(\mathbf{r})|^2 \cdot dV + \int_A \sigma(||,\perp)dA,$$
(2.9)

where F_{surf} contains a surface integral of $\sigma(||,\perp)$ from interfacial interactions over the substrate area A. V_{el} is the sample volume (extending out from the surface) over which the surface and electric field alignment forces are in competition, driving alignment perpendicular or parallel to the electric field, respectively. In the notation of Eq. 2.8, there is a critical field at which $\Delta F_{el} = -\Delta F_{surf}$ and the two contributions are balanced. Thurn-Albrecht et al. found that V_{el} was independent of the film thickness, indicating a finite boundary layer at the substrate at which F_{surf} is able to dominate F_{el} [32]. Slightly below the threshold field, the $C_{||}$ boundary layer coexists with $C_\perp$ in the rest of the film. Only above the threshold does the electric field dominate everywhere and produce "complete" alignment, i.e., $C_\perp$ microdomains that pass continuously from surface to substrate. Detailed theoretical studies have described the orientation of lamellar microdomains in symmetric BCP thin films taking into account the competition between electric (E) and surface fields [35, 44]. In the strong segregation regime, there are two critical fields E_1 and E_2 when there are strong, preferential interfacial interactions. For $E < E_1$, surface fields dominate and an $L_{||}$ morphology is favored, while for $E > E_2$, the electric field dominates and an $L_\perp$ is favored. For intermediate fields, $E_1 < E < E_2$, a mixed orientation of microdomains is predicted with $L_{||}$ near the interfaces and $L_\perp$ in the center of the film.

The electric field strength, dielectric contrast, interfacial energies, and film thickness, all play a role in achieving the desired alignment. Complete vertical alignment of domains can be achieved in two ways: (i) strengthening the electric field driving force by increasing the field strength and the dielectric contrast, and/or (ii) reducing the surface fields by making the substrate "neutral" to both blocks. Xu et al. found that in a lamellae-forming PS-b-PMMA copolymer, complete alignment in a 40 V/μm field could only be achieved once surface interactions were neutralized with a random copolymer surface-grafted brush [38].

The potential effectiveness of electric field BCP alignment of the C phase is demonstrated in Fig. 2.7 comparing two cylinder-forming poly(4-fluorostyrene-*b*-poly(lactide) (PFS-*b*-PLA) films thermally annealed under a 120 V μm^{-1} electric field. The film is heated above the glass transition (T_g) temperatures of both blocks while the field is applied, before cooling below T_g to freeze-in the microstructure with field still in place (experimental methods are further discussed in Section 2.4.4.2). In this case, the minority PLA domains were selectively etched away after alignment (see Section 2.3.1) in order to provide imaging contrast for scanning electron microscopy (SEM). A well-ordered array of hexagonally packed pores is visible at the surface, suggesting that the PLA cylinders are vertically aligned. The cross-sectional view confirms that the cylinders span the full film thickness. In contrast, a film thermally annealed without an electric field adopts a $C_{\parallel}$-lying cylinder morphlogy throughout the film as a result of the energetic preference of the PFS phase to contact the substrate (Fig. 2.7b,d).

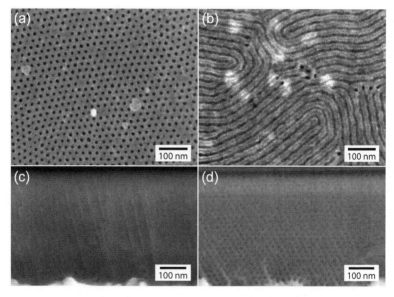

Figure 2.7 Electric field alignment of cylinder-forming PFS-*b*-PLA copolymer films. (a,c) Surface and cross-sectional SEM images of a cylinder-forming PFS-*b*-PLA film thermally annealed under a 120 V μm^{-1} electric field followed by selective removal of the minority PLA phase. (b,d) Equivalent images of a film annealed with no electric field.

2.4 Porous Block Copolymer Film Templates

The key to electrochemical replication of block copolymer film morphologies lies in producing a porous template coating on a suitable conducting substrate. If an electrolyte, containing precursor ionic species, is able to reach this underlying working electrode, electrochemical synthesis of new material can initiate here and grow upward through the pore structure. In essence, this requires that two processing stages can be successfully completed:

(i) Alignment of the copolymer microphase such that the minority phase forms continuous pathways from the surface to the substrate
(ii) Complete selective removal of the minority component without compromising the majority phase

Hereafter, electrochemical growth of new material in the pores of the film is reliant on optimizing many of the same parameters demanded by other nanoporous electrode structures, such as anodized alumina and track-etched membranes [45].

Stage I must be addressed using one or more of the alignment methods discussed in Section 2.3.1. The most general approach to achieving stage II is to work with selectively sacrificial block copolymers [46, 47], in which one component is susceptible to selective etching (potentially a wet chemical, photochemical, or gas phase reaction). The first nanoporous polymer materials made in this way were reported by Nakahama in 1988 using a poly(styrene-*b*-isoprene) combination [48]. There are now at least four copolymer systems amenable to selective etching of a single component to produce mesoporous materials: hydrolysis of polyesters [49], hydrofluoric acid cleavage of akylsilxoanes [50], ozonolysis of polydienes [51, 52], and UV-etching of poly(methylmethacrylate) [16]. A selective etching functionality can also take the form of a single cleavable linkage in the copolymer, such that the one phase is freed from the other and can be subsequently removed by selective solvents [53].

A summary of the reported, selectively sacrificial diblock copolymers and relevant etching methods used to produce porous materials is shown in Table 2.1. Rather few of these systems have yet been explored as electrochemical film templates, largely owing to difficulties in microphase alignment. A substrate-wetting layer as thin as a single copolymer domain (≤10 nm) can be sufficient to effectively block electronic contact with the working electrode and prevent replication [54].

Table 2.1 Reported selectively sacrifical diblock copolymer materials and etching methods

Sacrificial component	Etching method	Matrix material	Geometry[a]	Morphology[b]	Ref.
PMMA	UV (254 nm) irradiation and acetic acid rinse	PS	F	C*	[55]
		PS	F	L	[38]
		P2VP	F	C*	[57]
PLA	0.05 M NaOH$_{(aq)}$ at RT	PS	B	C	[49]
		PS	B	G	[58]
		PS	F	C*	[59, 60]
		PFS	F	G*	[61]
		PFS	F	C*	[54]
		PCHE	B	C*	[62]
PEO	Conc HI$_{(aq)}$	PS	B	G	[58, 63]
PI	Ozonolysis	PS	B	G	[64, 65]
	UV irradiation	PS	B	G	[66]
	1,4-diiodobutane + ozonolysis	P2VP	B	G	[67, 68]
	UV irradiation in 2% ozone	PMDSS	B	G	[69]
PMDS	Hydrofluoric acid	PS	B	G	[50]

[a]F = Film, B = Bulk samples.
[b]C = Cylindrical, G = Gyroid, L = Lamellar.
*Copolymer templates replicated electrochemically or used as an electrochemical etching mask.
PMMA = poly(methylmethacrylate), PLA = poly(lactide),
PS = poly(styrene), PFS = poly(4-fluorostyrene),
PEO = poly(ethyleneoxide), P2VP = poly(2-vinylpyridine),
PI = poly(isoprene), PMDSS = poly(pentamethyldisilylstyrene),
PDMS = poly(dimethylsiloxane), PCHE = poly(cyclohexylethylene).

So far the most widely studied copolymers studied as precursors for porous electrochemical film templates are poly PS-*b*-PMMA and PFS-*b*-PLA.

Poly(styrene)-*b*-(methylmethacrylate) (PS-*b*-PMMA): The first system to be used to form a porous electrochemical template by Thurn-Albrecht et al. in 2000 remains the most widely studied copolymer film template [55]. PMMA is removed by photodegradation under 254 nm UV irradiation (25 J/cm^2 dosage for a 1 μm thick film) and the degradation products rinsed away in a selective solvent such as acetic acid. The majority polystyrene component undergoes a cross linking reaction during UV exposure that can render the template insoluble in organic solvents.

Poly(4-fluorostyrene-*b*-poly(lactide) (PFS-*b*-PLA): Poly(lactic acid) presents a particularly interesting sacrificial copolymer component as it is very simple to degrade under mild chemical conditions—namely soaking the material in a 0.05 M NaOH$_{(aq)}$ solution [49]. In PFS-*b*-PLA, the fluorinated styrene majority block is intended to improve the dielectric contrast between the two blocks for ease of electric field alignment in comparison to the more widely available poly(styrene-*b*-poly(lactide) (PS-*b*-PLA) [56].

2.4.1 Nondegradative Routes to Porous Templates

In films thin enough to comprise only a monolayer of BCP domains, porous structures can be achieved without necessarily chemically degrading one component. One possibility is to mix a low molecular weight additive with the copolymer, chosen to segregate selectively into one phase. After microphase separation (also typically modified by the presence of a further component) the molecule can be released from the structure by soaking in a selective solvent for the host copolymer block. The copolymer microphase is not destroyed providing that the majority component remains insoluble. Two well-studied systems of this type are poly(vinylpyridine) (PVP) containing copolymers hosting either 2-(4-hydroxybenzeneazo) benzoic acid (HABA) or 3-pentadecylphenol (PDP) molecules [70, 71]. To date, this method has only been applied in very thin layers (~40 nm) where selected solvent interactions can switch orientation from lying to standing cylinders. Sidorenko et al. used these films as electrochemical templates for 40 nm long standing nickel nanorod arrays some 8 nm in diameter (Fig. 2.8).

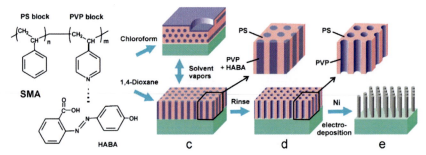

Figure 2.8 Porous electrochemical templates from copolymer (PS-*b*-PVP): small molecule (HABA) assembly. Reproduced with permission from Ref. [71].

The same selective dissolution method can in fact be used even without an additive. When the minority component is selectively solvated then dried, the chains are left coating the matrix phase of the majority matrix. In an ultrathin film, the chains also migrate to cover the free surface on drying, leaving an empty pore in what was the center of the minority domain. Porous arrays up to around 30 nm thick based on PS-*b*-PMMA [72] (trifluorotoluene as selective PMMA solvent), PS-*b*-PVP [73] and PMMA-*b*-PVP [57] (ethanol as PVP solvent) have been created in this way.

2.4.2 Accessible Pore Sizes

The range of feature sizes accessible using porous block copoylmer templates is largely determined by the degree of polymerization (N) of the copolymer, scaling as $N^{2/3}$ in the strong segregation regime (see Section 2.2). A lower limit is set by the thermodynamics of microphase separation requiring that the degree of segregation (χN) is around 10.5 or greater (smaller than this and the copolymer will form a single homogeneous phase). On the other hand, good vertical ordering of the microphase becomes progressively more difficult with increased N. While the driving force for local phase separation is stronger, this requires only short-range rearrangements of the chains. Alignment of the domains, however, requires coordinated motions of many chains—a diffusive process that becomes prohibitively slow with increasing N. High-molecular-weight copolymer films are trapped in a nonequilibrium state reflected in poor ordering. In practice, Xu et al. found that standing cylindrical pores between 14 and 50 nm in diameter with center-to-center distances of 24 to 89 nm could be acheived in PS-*b*-PMMA templates by varying the molecular weight between 42 and 295 Kg mol^{-1} (Fig. 2.9).

Figure 2.9 Cylindrical pore size range in electric field aligned PS-*b*-PMMA films. Tapping-mode atomic force microscopy (phase contrast) images of 14 to 50 nm diameter pores with center-to-center distances between 24 and 89 nm in thin PS-*b*-PMMA films with molecular weight ranging from 42 to 295 Kg mol^{-1}. Image size 1 μm × 1 μm. Reproduced with permission from Ref. [74]. Copyright Wiley-VCH Verlag GmbH & Co. KGaA.

2.4.3 Template Stability

Porous polymer templates are often soluble in common organic solvents, making them incompatible with many organic-based electrolytes. In some cases, this problem can be avoided by cross-linking the template, for example by UV-cross-linking of polystyrene. Aqueous electrolytes normally present no such solubility problem but must often contain additives such as alcohols to reduce their surface tension and ensure wetting of the internal pore structure [55]. Water-wettable pores can be achieved by chemical modification, for example using triblock copolymers with a hydrophobic central block that is left lining the pore surface [75].

The temperature stability of the template is dictated largely by the polymer glass transition temperature, which for polystyrene is found around 100°C (again further stability can be achieved using cross-linking). Polymer substrates are less robust than many inorganic nanostructures (and therefore more sensitive to extreme electrochemical deposition conditions). This apparent drawback becomes a significant advantage for applications in which freestanding electrodeposited nanostructures are desirable, for example, the patterning of high-surface-area capacitive or catalytic coatings, batteries, and photovoltaics [76, 77]. Non-cross-linked polymers can simply be rinsed away in an organic solvent. Alternatively, dry etching by exposure to UV-ozone or moderately high temperature (300–400°C in air) ensures that the polymer is degraded without exposing the templated nanostructures to a drying liquid phase. Capillary bridges formed by the receding liquid exert a powerful attractive force between adjacent objects, and often lead to collapse of freestanding nanostructures standing array. Relatively elaborate processing, such as supercritical drying, is required to keep these structures intact.

2.4.4 Nanowire Replication: Cylinder-Forming BCP Templates

2.4.4.1 In-plane nanowires

The extremely high aspect ratio of the cylindrical microphase makes it a natural candidate for templating arrays of nanowires. In-plane wires can be patterned from very thin films (comparable to domain spacing) containing a simple monolayer of lying cylinders. Park et al. demonstrated this approach with a P2VP-*b*-PMMA copolymer film on a Si substrate first coated with a thin, thermally evaporated Ag layer [57]. The resulting template (PMMA UV photodegraded, or alternatively P2VP selectively dissolved) was used as a mask to guide the electrochemical etching (0.7 V vs. Pt counterelectrode in $1\%_{vol}$ $HNO_{3(aq)}$) of the exposed

Ag substrate (Fig. 2.10). Electrodeposition rather than etching could of course be used to produce conducting or semi-conducting wires in a similar manner.

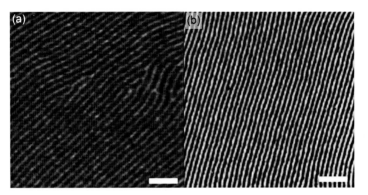

Figure 2.10 Electrochemical fabrication of in-plane nanowires. (a) 27 nm thick porous PS-*b*-PVP template after selective dissolution of PVP in ethanol. (b) Masked electrochemical etching of Ag silver nanowires produced from the template in (a). Scale bar 200 nm. Reproduced with permission from Ref. [57]. Copyright Wiley-VCH Verlag GmbH & Co. KGaA.

2.4.4.2 Standing nanowires arrays

Templating vertical nanowire arrays is generally more challenging than in-plane wires since the cylindrical microphase must first be persuaded to stand vertically on the substrate (see Section 2.3.1) to produce film-spanning porosity on removal of the minority domains. The method of applying an electric field across a substrate-supported film devised by Thurn-Albrecht et al. remains the most reliable: the copolymer film is incorporated into a simple parallel-plate capacitor geometry while heating above the glass transition temperature of both constituent blocks (shown schematically in Fig. 2.11). The copolymer film is coated on a conducting substrate (for example, ITO glass or highly doped silicon wafers with a thermally evaporated gold layer (~50 nm)), which forms one electrode contact. A 25 µm thick polyimide foil (Kapton CR, DuPont) with a 50 nm thermally evaporated Au layer on the reverse side makes a convenient counterelectrode and dielectrically strong insulating layer for the capacitor (to prevent breakdown at high fields). Conformal contact with the polymer layer is acheived using a thin layer (~2 µm or more) of cross-linked poly(dimethylsiloxane) rubber (PDMS, Sylgard 184, Dow Corning) coated on the front side of the Kapton sheet. A constant voltage of up to 4 kV is applied across the assembled layer while heating above the glass transistion temperature of both components, and removed after cooling to room temperature (where the microphase is frozen in place in the glassy polymer matrix). The counterelectrode is peeled away from the copolymer film, allowing subsequent etching of the minority component.

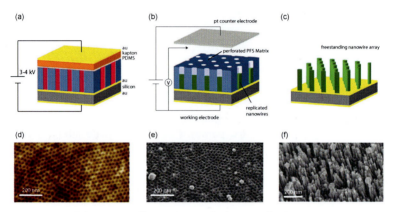

Figure 2.11 Schematic illustration of freestanding nanowire array fabrication from electric-field-aligned, sacrificial cylinder-forming block copolymer films. (a) Annealing of copolymer film above glass transition temperatures in a parallel-plate capacitor geometry with applied electric field. (b) Voids left by removal of the minority phase are electrochemically replicated from the underlying conducting substrate. (c) The supporting template is removed by UV ozone etching to leave a freestanding nanowire array. (d) Tapping mode atomic force microscopy (height image) of a porous PFS template after PLA removal. (e) Surface SEM of the electrochemically filled template. (f) Freestanding nanowires (Cu_2O/Cu mixed phase) after UV ozone etching of PFS matrix. Reproduced from Ref. [54] by permission of the Royal Society of Chemistry.

The effectiveness and high fidelity of electrochemical replication possible using this method is demonstrated in Fig. 2.12. In this case a PFS-b-PLA template was aligned at 180°C for 35 h with an applied field of 130 V μm^{-1} and the PLA component subsequently etched in a 0.05 M NaOH$_{(aq)}$ solution. The template is shown (Fig. 2.12a) after electrostatic deposition of Pt metal at +0.1 V in a standard three-electrode cell (Ag/AgCl reference electrode) and platinum wire counterelectrode. The electrolyte was a 20 mM hexachloroplatinic acid solution containing 20%$_{vol}$ methanol [78].

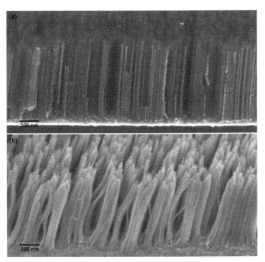

Figure 2.12 Preparation of free-standing platinum nanowires. (a) Cross-sectional view of a platinum-filled template. The copolymer template was annealed at 180°C for 35 h with an applied field of 130 V μm^{-1}. (b) Free-standing platinum nanowires after UV-ozone etching of supporting PFS template. There is partial buckling of individual wires and some degree of aggregation of neighboring wires after removal of the support. Viewing angle is 45°.

Removal of the polymer matrix in Fig. 2.12b was achieved by UV-ozone etching after electrochemical deposition. At such high aspect ratios (diameter 12 nm, length 800 nm) the wires are somewhat unstable even after dry etch processing. A much greater stability problem is encountered when the array is exposed to processing with a liquid phase. Figure 2.13 shows how rinsing away the polymer template in organic solvent and simply allowing the film to dry leads to extensive collapse of the array, most likely caused by capillary forces in the liquid solvent during drying.

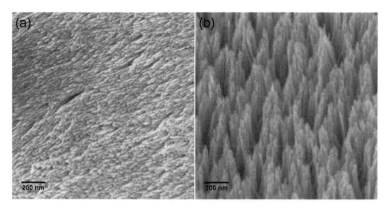

Figure 2.13 Effect of template degradation processing on Pt nanowire stability (SEM micrographs). (a) Removal by rinsing in organic solvent and drying can lead to complete collapse of the nanowire array. (b) Dry etching by UV-ozone or O_2 plasma etching leaves the wires standing. Viewing angle 45°.

2.4.4.3 Polymeric nanowire replication

The great attraction of electrochemical deposition as a means of BCP replication is the broad range of potential filling materials. Russell et al. demonstrated deposition of conducting polypyrrole (PPy) nanowires into the pores of a 100 nm thick PS-*b*-PMMA template on ITO glass using lithium perchlorate in propylene carbonate as a supporting electrolyte [79]. The template was vertically aligned on a polymer brush neutralized substrate by the addition of low molecular weight PMMA homopolymer (Section 2.3.1). Open pores were formed by rinsing away PMMA homopolymer in a selective solvent (acetic acid) such that the polystyrene matrix remained non-cross-linked and could be removed by simple dissolution to produce freestanding polymer nanowires (Fig. 2.14).

Li et al. produced polyaniline (PANI) nanowires by electropolymerization in the pores of a 22 nm thick PS-*b*-P2VP nanoporous film on a gold substrate. Growth nucleated only in the pores created after selective dissolution of the minority P2VP chains in ethanol. Nanowires with diameters of 30–40 nm continued to grow out of the template where they entangled to form a highly porous film.

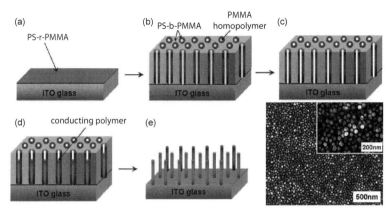

Figure 2.14 Schematic fabrication of conducting PPy polymer nanowires from PS-*b*-PMMA templates on neutralized ITO glass substrates. Reproduced with permission from Ref. [79].

2.4.5 Combination with Top-Down Lithography

Nanometer-scale patterning using block copolymer microphases can easily be integrated with micrometer scale top-down lithographic techniques. The most straightforward method is simple patterning of the conducting working electrode before coating the copolymer film. For example, Tseng et al. patterned a conducting ITO substrate supporting a poly(styrene-*b*-poly(L-lactide)) film and electrodeposited Ni metal nanowires only from these conduction regions [59]. This method is well suited to polymers because of the ease of solution coating the template film over general patterned substrate geometries.

Photodegradable components such as PMMA offer the possibility of being selectively etched away from regions exposed using standard photomasks or directed electron beam lithography. Patterning in this way adds another degree of freedom to that offered by patterning the working electrode itself. Bal et al. demonstrated this process using 50 $\mu C/cm^2$ electron beam exposure of an electric field aligned PS-*b*-PMMA template, followed by "developing" in acetic acid [80]. Cobalt nanowires were electrodeposited in the exposed pores on top of a patterned Au substrate (Fig. 2.15).

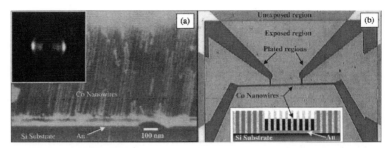

Figure 2.15 (a) Cross-sectional SEM image of electrodeposited Co nanowires in a porous PS template (small angle X-ray scattering pattern inset). (b) Optical micrograph of a four-probe patterned Au substrate after Co elecrodeposition (schematic cross-section inset). Reprinted with permission from Ref. [80]. Copyright 2002, American Institute of Physics.

2.4.6 Bicontinuous Gyroid Copolymer Templates

The technological challenges associated with vertical alignment of cylinder-forming templates can be avoided by targeting the gyroid (G) phase, comprising two interwoven continuous networks of the minority component held in a matrix phase of the majority block [81]. Since *both* minority and majority phases are continuous in all three dimensions, the structure is termed *bicontinuous*. If surface reconstruction (Section 2.3) can be avoided, templates made from the gyroid phase are spontaneously porous upon removal of the minority networks without any further alignment.

An approximation to the copolymer gyroid morphology is based on the "Schoen G" or "Gyroid" surface—a triply periodic, infinitely extending minimal surface (i.e., has zero mean curvature) that divides space into two continuous labyrinthine volumes without intersections. When this surface is given a finite thickness by simultaneously expanding two surfaces of uniform thickness on either side, shown in Fig. 2.16a, space is divided into three nonintersecting volumes. The "central" volume, centered on the original gyroid surface, forms a matrix through which run two distinct intertwining networks, each periodic in all three directions. The resultant morphology constitutes the constant thickness (CT) model for the G phase [82]. The BCP G phase is often termed a *double gyroid*, formed upon segregation of the majority block into a central matrix volume making up roughly 60% of space, with the minority component comprising both of the two network phases. A model for a G phase unit cell with cubic (Ia̅3d) symmetry is shown in Fig. 2.16b with the matrix phase removed and the two networks shaded blue and red for clarity.

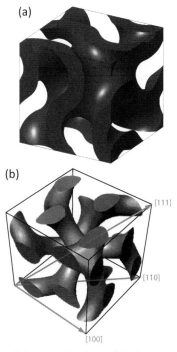

Figure 2.16 Models of the gyroid phase. (a) A constant thickness gyroid model surface. (b) A model for the cubic G phase block copolymer phase unit cell. The matrix phase is removed and the two network phases colored blue and red.

The gyroid phase was first identified in diblock copolymers by two groups in 1994 [82, 83]. Following these first investigations the gyroid phase has been observed in AB [50, 84, 85] and ABA [86–88] block-copolymers, blends of two AB diblocks [58], and AB diblocks blended with homopolymers [52, 67] or small plasticizing molecules [89]. A comprehensive review can be found in Ref. [81].

The fascinating bicontinuous topology of the gyroid phase makes it an irresistible target for replication in functional materials for applications in nanotechnology. While a of number porous gyroid materials obtained from sacrificial block copolymers have been identified in bulk (Table 2.1), successful replication of gyroid film templates has only been rather recently achieved

[61, 69, 90]. The first electrochemical replication, in 2008, was actually achieved using a EO_{17}-PO_{12}-C_{14} surfactant templated silica film [90]. Further exploration and applications of replicated gyroid networks remains a current topic of great interest.

2.4.6.1 Viewing the porous gyroid morphology

The porous gyroid morphology refers to the structure left by removal of the minority polymer network domains (the inverse is often referred to as the relief gyroid morphology). Whether viewed in bulk samples or thin films, the multitude of surface morphologies presented by the porous gyroid, visible under high resolution electron microscopy, can be rather confusing. A better understanding can be obtained by comparison to a simulation of the constant thickness gyroid morphology. In practice a trigonometric level set function is used as an approximation of the gyroid minimal surface [65, 91]:

$$f(x,y,z) = \sin\left(\frac{2\pi x}{L}\right)\cos\left(\frac{2\pi y}{L}\right) + \sin\left(\frac{2\pi y}{L}\right)\cos\left(\frac{2\pi z}{L}\right) + \sin\left(\frac{2\pi z}{L}\right)\cos\left(\frac{2\pi x}{L}\right)$$
$$= 0, \qquad (2.10)$$

where L is the cubic unit cell repeat distance. The surface defined by Eq. 2.10 divides space into two equal volumes. Computer simulation[3] of this surface is used to construct the two related level set surfaces representing the interfaces with the minority phase (so-called "skeletal surfaces"). The cross-sectional representations of this structure, defined by projections along the desired lattice vectors, are computed by the intersection of the surfaces with the desired plane. Figure 2.17 shows simulated sections normal to the (211) lattice vector taken at different heights through the unit cell. This face has the "double-wave" pattern characteristic of the gyroid structure. The series in effect views thin slices through the unit cell at various distances along the (211) lattice vector.

---
[3] With thanks to Dr. Gilman Toombes, based on the work of Jim Hoffman http://www.msri.org/about/sgp/jim/geom/level/skeletal/index.html.

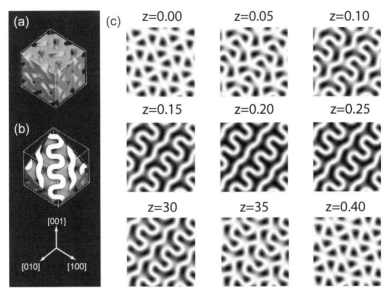

Figure 2.17 Simulated section views of (211) lattice planes. (a) A (2 × 2) standard cubic unit cell of the gyroid structure with the minority phases removed. (b) A section perpendicular to the (211) lattice vector shows the characteristic "double wave" pattern. The $(1\bar{1}1)$ direction runs parallel to the waves. (c) Simulated cross sections with the matrix phase bright and voids left by removal of the minority phase black. Each section has a thickness of 5% of the repeat distance along (211) and is taken at a height z (as a fraction of the total lattice vector). Images (a) and (b) are reproduced with permission from Ref. [65].

2.4.6.2 Replication of gyroid network arrays

Figure 2.18a,b shows SEM micrographs in both surface and cross-sectional projections of a porous template made from a gyroid-forming PFS-*b*-PLA copolymer with the PLA network phase removed at RT after thermal annealing at 180°C. The (211) lattice plane is seen clearly at the free surface. Electrochemical synthesis of hydrated Ti(IV) oxide was performed from a 0.2M aqueous $TiCl_3$ at pH 2.7 using a protocol developed by Kavan et al. [92], with the addition of 20%$_{vol}$ methanol to ensure wetting of pores. The filled templates were subsequently heated to 500°C for 2 h. This treatment served two purposes: to crystallize the deposited material to form nanocrystalline TiO$_2$ and to burn away the polymer template. The resulting films were transparent and optically smooth. Such uniform replication reflects the uniform cross-film percolation of the pore networks from surface to substrate.

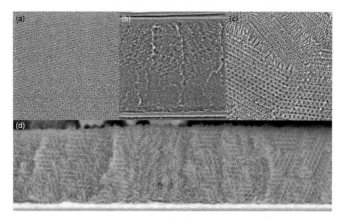

Figure 2.18 Electrochemical replication of Ti(IV) oxide in porous gyroid templates. (a) Surface and (b) cross-sectional SEM images of porous gyroid PFS templates after removal of the minority PLA component. (c) Surface and (d) cross-sectional SEM images of the Ti(IV) oxide inverse gyroid array by uniformly filling the template voids from the conducting substrate by electrochemical deposition and dissolving the surrounding template in organic solvent. (a,b,c) Image size 1 μm, (d) image width 3 μm. Reproduced with permission from Ref. [61].

An indication of the extremely high fidelity of electrochemical gyroid replication is given by comparison to computer-simulated sections of the BCP template before deposition, and of the replicated material after final removal of the template (Fig. 2.19). A gyroid simulation with a unit cell dimension of 47 nm was used to match these SEM images.

Over macroscopic-length scales, the order and orientation of template and replicate was probed using grazing incidence small angle X-ray scattering (GISAXS) for (I) the porous polymer template, (II) the electrochemically filled template, and (III) the heat annealed and contracted array. The patterns for all three stages could be indexed to a cubic I$\bar{a}$3d structure oriented with the (211) lattice vector perpendicular to the substrate. The fitted structures have a uniaxial compression perpendicular to the substrate, as summarized in Table 2.2. Remarkably, the periodic ordering is preserved after TiO_2 crystallization despite an accumulated compression of 52%.

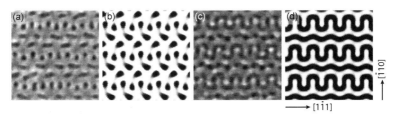

Figure 2.19 Surface SEM images of porous and relief (211) projections. (a) SEM image of a porous polymer template and (b) the simulated (211) porous projection with a unit cell dimension of 47 nm. (c) SEM image of replicated Ti(IV) oxide nanostructure electrochemically deposited into the template followed by dissolution of the polymer and (d) comparison to the (211) simulated relief projection. Reproduced with permission from Ref. [61].

Table 2.2 Gyroid structural compression calculated from GISAXS fitting to a uniaxially compressed cubic $I\bar{a}3d$ structure

	Unit cell (nm)	Incremental compression (%)	Accumulated compression (%)
Bulk PFS-b-PLA	49.0	0	0
Voided PFS template film	50.5	9	9
Hydrated Ti(IV) oxide film	50.0	13	21
Anatase titania array	50.5	39	52

Both the orientation and high-temperature contraction confirmed by scattering data can be observed directly in real space by SEM imaging. Figure 2.20a,d shows low-magnification SEM cross sections of a Ti(IV) oxide-filled template before and after annealing at 500°C with a corresponding compression of ~40%. High magnification views of the array after UV-ozone (low temperature) removal of the PFS template (Fig. 2.20b), or high temperature annealing (Fig. 2.20e), allow comparison with the simulated compression from the cubic lattice. Simulated projection matches to the nonannealed and heat annealed sections show a 14% and 48% vertical compression, respectively, both aligned with the (211) lattice direction.

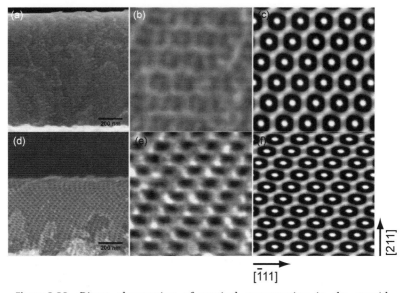

Figure 2.20 Direct observation of vertical compression in the gyroid structure after 500°C crystallization of Ti(IV) oxide into anatase TiO$_2$. (a, d) Cross-sectional views of the gyroid template electrochemically filled with Ti(IV) oxide before and after annealing at 500°C for 2 h. The film thickness compression in (d) is ~40%. (b) Magnified view of the network after low temperature UV-ozone template removal. (e) Magnified view of the compressed network in (d). (c) Simulated (0$\bar{1}$1) projection with 14% compression in (211) direction. (f) Simulated (0$\bar{1}$1) projection with 50% compression in (211) direction. Image size in (b-f) 200 nm, 50 nm simulation unit cell size. Reproduced with permission from Ref. [61].

The source of compression in the replicated network is a volume contraction of the electrochemically deposited material on heating. This is confirmed by thickness measurements of flat films (Fig. 2.22) before and after high temperature processing. Since the film cannot shrink in the lateral direction (without fracturing from the substrate), all compression is taken up by vertical compression.

Complete filling of the template was acheived in a standard manner by monitoring the electrochemical cell current under potentiostatic operation as shown in Figs. 2.22 and 2.23.

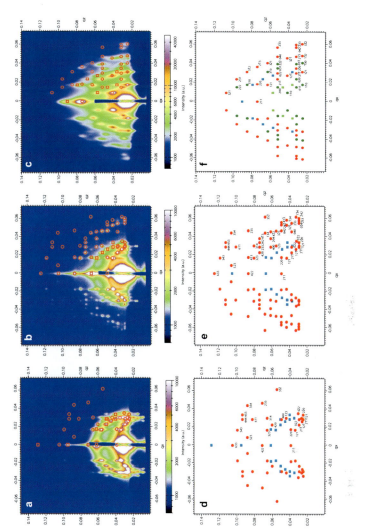

Figure 2.21 Characterization of the gyroid morphology by GISAXS. Experimental GISAXS pattern and indexation[93] of a porous PFS template after removal of the minority component (a, d), nanostructured hydrated Ti(IV) oxide films with the PFS block still in place (b, e) and annealed titania network array (c, f). The patterns were analyzed using the distorted wave Born approximation to account for scattering of both the direct and reflected beam.[94,95] ● and ■ denote the diffraction peaks due to scattering of the direct and reflected X-ray beams, respectively. ● and ■ indicate "forbidden" reflections. Only the diffraction peaks due to the direct beam scattering are indexed and only observed peaks are marked. Figure reproduced with permission from Ref. [61].

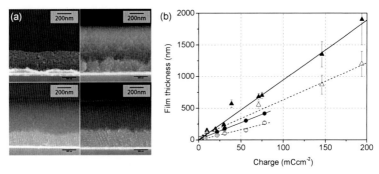

Figure 2.22 Ti(IV) oxide electrodeposition thickness calibration measurements and film contraction on high temperature annealing. (a) Example series of cross-sectional SEM images taken of gyroid film replication with 5, 10, 20, 30 mCcm^{-2} charge passed through the porous template in order to calibrate the deposition thickness. The first image is of a film heated to 500°C, showing additional material left from the empty template region. (b) Plot of deposition thickness (d) versus charge (Q) for nontemplated deposition before heating (●) and after annealing at 500°C (○). The deposition efficiency ($\eta = d/Q$) is 5.2 ± 0.2 nm/mCcm^{-2} with a 37 ± 6% compression on heating. The equivalent plot for gyroid templated arrays (▲, △) gives a deposition efficiency of 10 ± 1 nm/mCcm^{-2} and a 40 ± 12% compression on heating. The straight lines are linear fits to the data.

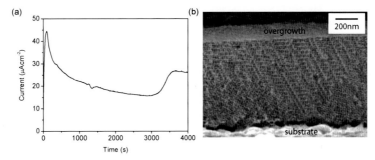

Figure 2.23 Example of template filling by monitoring electrochemical current. (a) Anodic current density in a 780 nm template. The current increases by a factor of ~1.8 on filling of the polymer template (accounting for the steady fall in current with a single exponential decay fit). (b) Cross section of templated array after removal of PFS matrix by UV-ozone etching shows the uniform layer of overgrowth over the template surface.

Electrochemical replication of gyroid templates can be extended to rather thick layers (in comparison to the characteristic unit cell dimension); the network voids retain uniform porosity from the film surface to the underlying substrate for arrays at least several microns thick. The remarkable continuity of the network through thick template layers is seen in Fig. 2.24, showing the cross section of a ~4 μm thick replicated network. Clearly the array grows uniformly through the template, and grains of ~1 μm are visible, with complete connectivity. The implication is that all parts of the pore network are in electrical contact with the underlying substrate throughout replication. By extension, the template matrix phase is also fully continuous, so that any material later infiltrated into the array will have complete access to every part of the network.

Figure 2.24 SEM image of thick gyroid template replication in TiO_2. A ~4 μm thick TiO_2 array demonstrates that the gyroid template remains fully porous in much thicker films, enabling deposition of thick arrays which grow uniformly upward from the underlying substrate. Reproduced with permission from Ref. [61].

The thickness of electrochemically grown Titania gyroid arrays is limited by mechanical stability during high temperature processing. Macroscopic signs of cracking appear for templates around 900 nm and thicker. Kavan et al. reported cracking of nonstructured electrodeposited Ti(IV) oxide layers above

250 nm [92]. Eventually the films crack and buckle from the substrate, as shown in Fig. 2.25, and complete delamination begins above this thickness. Conceptually, stable thick layers would be possible by increasing the density of electrodeposited material, ideally synthesizing a fully crystalline material, such that volume contraction is reduced or completely eliminated.

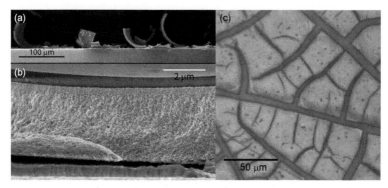

Figure 2.25 Film failure by cracking during high temperature annealing. (a,b) Low and high magnification SEM cross section views of 4 µm thick TiO_2 gyroid arrays after 500°C annealing for 2 h. Lateral stress is relieved by cracking and buckling of ~50–100 µm pieces of the array which delaminate from the substrate. The failure plane is at the interface with the substrate. (c) Cracking and partial delamination of a 1.5 µm thick array after annealing.

2.5 Applications: The Bulk Heterojunction Solar Cell

The ability to pattern ordered nanometer-scale structures into semiconducting composite materials is of great interest for emerging solar energy technologies based on so-called excitonic photovoltaic systems [76]. These multicomponent devices rely on a donor–acceptor interface or *heterojunction* as a means of generating free charges from otherwise energetically bound photoexcited electron-hole pairs (excitons). Unfortunately, excitons in many interesting materials (including many organic molecular systems) diffuse only a short distance (of the order 10 nm) before recombination of the charge pair. As such, only excitons generated within 10 nm of the heterojunction interface have the possibility to contribute to photocurrent in the cell. Since a viable solar cell must also be thick enough to absorb most of the incident sun light, in planar organic films a significant fraction of excitons decay prior to reaching the etrojunction where they are capable of ionising and generating free charges. The idea of a *bulk heterojunction* is to distribute the donor–acceptor interface throughout the active layer, while simultaneously maintaining free charge extraction pathways outwards through both material phases. This concept calls for an intimate vertical interdigitation of two materials on the 10 nm length scale.

A great number of donor–acceptor material combinations, including organic, inorganic, and hybrid composites, are now being explored as the basis for solar cells. Often simple mixing of the two components, and relying on a degree of spontaneous macrophase separation during processing, is enough to achieve a fairly efficient bulk heterojunction architecture. However, optimizing an intimate and distributed interdigitation on the right length scale in a film in this way is extremely difficult and varies enormously from system to system and batch to batch. Hence the attraction of strategies in which an idealized morphology can be templated into a great many donor–acceptor combinations which can then be chosen for their interesting intrinsic material properties such as light absorption, molecular energy levels, and ease of charge transport.

A concerted exploration of the concept of patterning semiconductor composite films using BCP templates (replicated electrochemically or otherwise) should be possible now that suitable copolymers are more widely available, and the required processing is better understood. Evaluation and comparison of the relative effectiveness of each microphase morphology has already started, using for example the dye-sensitized solar cell.

2.5.1 Block Copolymers in the Dye-Sensitized Solar Cell

Dye-sensitized solar cells (DSCs) are a paricularly successful example of a bulk heterojunction cell architecture. A wide bandgap inorganic semiconductor (typically a metal oxide) is sensitized to the solar spectrum by attaching a surface-adsorbed monolayer of an organic or organic-metal complex dye [96, 97]. If the redox potential of the photoexcited dye lies above the conduction band edge of the inorganic semiconductor then an electron may be injected into the layer. The oxidized dye is regenerated by electron transfer from a surrounding donor species. The first, and still most successful version of this concept (though there are now many variations) uses a donor species dissolved in a liquid electrolyte, usually an organic or ionic liquid solvent containing the I^-/I_3^- redox couple [98]. The oxidized donor diffuses away to a counterelectrode where it is subsequently reduced to complete the cell circuit.

The problem of exciton diffusion is avoided since all excited states are generated directly at a charge-separating interface. However, the confinement of the cell's absorbing component to a single interfacial monolayer makes it very difficult to achieve significant light absorption in a flat layer. The bulk heterojunction concept again offers a solution in which a highly structured (or porous) inorganic layer provides the massively increased surface required to load with dye. Using a typical transition metal complex dye, a surface area enhancement of around 1000 over the flat surface equivalent is needed to achieve significant (*ca.* 90%) light harvesting. Electron extraction occurs via the diffusion of electrons through the nanocrystalline inorganic layer [99], again requiring a fully continuous network structure. This well-studied and modeled system provides an ideal testing ground for probing both the device performance of BCP morphologies and the dependence of important optoelectronic characteristics such as charge transport and recombination rates on the semiconductor topology.

Figure 2.26a–f summarizes the templated BCP morphologies accesible using electrochemical replication of TiO_2 detailed in Section 2.4.6.2. In addition to standing nanowires and the gyroid network, a disordered mesoporous layer of thermally sintered, 20 nm TiO_2 nanoparticles is taken as comparison, representing the current state of the art inorganic nanostructure in DSCs. The predicted geometric surface areas (Table 2.3) expressed as the surface enhancement (or *roughness factor (RF)*) per micron in array thickness, are comparable for the three arrays; 100, 125, and 120 for wires, gyroid and nanoparticles, respectively. We expect the primary difference between the morphologies to lie

Table 2.3 Calculated and measured roughness factors for each array morphology and photovoltaic performances in 1.2 μm liquid electrolyte cells under AM1.5 illumination

Morphology	RF_{calc}[a]	RF_{meas}	Porosity (%)	J_{sc} (mAcm^{-2})	V_{oc} (mV)	FF	η(%)
Wires	100	60	65	1.79	845	0.71	1.07
Gyroid	125	130	60	5.83	713	0.71	2.97
Particles	120	116[100]	60	5.42	785	0.63	2.67

[a]Model calculations using "smooth" geometric constructions. Wires: 12 nm diameter, 21 nm centre-to-centre hexagonal standing rods; cubic gyroid: 50 nm unit cell, consisting of cylindrical rods of 12 nm diameter; nanoparticles: 20 nm diameter spheres, 60% porosity with point-like inter-particle contact.

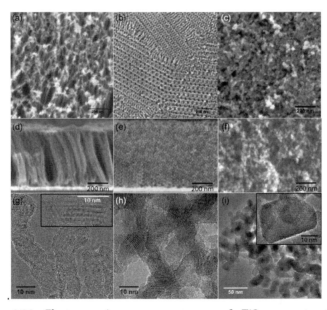

Figure 2.26 Electron microscopy summary of TiO$_2$ nanostructures. (a–f) Top and cross-sectional SEM perspectives of standing nanowire, gyroid network and sintered nanoparticle mesoporous layers. (g–i) Corresponding high resolution TEM images. Reproduced with permission from Ref. [56].

rather in the dimensionality of charge transport out of the array. Standing wires offer direct and, in principle, independent extraction pathways while gyroid and nanoparticle layers privide an ordered or disordered 3D network of routes, respectively. The TiO$_2$

crystal phase (indexed in all cases to anatase) and characteristic grain size in each of the three morphologies is shown by high resolution transmission electron microscopy (Fig. 2.26). In both electrochemically replicated morphologies the grain size is of the same order as the structural confinement (~10 nm). Crystallite dimensions in nanowire templated structures may exceed this confinement length along the wire axis in some places (Fig. 2.26g, inset). The nanoparticle network (Fig. 2.26c) is made up of ~20 nm randomly contacted single-crystal nanoparticles.

The absorption spectra of the dye-sensitized arrays before cell construction is shown in Fig. 2.27a. Both the gyroid and wire spectra show vertical offsets caused by scattering. Comparing the (offset corrected) peak optical density of the arrays at 800 nm, the nanoparticle and gyroid have comparable degrees of dye-loading while the nanowire array has only half the peak absorbance, indicating a considerably lower accessible area in the device for dye uptake. The spectral response, that is, the wavelength-dependent conversion efficiency of incident photons to extracted charge, of cells based on the three arrays are shown in Fig. 2.27b. At these low light intensities (~0.1 mW cm^{-2}), the nanoparticle film shows the highest peak external quantum efficiency (EQE) of 34% compared with 28% for the gyroid array, while the nanowire cell produces a peak EQE of only 9%.

Solar cell current-voltage curves under simulated AM1.5 solar illumination at 100 mW cm^{-2} are shown in Fig. 2.27c and summarized in Table 2.2. The gyroid, nanoparticle and nanowire cells have power conversion efficiencies of 3.0, 2.7 and 1.1%, respectively. The relatively weak adsorption, low EQE and power conversion efficiencies of the nanowire devices are consistent with low dye loading. Indeed, the electrochemically determined surface area of arrays replicated in Pt (Table 2.3) is only 60% of the value expected for a perfect standing array. Low power conversion efficiency could also arise from physical damage to a wire during processing which would, for instance, prevent collection of charge above the level of a fracture. This is not generally the case for network structures in which alternative extraction routes remain. The improved performance of both nanoparticle and gyroid devices over nanowires is most likely due to greater structural stability of the network structure.

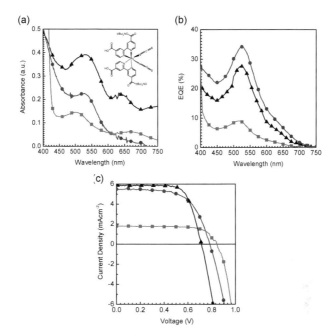

Figure 2.27 DSC solar cell characteristics. 1.2 μm thick nanowire (■), 1.2 μm thick gyroid (▲), and 1.4 μm thick nanoparticle (●) TiO$_2$ arrays. (a) Absorption spectra after sensitization with N719 (inset). (b) Spectral response of liquid electrolyte dye-sensitized solar cells. (c) Current–voltage curves under simulated AM1.5 100 mW cm^{-2} solar illumination. Reproduced with permission from Ref. [56].

The dynamics of photogenerated electrons in the solid semiconductor phase has a strong bearing on the operational efficiency of the solar cell. Ideally, charge should be able to escape the device as quickly as possible, leaving as short a time as possible to undergo recombination events with species in the surrounding electrolyte. Under shortcircuit conditions, measuring the transient current decay time following an incident light pulse gives a measure of the time taken for electrons to diffuse out of the array. Assuming the transport mechanism to be one of multiple trapping and retrapping events, one would expect a one-dimensional diffusion to progress three times faster than a three-dimensional diffusion. However, this argument only applies if the confinement lengthscale is on the same order as the distance between charge trapping sites. Charge collection times measured in wire arrays are approximately twice as fast as in gyroid arrays of the same thickness (Fig. 2.28). That a full threefold improvement was not observed could be a result of interwire electrical contacts, or a density of traps greater than one per nanocrystal, such that a wire cannot be reasonably modeled as an ideal one-dimensional sequence of trap sites.

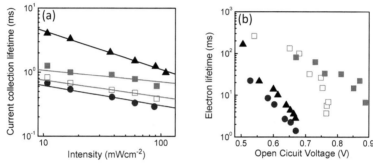

Figure 2.28 Transport and recombination characteristics. (a) Current collection lifetimes for the samples in Fig. 2.27, 1.2 μm thick nanowire (■), 1.2 μm thick gyroid (▲) and 1.4 μm thick nanoparticle (●) TiO$_2$ arrays. A second nanowire-array data set (□) illustrates the variation in transport response from this type of device. (b) Recombination lifetimes measured at open-circuit for the arrays shown in (a) as a function of cell voltage. Reproduced with permission from Ref. [56].

Recombination of photogenerated charge must be minimized in order to achieve efficient power conversion efficiency. The rate at which electrons in the semiconductor are lost to recombination events can be measured by tracking the decay of the open circuit voltage following an incident light pulse. Figure 2.28 shows this measurement for steadily increasing background light intensities (that is, for steadily increasing background charge density in the semiconductor, reflected in the increasing cell voltage). Here it was found that the recombination rates in nanowire devices are up to an order of magnitude slower than in gyroid or nanoparticle arrays. This trend might reflect a tendency of trap sites in the electrochemically templated structures to be found inside the scaffold rather than at the surface, where the charge is phyiscally closer to the electrolyte. Additionally, the reduced surface area resulting from interwire contacts would also reduce the number of such surface trap sites accessible from the electrolyte.

2.6 Concluding Remarks

Electrochemical deposition has been shown to be a rather effective method of transferring model BCP morphologies into functional materials. The first applications of which in real devices already show encouraging optoelectronic properties and performances competitive with well-developed state-of-the-art mesostructures. However, it is likely that the real benefits of this technique, that is, the enormous diversity of accessible functional materials and nanostructured composites, will only emerge with further exploration of previously inaccessible material systems. The current development and availability of BCP templating technology itself is an important step toward enabling this goal.

References

1. I. W. Hamley, *The Physics of Block Copolymers* (Oxford University Press, 1998).
2. F. S. Bates and G. H. Fredrickson, "Block Copolymers—Designer Soft Materials," *Physics Today.* **52(2)**, 32–38 (1999).
3. R. A. L. Jones, *Soft Condensed Matter* (Oxford University Press, 2002).
4. L. Leibler, "Theory of Microphase Separation in Block Copolymers," *Macromolecules.* **13**, 1602–1617 (1980).
5. F. S. Bates and G. H. Fredrickson, "Block Copolymer Thermodynamics: Theory and Experiment," *Annu. Rev. Phys. Chem.* **41**, 525–557 (1990).
6. E. W. Cochran, C. J. Garcia-Cervera, and G. H. Fredrickson, "Stability of the Gyroid Phase in Diblock Copolymers at Strong Segregation," *Macromolecules.* **39**, 2449–2451 (2006).
7. S. B. Darling, "Directing the Self-Assembly of Block Copolymers," *Prog. Polym. Sci.* **32**, 1152–1204 (2007).
8. R. A. Segalman, "Patterning with Block Copolymer Films," *Mater. Sci Eng. R.* **48**, 191–226 (2005).
9. T. P. Russell, G. Coulon, V. Deline, and D. C. Miller, "Characteristics of the Surface-Induced Orientation for Symmetric Diblock PS/PMMA Copolymers," *Macromolecules.* **22**, 4600–4606 (1989).

10. S. H. Anastasiadis, T. P. Russell, S. K. Satija, and C. F. Majkzak, "Neutron Reflectivity Studies of the Surface-Induced Ordering of Block Copolymer Films," *Phys. Rev. Lett.* **62**, 1852–1855 (1989).
11. T. Xu, C. J. Hawker, and T. P. Russell, "Interfacial Interaction Dependence of Microdomain Orientation in Diblock Copolymer Films," *Macromolecules.* **38**, 2802–2805 (2005).
12. P. Mansky, Y. Liu, E. Huang, T. P. Russell, and C. Hawker, "Controlling Polymer-Surface Interactions with Random Copolymer Brushes," *Science.* **275**, 1458–1460 (1997).
13. J. Heier, E. J. Kramer, S. Walheim, and G. Krausch, "Thin Diblock Films on Chemically Heterogeneous Surfaces," *Macromolecules.* **30**, 6610–6614 (1997).
14. P. Mansky, T. P. Russell, C. J. Hawker, M. Pitsikalis, and J. Mays, "Ordered Diblock Copolymer Films on Random Copolymer Brushes," *Macromolecules.* **30**, 6810–6813 (1997).
15. A. Knoll, A. Horvat, K. S. Lyakhova, G. Krausch, G. J. A. Sevink, A. V. Zvelindovsky, and R. Magerle, "Phase Behaviour in Thin Films of Cylinder-Forming Block Copolymers," *Phys. Rev. Lett.* **89**, 035501 (2002).
16. T. Thurn-Albrecht, R. Steiner, J. DeRouchey, C. M. Stafford, E. Huang, M. Bal, M. Tuominen, C. J. Hawker, and T. P. Russell, "Nanoscopic Templates from Oriented Block Copolymer Films," *Adv. Mater.* **12**, 787–791 (2000).
17. H. W. Li and W. T. S. Huck, "Ordered Block-Copolymer Assembly Using Nanoimprint Lithography," *Nano. Lett.* **4**, 1633–1636 (2004).
18. U. Jeong, D. Ryu, D. Kho, J. Kim, J. Goldbach, D. Kim, and T. P. Russell, "Enhancement in the Orientation of the Microdomain in Block Copolymer Thin Films on the Addition of Homopolymer," *Adv. Mater.* **16**, 533–536 (2004).
19. D. Sundrani, S. B. Darling, and S. J. Sibener, "Hierachical Assembly and Compliance of Aligned Nanoscale Polymer Cylinders in Confinement," *Langmuir.* **20**, 5091–5099 (2004).
20. V. P. Chuang, J. Y. Cheng, T. A. Savas, and C. A. Ross, "Three-Dimensional Self-Assembly of Spherical Block Copolymer Domains into V-Shaped Grooves," *Nano Lett.* **6**, 2332–2337 (2006).
21. S. O. Kim, H. H. Solak, M. P. Stoykovich, N. J. Ferrier, J. J. de Pablo, and P. F. Nealey, "Epitaxial Self-Assembly of Block Copolymers on Lithographically Defined Nanopatterned Substrates," *Nature.* **424**, 411–414 (2003).

22. M. P. Stoykovich, M. Müller, S. O. Kim, H. H. Solak, E. W. Edwards, J. J. de Pablo, and P. F. Nealey, "Directed Assembly of Block Copolymer Blends into Nonregular Device-Oriented Structures," *Science.* **308**, 1442–1446 (2005).

23. C. Tang, A. Tracz, M. Kruk, R. Zhang, D. M. Smilgies, K. Matyjaszewski, and T. Kowalewski, "Long-Range Ordered Thin Films of Block Copolymer Prepared by Zone-Casting and Then Thermal Conversion into Ordered Nanostructured Carbon," *JACS.* **127**, 6918–6919 (2005).

24. D. E. Angelescu, J. H. Waller, D. H. Adamson, R. A. Register, and P. M. Chaikin, "Enhanced Order of Block Copolymer Cylinders in Single-Layer Films Using a Sweeping Solidification Front," *Adv. Mater.* **19**, 2687–2690 (2007).

25. Z. Lin, D. H. Kim, X. Wu, L. Booshahda, D. Stone, L. LaRose, and T. P. Russell, "A Rapid Route to Arrays of Nanostructures in Thin Films," *Adv. Mater.* **14**, 1373–1376 (2002).

26. S. H. Kim, M. J. Misner, T. Xu, M. Kimura, and T. P. Russell, "Highly Oriented and Ordered Arrays from Block Copolymers via Solvent Evaporation," *Adv. Mater.* **16**, 226 (2004).

27. I. W. Hamley, "Structure and Flow Behaviour of Block Copolymers," *J. Phys. Condens. Matter.* **13**, R643–R671 (2001).

28. M. A. Villar, D. R. Ania, and E. L. Thomas, "Study of Oriented Block Copolymer Films Obtained by Roll-Casting," *Polymer.* **43**, 5139–5145 (2002).

29. B. J. Dair, A. Avgeropoulos, N. Hadjichristidis, M. Capel, and E. L. Thomas, "Oriented Double Gyroid Films via Roll Casting," *Polymer.* **41**, 6231–6236 (2000).

30. D. E. Angelescu, J. D. Waller, D. H. Adamson, P. Deshpande, S. Y. Chou, R. A. Register, and P. M. Chaikin, "Macroscopic Orientation of Block Copolymer Cyclinders in Single-Layer Films by Shearing," *Adv. Mater.* **16**, 1736–1740 (2004).

31. L. D. Landau, E. M. Lifshitz, and L. P. Pitaevskii, *Landau and Lifshitz Course of Theoretical Physics: Electrodynamics of Continuous Media*, 2 edition (Pergamon Press, 1984).

32. T. Thurn-Albrecht, J. DeRouchey, T. P. Russell, and H. M. Jaeger, "Overcoming Interfacial Interactions with Electric Fields," *Macromolecules.* **33**, 3250–3253 (2002).

33. K. Admundson, E. Helfand, D. D. Dais, X. Quan, S. S. Patel, and S. D. Smith, "The Effect of an Electric Field on Block Copolymer Microstructure," *Macromolecules.* **24**, 6546–6548 (1991).

34. K. Admundson, E. Helfand, X. Quan, and S. D. Smith, "Alignment of Lamellar Block Copolymer Microstructure in An Electric Field, 1. Alignment Kinetics," *Macromolecules.* **26**, 2698–2703 (1993).

35. Y. Tsori and D. Andelman, "Thin Film Diblock Copolymers in Electric Field: Transition from Perpendicular to Parallel Lamellae," *Macromolecules.* **35**, 5161–5170 (2002).

36. T. L. Morkved, M. Lu, A. M. Urbas, E. E. Ehrichs, H. M. Jaeger, P. Mansky, and T. P. Russell, "Local Control of Microdomain Orientation in Diblock Copolymer Thin Films with Electric Fields," *Science.* **273**, 931–933 (1996).

37. T. Thurn-Albrecht, J. DeRouchey, T. P. Russell, and R. Kolb, "Pathways Toward Electric Field Induced Alignment of Block Copolymers," *Macromolecules.* **35**, 8106–8110 (2002).

38. T. Xu, C. J. Hawker, and T. P. Russell, "Interfactial Energy Effects on the Electric Field Alignment of Symmetric Diblock Copolymers," *Macromolecules.* **36**, 6178–6182 (2003).

39. T. Xu, A. V. Zvelindovsky, G. J. A. Sevink, K. S. Lyakhova, H. Jinnai, and T. P. Russell, "Electric Field Alignment of Asymmetric Diblock Copolymer Films," *Macromolecules.* **38**, 10788–10798 (2005).

40. J. Y. Wang, T. Xu, J. M. Leiston-Belanger, S. Gupta, and T. P. Russell, "Ion Complexation: A Route to Enhanced Block Copolymer Alignment with Electric Fields," *Phys. Rev. Lett.* **96**, 128301 (2006).

41. T. Xu, J. T. Godbach, J. Leiston-Belanger, and T. P. Russell, "Effect of Ionic Impurities on the Electric Field Alignment of Diblock Copolymer Thin Films," *Polymer.* **282**, 927–931 (2004).

42. Y. Tsori, F. Tournilhac, D. Andelman, and L. Leibler, "Structural Changes in Block Copolymers: Coupling of Electric Field and Mobile Ions," *Phys. Rev. Lett.* **90**, 145504 (2003).

43. T. Xu, A. V. Zvelindovsky, G. J. A. Sevink, O. Gang, B. Ocko, Y. Zhu, S. P. Gido, and T. P. Russell, "Electric Field Induced Sphere-to-Cylinder Transition in Diblock Copolymer Thin Films," *Macromolecules.* **37**, 6980–6984 (2004).

44. G. C. Pereira and D. R. M. Williams, "Diblock Copolymer Melts in Electric Fields: The Transition from Parallel to Perpendicular Alignment Using a Capacitor Analogy," *Macromolecules.* **32**, 8115–8120 (1999).

45. C. R. Martin, "Nanomaterials: A Membrane Based Approach," *Science.* **266**, 1961–1966 (1994).

46. M. A. Hillmyer, "Nanoporous Materials from Block Copolymer Precursors," *Adv. Polym. Sci.* **190**, 137–181 (2005).
47. D. A. Olson, L. Chen, and M. A. Hillmyer, "Templating Nanoporous Polymers with Ordered Block Copolymers," *Chem. Mater.* **20**, 869–890 (2008).
48. J. Lee, A. Hirao, and S. Nakahama, "Polymerization of Monomers Containing Functional Silyl Groups. 5. Synthesis of New Porous Membranes with Functional Groups," *Macromolecules.* **21**, 274–276 (1988).
49. A. S. Zalusky, R. Olayo-Valles, J. H. Wolf, and M. A. Hillmyer, "Ordered Nanoporous Polymers from Polystyrene-Polylactide Block Copolymers," *J. Am. Chem. Soc.* **124**, 12761–12773 (2002).
50. S. Ndoni, M. E. Vigild, and R. H. Berg, "Nanoporous Materials with Spherical and Gyroid Cavities Created by Quantitative Etching of Polydimethylsiloxane in Polystyrene-Polydimethylsiloxane Block Copolymers," *J. Am. Chem. Soc.* **125**, 13366–13367 (2006).
51. J. S. Lee, A. Hirao, and S. Nakahama, "Polymerization of Monomers Containing Functional Silyl Groups. 5. Synthesis of New Porous Membranes with Functional Groups," *Macromolecules.* **21**, 274–276 (1988).
52. T. Hashimoto, K. Tsutsumi, and Y.Funaki, "Nanoprocessing Based on Bicontinuous Microdomains of Block Copolymers: Nanochannels Coated with Metals," *Langmuir.* **13**, 6869–6872 (1997).
53. M. Zhang, L. Yang, S. Yurt, M. J. Misner, J. T. Chen, E. B. Coughlin, D. Venkataraman, and T. P. Russell, "Highly Ordered Nanoporous Thin Films from Cleavable Polystyrene-Block-Poly(Ethylene Oxide)," *Adv. Mater.* **19**, 1571–1576 (2007).
54. E. J. W. Crossland, S. Ludwigs, M. A. Hillmyer, and U.Steiner, "Freestanding Nanowire Arrays from Soft-Etch Block Copolymer Templates," *Soft Matter.* **3**, 94–98 (2007).
55. T. Thurn-Albrecht, J. Schotter, G. Kästle, N. Emley, T. Shibauchi, L. Krusin-Elbaum, K. Guarini, C. T. Black, M. T. Tuominen, and T. P. Russell, "Ultrahigh-Density Nanowire Arrays Grown in Self-Assembled Diblock Copolymer Templates," *Science.* **290**, 2126–2129 (2000).
56. E. J. W. Crossland, M. Kamperman, M. Nedelcu, C. Ducati, S. Ludwigs, M. A. Hillmyer, U. Steiner, and H. J. Snaith, "Block Copolymer Morphologies in Dye-Sensitized Solar Cells: Probing the Photovoltaic Structure-Function Relation," *Nano Lett.* **9**, 2813–2819 (2009).

57. S. Park, B. Kim, A. Cirpan, and T. Russell, "Preparation of Metallic Line Patterns from Functional Block Copolymers," *Small.* **5**, 1343–1348 (2009).

58. H. Mao and M. A. Hillmyer, "Macroscopic Samples of Polystyrene with Ordered Three-Dimensional Nanochannels," *Soft Matter.* **2**, 57–59 (2006).

59. Y. Tseng, W. Tseng, C. Lin, and R. Ho, "Fabrication of Double-Length-Scale Patterns via Lithography, Block Copolymer Templating, and Electrodeposition," *Adv. Mater.* **19**, 3584–3588 (2007).

60. R. Olayo-Valles, M. S. Lund, C. Leighton, and M. A. Hillmyer, "Large Area Nanolithographic Templates by Selective Etching of Chemically Stained Block Copolymer Films," *J. Mater. Chem.* **14**, 2729–2731 (2004).

61. E. J. W. Crossland, M. Kamperman, M. Nedelcu, C. Ducati, U. Weisner, D. M. Smilgies, G. E. S. Toombes, M. A. Hillmyer, S. Ludwigs, U. Steiner, and H. J. Snaith, "A Bicontinuous Double Gyroid Dye-Sensitized Solar Cell," *Nano Lett.* **9**, 2807–2812 (2009).

62. J. Wolf and M. Hillmyer, "Ordered Nanoporous Poly(Cyclohexylethylene)," *Langmuir.* **19**, 6553–6560 (2003).

63. H. Mao and M. Hillmyer, "Nanoporous Polystyrene by Chemical Etching of Poly(Ethylene Oxide) from Ordered Block Copolymers," *Macromolecules.* **38**, 4038–4039 (2005).

64. E. S. M. Adachi, A. Okumura and T. Hashimoto, "Incorporation of Metal Nanoparticles into a Double Gyroid Network Texture," *Macromolecules.* **39**, 7352–7357 (2006).

65. T. Hashimoto, Y. Nishikawa, and K. Tsutsumi, "Identification of the 'Voided Double-Gyroid-Channel': A New Morphology in Block Copolymers," *Macromolecules.* **40**, 1066–1072 (2007).

66. A. M. Urbas, M. Maldovan, P. Derege, and E. L. Thomas, "Bicontinuous Cubic Block Copolymer Photonic Crystals," *Adv. Mater.* **14**, 1850–1853 (2002).

67. Y. N. A. Okumura and T. Hashimoto, "Nano-Fabrication of Double Gyroid Network Structure via Ozonolysis of Matrix Phase of Polyisoprene in Poly(2-Vinylpyridine)-Block-Polyisoprene Films," *Polymer.* **47**, 7805–7812 (2006).

68. M. Adachi, A. Okumura, E. Sivaniah, and T. Hashimoto, "Incorporation of Metal Nanoparticles into a Double Gyroid Network Texture," *Macromolecules.* **39**, 7352–7357 (2006).

69. V. Z. Chan, J. Hoffman, V. Y. Lee, H. Iatrou, A. Avgeropoulos, N. Hadjichristidis, R. D. Miller, and E. L. Thomas, "Ordered Bicontinuous

Nanoporous and Nanorelief Ceramic Films from Self Assembling Polymer Precursors," *Science.* **286**, 1716–1719 (1999).

70. R. Maki-Ontto, K. de Moel, W. de Odorico, J. Ruokolainen, M. Stamm, G. ten Brinke, and O. Ikkala, "Hairy Tubes: Mesoporous Materials Containing Hollow Self-Organized Cylinders with Polymer Brushes at the Walls," *Adv. Mater.* **13**, 117–121 (2001).

71. A. Sidorenko, I. Tokarev, S. Minko, and M. Stamm, "Ordered Reactive Nanomembranes/Nanotemplates from Thin Films of Block Copolymer Supramolecular Assembly," *JACS.* **125**, 12211–12216 (2003).

72. T. Xu, J. Stevens, J. Villa, J. Goldbach, K. Guarini, C. Black, C. Hawker, and T. Russell, "Block Copolymer Surface Reconstruction: A Reversible Route to Nanoporous Films," *Adv. Func. Mater.* **13**, 698–699 (2003).

73. X. Li, S. Tian, Y. Ping, D. Kim, and W. Knoll, "One-Step Route to the Fabrication of Highly Porous Polyaniline Nanofiber Films by Using Ps-*b*-PVP Diblcock Copolymers as Templates," *Langmuir.* **21**, 9393–9397 (2005).

74. T. Xu, H. Kim, J. DeRouchey, C. Seney, C. Levesque, P. Martin, C. Stafford, and T. Russell, "The Influence of Molecular Weight on Nanoporous Polymer Films," *Polymer.* **42**, 9091–9095 (2001).

75. J. Rzayev and M. A. Hillmyer, "Nanochannel Array Plastics with Tailored Surface Chemistry," *JACS.* **127**, 13373–13379 (2005).

76. B. A. Gregg, "Excitonic Solar Cells," *J. Phys. Chem. B.* **107**, 4688–4698 (2003).

77. M. Law, L. E. Greene, J. C. Johnson, R. Saykally, and P. Yang, "Nanowire Dye-Sensitized Solar Cells," *Nat. Mater.* **4**, 455–459 (2005).

78. A. M. Feltham and M. Spiro, "Platinized Platinum Electrodes," *Chem. Rev.* **71**, 177–192 (1971).

79. J. Lee, S. Cho, S. Park, J. Kim, J. Kim, J. Yu, Y. Kim, and T. Russell, "Highly Aligned Ultrahigh Density Arrays of Conducting Polymer Nanorods Using Block Copolymer Templates," *Nano Lett.* **8**, 2315–2320.

80. M. Bal, A. Ursache, M. Tuominen, J. Goldbach, and T. Russell, "Nanofabrication of Integrated Magnetoelectronic Devices Using Patterned Self-Assembled Copolymer Templates," *App. Phys. Lett.* **81**, 3479–3481 (2002).

81. A. J. Meuler, M. A. Hillmyer, and F. S. Bates, "Ordered Network Mesostructures in Block Copolymer Materials," *Macromolecules.* **42**, 7221–7250 (2009).

82. D. Hadjuk, P. E. Harper, S. M. Gruner, C. C. Honeker, G. Kim, E. L. Thomas, and L. J. Fetters, "The Gyroid: A New Equilibrium Morphology

in Weakly Segregated Diblock Copolymers," *Macromolecules.* **27**, 4063–4075 (1994).

83. M. F. Schulz, F. S. Bates, K. Almdal, and K. Mortensen, "Epitaxial Relationship for Hexagonal-to-Cubic Phase Transition in a Block Copolymer Mixture," *Phys. Rev. Lett.* **73**, 86–89 (1994).

84. G. Floudas, R. Ulrich, and U. Wiesner, "Microphase Separation in Poly(Isoprene-*b*-Ethylene Oxide) Diblock Copolymer Melts. i. Phase State and Kinetics of the Order-to-Order Transition," *J. Chem. Phys.* **110**, 652–663 (1999).

85. L. Sun, L. Zhu, Q. Ge, R. P. Quirk, C. Xue, S. Z. D. Cheng, B. S. Hsiao, C. A. Avila-Orta, I. Sics, and M. E. Cantino, "Comparison of Crystallization Kinetics in Various Nanoconfined Geometries," *Polymer.* **45**, 2931–2939 (2004).

86. B. J. Dair, C. C. Honeker, D. B. Alward, A. Avgeropoulos, N. Hadjichristidis, L. J. Fetters, M. Capel, and E. L. Thomas, "Mechanical Properties and Deformation Behaviour of the Double Gyroid Phase in Unoriented Thermoplastic Elastomers, *Macromolecules.* **32**, 8145–8152 (1999).

87. H. Jinnai, Y. Nishikawa, R. J. Spontak, S. D. Smith, D. A. Agard, and T. Hashimoto, "Direct Measurement of Interfacial Curvature Distributions in a Bicontinuous Block Copolymer Morphology," *Phys. Rev. Lett.* **84**, 518–521 (2000).

88. A. Nykaenen, M. Nuopponen, A. Laukkanen, S. P. Hirvonen, M. Rytela, O. Turunen, H. Tenhu, R. Mezzenga, O. Ikkala, and J. Ruokolainen, "Phase Behaviour and Temperature-Responsive Molecular Filters Based on Self-Assembly of Polystyrne-Block-Poly(*n*-Isopropylacrylamide)-Block Polystyrene," *Macromolecules.* **40**, 5827–5834 (2007).

89. S. Valkama, T. Ruotsalainen, A. Nykanen, A. Laiho, H. Kosonen, G. T. Brinke, O. Ikkala, and J. Ruokolainen, "Self-Assembled Structures in Diblock Copolymemrs with Hydrogen-Bonded Amphiphilic Plasticizing Compounds," *Macromolecules.* **39**, 9327–9336 (2006).

90. V. N. Urade, T. Wei, M. P. Tate, J. D. Kowalski, and H. W. Hillhouse, "Nanofabrication of Double-Gyroid Thin Films," *Chem. Mater.* **19**, 768–777 (2007).

91. G. E. S. Toombes, A. C. Finnefrock, M. W. Tate, R. Ulrich, U. Wiesner, and S. M. Gruner, "A Re-Evaluation of the Morphology of a Bicontinuous Block Copolymer-Ceramic Material," *Macromolecules.* **40**, 8974–8982 (2007).

92. L. Kavan, B. O'Regan, A. Kay, and M. Grätzel, "Preparation of TiO$_2$ (Anatase) Films on Electrodes by Anodic Oxidative Hydrolysis of TiCl$_3$," *J. Electroanal. Chem.* **346**, 291–307 (1993).
93. D. M. Smilgies and D. R. Blasini, "Indexation Scheme for Oriented Molecular Thin Films Studied with Grazing Angle-Incidence Reciprocal-Space Mapping," *J. Appl. Cryst.* **40**, 716–718 (2007).
94. P. Busch, M. Rauscher, D. M. Smilgies, D. Posselt, and C. M. Papadakis, "Grazing-Incidence Small-Angle X-Ray Scattering From Thin Polymer Films with Lamellar Structures—The Scattering Cross Section in the Distorted-Wave Approximation," *Appl. Crys.* **39**, 433–442 (2006).
95. M. Rauscher, T. Salditt, and H. Spohn, "Small-Angle X-Ray Scattering Under Grazing Incidence: The Cross Section in the Distorted-Wave Born Approximation," *Phys. Rev. B.* **52**, 16855–16863 (1995).
96. H. J. Snaith and L. Schmidt-Mende, "Advances in Liquid-Electrolyte and Solid-State Dye-Sensitized Solar Cells," *Adv. Mater.* **19**, 3187–3200 (2007).
97. M. Grätzel, "Photoelectrochemical Cells," *Nature.* **414**, 338–344 (2001).
98. B. O'Regan and M. Grätzel, "A Low-Cost, High-Efficiency Solar Cell Based on Dye-Sensitized Colloidal TiO$_2$," *Nature.* **353**, 737–739 (1991).
99. A. J. Frank, N. Kopidakis, and J. van de Lagemaat, "Electrons in Nanostructured TiO$_2$ Solar Cells: Transport, Recombination and Photovoltaic Properties," *Coord Chem. Rev.* **248**, 1165–1179 (2004).
100. S. Ito, P. Liska, P. Comte, R. Charvet, P. Pechy, U. Bach, L. Schmidt-Mende, S. M. Zakeeruddin, A. Kay, M. K. Nazeeruddin, and M. Grätzel, "Control of Dark Current in Photoelectrochemical (TiO$_2$/I$^-$-I$_3$) and Dye-Sensitized Solar Cells," *Chem. Commun.* 4351–4353 (2005).

Chapter 3

Synthesis of Organic Electroactive Materials in Ionic Liquids

Michal Wagner,[a] Carita Kvarnström,[b] and Ari Ivaska[a]

[a]*Process Chemistry Centre, c/o Laboratory of Analytical Chemistry, Åbo Akademi University, Biskopsgatan 8, 20500 Åbo/Turku, Finland*
[b]*Laboratory of Materials Chemistry and Chemical Analysis, Department of Chemistry, University of Turku, Vatselankatu 2, 20014 Turku, Finland*

ari.ivaska@abo.fi

3.1 Introduction

Electroactive materials are a wide group of compounds, ranging from conducting polymers and polymer nanocomposites to carbon nanotubes and fullerene-based materials. There are also inorganic materials that belong to this category of compounds, but we have limited our scope to conducting polymers that are materials with intrinsic conduction undergoing reversible redox reactions and to other nanocomposites based on organic material. Recently, extensive research has been focused on applications of organic electroactive materials in areas of advanced electronic devices, catalysis, and chemical sensors. The main limiting factor in development of such devices is connected with difficulties in

---

Electrochemical Nanofabrication: Principles and Applications (Second Edition)
Edited by Di Wei
Copyright © 2016 Pan Stanford Publishing Pte. Ltd.
ISBN 978-981-4613-86-6 (Hardcover), 978-981-4613-87-3 (eBook)
www.panstanford.com

the synthesis of the material, especially for large-scale production. In recent years, new molten salts that are in liquid state at room temperature, the so-called ionic liquids, have played an important role in science and technology. Their application in electrochemistry as electrolytes brought successful and promising preparation methods for the future. Hence, in this chapter, the focus is on the enhanced synthesis of organic electroactive materials in ionic liquids.

3.2 Structure and Properties of Ionic Liquids

In general, ionic liquids (ILs) are salts in molten state at room temperature and usually consist of a bulky organic cation and an inorganic anion. They have been known for almost 100 years since the work of Walden on electrical conductivity study of ethylammonium nitrate [1]. Another step in the development of molten salts was investigations on chloroaluminate $AlCl_4^-$ -based ILs [2, 3]. Applications of these ILs are, however, limited due to their water sensitivity. Since the beginning of the 1990s, air- and water-stable ILs have been synthesized. The recent years have brought intense attention to the physical and chemical properties of organic molten salts that are in liquid phase at room temperature and are stable in the presence of moisture as well [4–9]. The reason that these salts remain liquid at ambient temperature is threefold. Seddon and coworkers mention low symmetry of ions, which prevents packing of the lattice [10]. Other factors can be attributed to charge dissipation [7] and steric hindrance effects [11] in ILs.

ILs can be divided in water-immiscible and -miscible compounds. Hydrophobic ILs typically consist of anions such as PF_6^- or Tf_2N^- and the hydrophilic ILs of Cl^-, Br^-, BF_4^- or $AlCl_4^-$. According to Anthony et al., the water miscibility is affected strongly by the anion [12]. Also the cation has its impact on water affinity, since when the alkyl chain is elongated, miscibility of the ionic liquid is decreased.

From an electrochemical point of view the promising applications of ILs are in batteries [13, 14], capacitors [15–17], solar cells [18–20] and sensors [21–24]. Properties of ILs that are crucial for the above-mentioned applications are high ionic conductivity [5], high viscosity [5], low melting point (below

100°C) [6, 7], sufficient thermal stability [3], and a wide electrochemical window [8, 9]. Another advantage is that the organic structure of the ions gives possibility to almost unlimited combinations of anions and cations leading to various kinds of ILs by manipulating the size of both anionic and cationic parts giving the desired properties [25, 26]. The toxicity of ILs has been discussed by Matzke et al. [27]. They found some toxicity in the studied ILs, and therefore, the commonly accepted claim that ILs are nontoxic is not necessary always valid. However, ILs still have one big advantage over the conventional organic solvents: the almost negligible vapor pressure [28]. Figure 3.1 presents some of the most common ions in ILs used in electrochemistry.

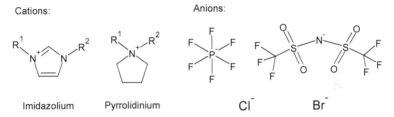

Figure 3.1 Structure of ionic liquids.

The electrochemical window of electrolytes is very crucial in electrochemical applications. ILs are electrochemically rather stable having an electrochemical window typically from 2 to even 6 V. The electrochemical stability of the ILs depends on redox reactions of both the anion and the cation. Anions such as Tf_2N^- oxidize at relatively high anodic potentials compared with, for example, BF_4^- resulting in extended stability of ILs containing Tf_2N^- [6, 29]. Also, tetraalkylammonium cation-based ILs show extended electrochemical window on the negative side of the potential scale [30]. Another important issue connected with the electrochemical properties of ILs is the influence of moisture. In chloride-based ILs, the addition of water produces HCl, which is then reduced further, greatly changing the electrochemical response of IL [31]. Also, the residual water can drastically decrease the electrochemical window. Shröder et al. studied the impact of water on cyclic voltammetry of the electrolytes [BMIM][BF$_4$] and [BMIM][PF$_6$] [9]. For both of those ILs, a decrease in the electrochemical window was observed with increasing concentration of water. Also, under exposure of [BMIM][PF$_6$] to water-saturated argon, the electrochemical window decreased from ca. 4 to 2 V within less than 1 h. Even ppm levels of water can affect the electrochemical response [32]. Hence, great care must be taken in synthesis of ILs as well as in their electrochemical studies. Viscosity of ILs ranges from ca. 30 cP to 600 cP [5]. These relatively high values of viscosities have strong influence on the conductivity of ILs and mass transport in them. For conventional electrolytes conductivity is proportional to the concentration of charge carriers. However, in ILs the apparent cation–anion interaction makes it difficult to exactly define the number of charge carriers. Hence, ILs have much lower ionic conductivities than could be expected when compared with some high-temperature molten salts [33]. High viscosities of ILs also result in relatively low self-diffusion and difficulties in determination of the exact transport numbers.

Table 3.1 Physical properties important for electrochemical applications of common ionic liquids (25°C)

Electrolyte	Abbreviation	Viscosity [cP]	Ionic conductivity [mS·cm^{-1}]	Electro-chemical window [V]
1-Butyl-3-methylimidazolium hexafluorophosphate	[BMIM][PF$_6$]	312 [36]	1.4 [36]	3.2 (Pt) [38]
Butylmethylpyrrolidinium bis(trifluoromethylsulfonyl)imide	[BMP][Tf$_2$N]	85 [39]	2.2 [39]	5.5 (GC) [37]
1-Butyl-3-methylimidazolium tetrafluoroborate	[BMIM][BF$_4$]	52 [6]	3.9 [6]	4.6 (Pt) [29]
1-Ethyl-3-methylimidazolium bis(trifluoromethylsulfonyl)imide	[EMIM][Tf$_2$N]	34 [6]	8.8 [6]	(GC) [40]

From an electrochemical point of view the properties such as extended potential window, relatively high viscosity, ionic conductivity, electrochemical, and thermal stability are important when ILs are used in electrochemistry. Since ILs have quite broad electrochemical windows combined with high stability

[34, 35], it is possible to use them at high potentials and to characterize doping abilities of synthesized films of conducting polymers (especially on the negative side of the potential scale). The relatively low nucleophilicity of ILs is also an advantage. In the synthesis of conducting polymers, the yield of polymerization is in some cases enhanced due to higher viscosities of ILs than of conventional solvents [51]. Table 3.1 summarizes some of the physical properties of ILs used in electrochemistry.

3.3 Electropolymerization of Conducting Polymers in Ionic Liquids

Synthesis of conducting polymers is usually conducted in organic solvents such as acetonitrile (ACN) or tetrahydrofuran (THF) [41–43]. However, these solvents are highly volatile and toxic. First attempts to use ILs for synthesis of conducting polymers were made in the 1980s and the beginning of the 1990s. The commonly known conducting polymers such as poly(paraphenylene) (PPP) [44, 45], polypyrole (PPy) [46], polyaniline (PANI) [47], and polythiophene (PTh) [48] were electrochemically polymerized in chloroaluminate ILs. Because of several factors, such as moisture sensitivity of chloroaluminates producing HCl that enhanced the decomposition of the formed film and difficulties in processability of the film, moisture-stable ILs based on PF_6^-, BF_4^-, or Tf_2N^- anions were introduced. Up to date PTh, PPy, PEDOT, PANI, and PPP have been synthesized in these ILs.

The synthesis of PPy in ILs was extensively studied by the Wallace group [49, 50]. They used [BMIM][PF_6] and [BMP][Tf_2N] in their polymerization studies. In the case of [BMIM][PF_6], the enhancement over the conventional organic solvents was connected with the increased electrochemical stability of the film during potential cycling and more distinct reduction peaks were observed. Figure 3.2 shows cyclic voltammograms of PPy films made in different ILs and propylene carbonate (PC). All the films are cycled in PC/TBAPF_6 electrolyte. It was found that films obtained in both studied ILs are electrochemically more active than the films made in PC. A possible explanation for the increased electrochemical activity may be connected with the higher amount of dopant ions in the films made in ILs than in PC/salt solution. Additionally films obtained in ILs were observed to be smoother and more compact when compared with films made in conventional organic solvents. Fuchigami and coworkers studied the use of [EMIM][OTf] (OTf = $CF_3SO_3^-$) in electrosynthesis of PPy [51]. Polymerization rate as well as the doping level (determined by elemental analysis) was improved when compared with the case when ACN was used as the solvent. The electrochemical polymerization process was also more reproducible in [EMIM][OTf] than in organic solvents as indicated by the smaller standard deviation of the thickness of the film obtained in [EMIM][OTf]. Recently Deepa et al. made morphological studies of PPy films electropolymerized in [EMIM][OTf] confirming the previous observations of decrease in polymer surface roughness [52]. The enhancement in polymerization yield and rate in [EMIM][OTf] may be explained by its high viscosity [53]. Since the diffusion of the formed radicals and the oligomers from the electrode surface is slower in highly viscous media, and therefore, the concentration of the reaction products at the electrode interface is increased, the material deposition is increased as well. A new

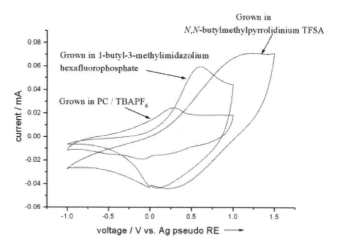

Figure 3.2 Comparison of the electrochemical activity of the polypyrrole films in PC/TBAPF$_6$ solution. Scan rate 100 mVs^{-1}. Reprinted from Ref. [50], Copyright (2004), with permission from Elsevier.

interesting method of PPy synthesis was presented by Han et al. where polyelectrolyte-functionalized IL (PEFIL) has been used [54]. PEFIL (imidazolium based) was functionalized with poly(styrenesulfonate). The reaction was performed in aqueous solution of the monomer. The surface of the working electrode was covered by highly viscous PEFIL, which assisted the polymerization of PPy by donating anions. This method provided the way for electrolyte free electrosynthesis of PPy films with sufficient electroactivity.

Synthesis and doping studies of PEDOT have been conducted by Damlin et al. with [BMIM][PF_6] and [BMIM][BF_4] as the solvents [55]. They noticed continuous polymer growth and the process had similar features as reported for electrosynthesis in organic solvents. The polymerization process was also studied by in situ attenuated total reflectance Fourier transform infrared (ATR-FTIR) spectroscopy. Figure 3.3 presents the changes in the IR spectra with oxidation potential, showing increase in the band intensities with increasing potential. The IR peaks observed are similar to the peaks of PEDOT obtained in organic solvents. However, during p-doping of the polymer at higher potentials, the intensity of the IR induced bands was increased compared with the peaks observed at the same doping potentials in conventional organic solvents. Wagner et al. studied electropolymerization of PEDOT in [BMP][Tf_2N], [EMIM][Tf_2N] and ACN [56]. They found that the cation of the IL used has an effect on the polymerization yield since the rate of the synthesis of PEDOT was faster with $EMIM^+$ as the cation. The authors also made the conclusion that the $Tf2N^-$ anion has an effect on the film morphology in terms of ion incorporation. Randriamahazaka et al. found that the Tf_2N^- anion was trapped in PEDOT film when it was p-doped in IL mixed with Li[Tf_2N] [57]. The system preserved electroneutrality by the simultaneous incorporation of cations in the film. Impedance and EQCM studies conducted by Ispas et al. showed as well that during the potential scanning part of the anions were trapped in the film [58]. PEDOT has also been synthesized in polymeric IL by Kim et al. [59]. Organic dispersion of PEDOT was done with the use of polymeric imidazolium-based IL with Br^- as the anion. The polymeric IL acted as a polymerization template. The resulting polymer complex had both better electroactivity and controlled structure than PEDOT made without polymeric IL.

The possible mechanism of polymerization includes ammonium persulfate reaction with polymeric IL and persulfate oxidation of EDOT leading to nonaqueous dispersions of PEDOT. Döbbelin et al. reported electrosynthesis of PEDOT derivatives that were functionalized with imidazolium IL [60]. Three different anions: BF_4^-, Tf_2N^- and PF_6^- were tested. It was found that by changing the anion the hydrophobicity of PEDOT derivative can be tuned and was verified by contact angle measurements.

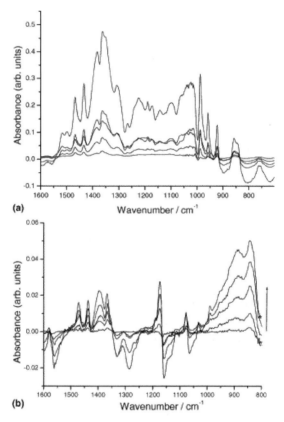

Figure 3.3 Changes in the in situ FTIR–ATR spectra obtained in the wavenumber range 1600–700 cm^{-1} during electropolymerization of 0.1M EDOT in the ionic liquid [BMIM][BF$_4$]. Potential cycling was performed between –0.95 and +1.3 V using a 50 mV/s scan rate. IR spectra were recorded at the end of each scan at about –0.95 V. Reprinted from Ref. [55], Copyright (2004), with permission from Elsevier.

Polymerization of PEDOT derivatives was conducted on ITO in dichloromethane, and it was found that polymerization rate was higher in the BF_4^--based derivative than in the PF_6^-, and Tf_2N^- derivatives.

Pringle et al. used [EMIM][Tf$_2$N] and [BMP][Tf$_2$N] in polymerization of PTh [61]. They studied polymerization of three different monomers: thiophene, bithiophene, and terthiophene. All the three monomers were polymerized both in [EMIM][Tf2N] and in [BMP][Tf$_2$N]. The polymers showed enhanced electrochemical activity. The polymer made in [BMP][Tf$_2$N] had, however, a smoother surface than the polymers electrosynthesized in the imidazolium-based IL. The polymer growth rate was faster in [EMIM][Tf$_2$N] since it is less viscous than [BMP][Tf$_2$N] and was explained by faster reaction kinetics at the electrode. The type of the monomer also influenced the structure of the final PTh. It was observed that with increasing monomer size the relative conjugation length as well as the impact of the IL on the electropolymerization process decreased.

The Fuchigami group electrosynthesized PANI in [EMIM][OTf] [62]. Since the synthesis of this polymer is usually made in acidic aqueous solutions [63], the IL was mixed with CF_3SO_3H, which acts as proton donor in polymerization. Figure 3.4 shows cyclic voltammograms of the growing PANI film in both aqueous and IL solutions. The polymerization rate was found to be much slower in [EMIM][OTf] than in aqueous solution, and therefore, increased concentration of the monomer was used. Also the current response during potential scanning was lower in IL than in the aqueous electrolyte. The films made by potential cycling in [EMIM][OTf] are, however, smoother than the films made in aqueous solution and they also exhibit higher conductivities. Li et al. synthesized the IL 1-ethylimidazolium trifluoroacetate ([EIM][TFA]) with high yield and used it as the electrolyte in polymerization of PANI [64]. Polymerization rate in this IL was higher than in sulfuric acid. Additionally, the films made in [EIM][TFA] had high electrocatalytic activity to oxidation of formic acid. Cycling the PANI film in this IL for 300 scans did not show any significant decrease in the electrocatalytic activity of the film. Morphological studies showed that the films made in [EIM][TFA] were smoother than the films made in sulfuric acid. The authors claim that the enhancement in polymerization is due to the high viscosity of the used ionic liquid. Polymerization of PANI in ionic liquids was also made by Wei et al. [65]. Nanotubular structures of PANI were obtained on modified ITO glass by using [BMIM][PF$_6$] mixed with 1M trifluoroacetic acid. The electrosynthesis resulted in the emerladine form of PANI. Potential cycling to 50 scans indicate continuous growth of PANI film in IL acidic solution and the SEM analysis show tubular structure of PANI with a diameter of ca. 120 nm.

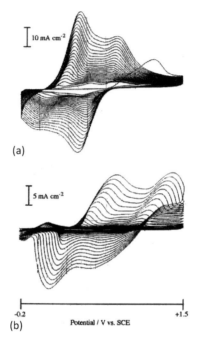

Figure 3.4 Cyclic voltammograms in the course of polymerization of aniline for 25 potential scan cycles. Electrolytic solution: (a) 0.5 M aniline in 1.0 M CF$_3$SO$_3$H + H$_2$O; (b) 1.0 M aniline in 1.0 M CF$_3$SO$_3$H + [EMIM][OTf]. Reprinted from Ref. [62], Copyright (2003), with permission from Elsevier.

Conducting polymers, based on polyphenylenes have intensively been studied due to their interesting electroluminescence property [66, 67]. However, PPP is unstable and also the synthesis is problematic due to the humidity sensitivity of the polymerization process. To overcome these problems, ILs can be applied. Endres and coworkers used moisture-stable ionic liquids for electrosynthesis of benzene [68, 69]. They used 1-hexyl-3-methylimidazolium tris(pentafluoroethyl) trifluorophosphate in electrosynthesis of PPP. The EQCM studies in that work showed incorporation of anions from the IL and a significant cation exchange in order to preserve electroneutrality in the film, resulting in porous PPP structures. Electrochemical polymerization of PPP from biphenyl in ILs was conducted by Wagner et al. [70]. Two common ionic liquids: [BMIM][PF$_6$] and [BMP][Tf$_2$N] have been used as the solvent as well as ACN + TBAPF$_6$ for comparison. Polymerization was greatly improved in [BMIM][PF$_6$]. Results from doping studies showed that n-doping in [BMP][Tf$_2$N] exhibited the best electrochemical response. In addition, the use of ionic liquids enabled spectroscopic studies of PPP films at higher doping levels in contrast to the conventional organic electrolyte.

3.4 Synthesis of Polymer Composites and Carbon-Based Nanomaterials in Ionic Liquids

Functionalized single-walled carbon nanotubes (SWNTs) have attracted attention because of their promising applications in photovoltaic devices [71–73]. Since SWNTs are n-type materials and PANI is both stable and has sufficient electron-donating properties, electrochemical functionalization of carbon nanotubes with PANI has been studied [74–76]. In order to achieve functionalized SWNTs in larger quantities, Wei et al. used [BMIM][PF$_6$] mixed with 1 M trifluoroacetic acid [77]. SWNTs were mixed and grind with [BMIM][PF$_6$] containing the monomer and the acid. The electrochemical polymerization was done by potential scanning with 300 cycles on modified ITO substrate. After 50 cycles, a growth of PANI around the nanotubes was observed. Figure 3.5 presents SEM images of SWNTs and PANI wrapped around them. The possible mechanism of functionalization is also shown in Fig. 3.5. Charged radicals produced during polymerization can make covalent bonding with the surface of the nanotube [78]. This suggestion was confirmed by changes in the carbon D and G band intensities in the Raman spectra. IL provided a good environment for obtaining sufficient quantities of PANI/SWNTs composites, since SWNTs dissolve more easily in

ILs than in aqueous or conventional organic solvent. This can probably be attributed to cation/π-electron interactions [79]. Also, the high dielectric constant of ionic liquids may be an important factor in dispersibility of nanomaterials since it leads to shielding of the van der Waals interactions.

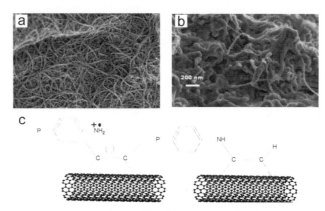

Figure 3.5 SEM images and schematic representation of SWNTs modification: (a) purified SWNTs, (b) SWNTs modified with PANI (after 300 cycles), and (c) possible mechanism of polymerization. *Note*: The images were changed with respect to the original use of pictures in the publication. Reprinted from Ref. [77], Copyright (2007), with permission from Elsevier.

Wei et al. synthesized PANI-gold composite (PANI:Au) nanoparticles in ionic liquid media for a nonvolatile memory device [80]. The PANI:Au particles were synthesized by cyclic voltammetry in [EMIM][Tos] (tosylate) under acidic conditions. The resulting composite could directly be electrodeposited on modified ITO substrate to construct an active layer in the memory device, simplifying, therefore, the processing procedure considerably. The device was found to be stable in air, and it exhibited bistability and negative differential resistance. Reproducibility of these two parameters is of great importance in organic electronics. The obtained PANI:Au single-layer device was tested after 8 weeks and the device still exhibited sufficient current–voltage characteristics.

Pringle et al. used [MEIM][Tf$_2$N] (1-methyl-3-ethylimidazolium) in chemical synthesis of PPy and PTh. Different chemical oxidants (gold chloride, Fe tosylate, silver nitrate, and Fe(ClO$_4$)$_3$) were used as dopants [81]. It was found that gold chloride suits best for fabrication of conducting polymer nanoparticles. The size of PTh and PPy particles were from 100 to 500 nm with conductivity ranging from 1 to 3 mS cm^{-1}. Recently Kim et al. made PPy nanostructures in the so-called magnetic IL [82]. They used [BMIM][FeCl$_4$], which is active under magnetic field and can be considered a paramagnetic compound [83]. PPy particles were made by self-assembled method where [BMIM][FeCl$_4$] was mixed with monomer and a magnetic field was applied resulting in PPy precipitate. Magnetic IL acted threefold: as dopant, catalyst, and solvent. The self-assembled particles were spheres of the size ranging from 50 to 100 nm with high electrical conductivities. Also in a recent paper Pringle et al. utilized [EMIM][Tf$_2$N] for synthesis of PPy-gold and -silver composites with the use of gold chloride and silver nitrate [84]. The use of these oxidants enabled a simple one-step fabrication of nanocomposites and IL assisted incorporation of metallic gold and silver into the polymer structure.

Modification of carbon nanotubes (CNTs) with metal nanoparticles (NPs) to obtain nanohybrids opens new possibilities in their applications for catalysis. The studies on CNTs decorated with NPs were focused on tuning the size and the distribution of NPs [85, 86]. Integrated NPs-CNTs materials have been synthesized by several chemical and physical methods involving chemical vapor deposition and laser ablation [87–89]. It was found, however, that in these methods the size and particle distribution were difficult to control. Recently, Hong and coworkers applied ionic liquids into the synthesis of CNT-nanohybrids [90]. They developed an IL-assisted sonochemical method (ILASM) by which it is possible to fabricate nanohybrids with controlled integration of NPs (e.g., Pt, Au, and Pd). The multiwalled carbon nanotubes (MWNTs) were mixed and grinded with [BMIM][BF$_4$]. The IL is self-assembled by π–π stacking interaction onto the MWNTs and acts as active sites for NPs. In Fig. 3.6, the synthesis process is presented. After mechanical grinding the NP precursors are interacting with BF$_4^-$ anions and after sonication the NPs are deposited onto the surface of MWNTs. The most important finding was the fact that the size and distribution of the used particles can relatively easy be controlled by changing the sonication time and the precursor concentration.

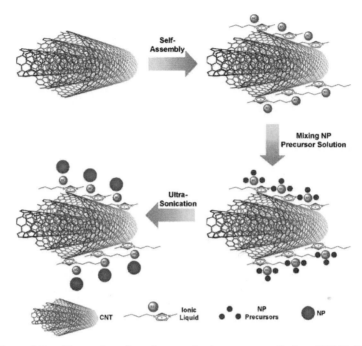

Figure 3.6 Illustration for the synthetic process of the CNT/IL/NP hybrid via ILASM. Park, H. S., Choi, B. G., Yang, S. H., Shin, W. H., Kang, J. K., Jung, D., and Hong, W. H. Ionic-liquid-assisted sonochemical synthesis of carbon-nanotube-based nanohybrids: control in the structures and interfacial characteristics [90]. Copyright Wiley-VCH Verlag GmbH & Co. KGaA. Reproduced with permission.

An interesting approach in generation of graphene sheets from graphite electrode was presented by Liu et al. [91]. The synthesis was conducted in ionic liquid environment. In the past, carbon nanoparticles have been exfoliated from graphite mainly by laser ablation [92] or by electrochemical methods [93]. ILs may be a promising and inexpensive alternative to make graphene sheets. The method developed by Liu et al. [91] is a one-step process and can be used in large-scale production of graphene. They used a mixture containing [BMIM][PF$_6$] and 10 mL of water and a potential of 15 V was applied between the two graphite electrodes (Fig. 3.7). The resulting precipitate consisted of exfoliated graphene sheets of ca. 500 nm in size. This method has the advantage over the chemical method to produce graphene films from graphene oxide derivatives due to the fact that it provide less defects, hence resulting in the increase in charge carrier mobility [94].

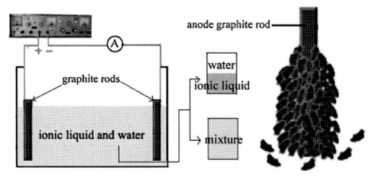

Figure 3.7 Experimental setup diagram (left) and exfoliation of the graphite anode (right). Liu, N., Luo, F., Wu, H., Liu, Y., Zhang, C., and Chen, J. One-step ionic-liquid-assisted electrochemical synthesis of ionic-liquid-functionalized graphene sheets directly from graphite [91]. Copyright Wiley-VCH Verlag GmbH & Co. KGaA. Reproduced with permission.

Lu et al. made carbon nanoparticles and graphene from graphite electrode in ILs [95]. The scope of their research was to make fluorescent nanomaterials that could be applied in bioimaging [96]. In their work, [BMIM][BF$_4$] was mixed with water and used to produce carbon nanoparticles (graphene sheets) in similar manner as in the studies conducted by Liu et al. [91]. The schematic presentation of their exfoliation process is shown in Fig. 3.8. They proposed a strong influence of water on the exfoliation mechanism since the increase in the water content in IL was found to result in a decrease in the activation energy required for graphite separation. As already stated, addition of water (also present as impurity) decreases electrolyte resistance and results in a decrease in the potential window. The hydroxyl and oxygen radicals formed during application of relatively high voltages (ranging, for instance, from 1.5 to 8 V; depending on the water content) facilitate both expansion (intercalation of BF_4^- anion) and corrosion of the graphite electrode. The resulting precipitate is composed of graphene sheets. It was shown that by changing the water/IL ratio several, nanomaterials can be made in different shapes and sizes. By controlling this ratio, it was possible to generate fluorescent nanomaterials to be used in the ultraviolet and visible range.

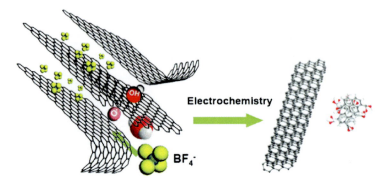

Figure 3.8 Illustration of the exfoliation process showing the attack of the graphite edge planes by hydroxyl and oxygen radicals, which facilitate the intercalation of the BF_4^- anion. Oxidative cleavage of the expanded graphite produces graphene nanoribbons. Reprinted from Ref. [95], Copyright (2009), with permission from Elsevier.

3.5 Conclusions

Use of ILs in the electrochemical synthesis of organic electroactive materials opens up new possibilities to improve both quality and quantity of the produced material. Production of nanomaterials can even be made in one step since ILs in many cases act as both solvent and activator in the synthesis. Ionic conductivity of ILs is sufficient for electrochemistry and the anions are incorporated into the polymer film during polymerization to preserve electroneutrality. Both the yield and the rate of polymerization for some conducting polymers can greatly be improved when performed in ILs. All studied electrochemical synthesis of conducting polymers except of PANI resulted in a higher material deposition in ILs than in conventional electrolytes. The quality of polymer films made in ILs is also enhanced, showing higher electrical conductivity of the polymer. ILs provide extended electrochemical windows. Hence, it is possible to dope the polymers at potential levels unavailable in conventional solvents. Also films made in ILs (e.g., PPP) exhibit enhanced stability. In IL-functionalized conducting polymer, the properties of the film may be changed by varying the type of anion. Polymerization of conducting polymers in viscous polymeric ILs provides a possibility of synthesis without any supporting electrolyte, since polymeric IL is both the polymerization template and the doping agent. Studies in IL-assisted synthesis of conducting polymer/noble particle composites have shown that these methods are promising in catalytic and electronic applications. Since carbon nanotubes and graphene sheets are easily dispersible in ILs, improvement in the quantity of the material production during synthesis can be expected. The size and distribution of metal nanoparticles can be controlled by changing water content in ionic liquid, sonication time, or applied potential. The fast development of ionic liquids in electrochemical applications is connected with their tunable properties that can meet different requirements of materials synthesis. This trend can be predicted to continue also in the future.

References

1. Walden, P. (1914). Ueber die molekulargroesse und elektrische leitfaehigkeit einiger geschmolzenen salze. *Bulletin de l'Academie Imperiale des Sciences de St. Petersbourg*, pp. 405–422.
2. Chum, H. L., Koch, V. R., Miller, L. L., and Osteryoung, R. A. (1975). Electrochemical scrutiny of organometallic iron complexes and hexamethylbenzene in a room temperature molten salt. *J. Am. Chem. Soc.*, **97**, pp. 3264–3265.
3. Wilkes, J. S., Levisky, J. A., Wilson, R. A., and Hussey, C. L. (1982). Dialkylimidazolium chloroaluminate melts: a new class of room-temperature ionic liquids for electrochemistry, spectroscopy and synthesis. *Inorg. Chem.*, **21**(3), pp. 1263–1264.
4. Fredlake, C. P., Crosthwaite, J. M., Hert, D. G., Aki, S. N. V. K., and Brennecke, J. F. (2004). Thermophysical properties of imidazolium-based ionic liquids. *J. Chem. Eng. Data.*, **49**, pp. 954–964.
5. Galinski, M., Lewandowski, A., and Stepniak, I. (2006). Ionic liquids as electrolytes. *Electrochim. Acta*, **51**, pp. 5567–5580.
6. Fuller, J., Carlin, R. T., and Osteryoung, R. A. (1997). The room temperature ionic liquid 1-ethyl-3-methylimidazolium tetrafluoroborate: electrochemical couples and physical properties. *J. Electrochem. Soc.*, **144**, pp. 3881–3886.

7. Bonhote, P., Dias, A.-P., Papageorgiou, N., Kalyanasundaram, K., and Grätzel, M. (1996). Hydrophobic, highly conductive ambient-temperature molten. *Inorg. Chem.*, **35**, pp. 1168–1178.
8. Hultgren, V. M., Mariotti, A. W. A., Bond, A. M., and Wedd, A. G. (2002). Reference potential calibration and voltammetry at macrodisk electrodes of metallocene derivatives in the ionic liquid [BMIM][PF_6]. *Anal. Chem.*, **74**, pp. 3151–3156.
9. Schröder, U., Wadhawan, J. D., Compton, R. G., Marken, F., Suarez, P. A. Z., Consorti, C. S., de Souza, R. F., and Dupont, J. (2000). Water-induced accelerated ion diffusion: voltammetric studies in 1-methyl-3-[2,6-(S)-dimethyllocten-2-yl]imidazolium tetrafluoroborate, 1-butyl-3-methylimidazolium tetrafluoroborate and hexafluorophosphate ionic liquids. *New J. Chem.*, **24**, pp. 1009–1015.
10. Deetlefs, M., Seddon, K. R., and Shara, M. (2006). Predicting physical properties of ionic liquids, *PCCP*, **8**, pp. 642–649.
11. Larsen, A. S., Tham, F. S., and Reed, C. A. (2000). Designing ionic liquids: imidazolium melts with inert carborane anions. *J. Am. Chem. Soc.*, **122**, 7264–7272.
12. Anthony, J. L., Maginn, E. J., and Brennecke, J. F. (2001). Solution thermodynamics of imidazolium-based ionic liquids and water. *J. Phys Chem. B*, **105**, 10942–10949.
13. Fung, Y. S., and Zhu, D. R. (2002). Electrodeposited tin coating as negative electrode material for lithium-ion battery in room temperature molten salt. *J. Electrochem. Soc.*, **149**, pp. A319–A324.
14. Shobukawa, H., Tokuda, H., Susan, M. A. B. H., and Watanabe, M. (2005). Ion transport properties of lithium ionic liquids and their ion gels. *Electrochim. Acta*, **50**, pp. 3872–3877.
15. Stenger-Smith, J. D., Webber, C. K., Anderson, N., Chafin, A. P., Zong, K., and Reynods, J. R. (2002). Poly(3,4-alkylenedioxythiophene)-based supercapacitors using ionic liquids as supporting electrolytes. *J. Electrochem. Soc.*, **149**, pp. A973–A977.
16. Sato, T., Masuda, G., and Takagi, K. (2004). Electrochemical properties of novel ionic liquids for electric double layer capacitor applications. *Electrochim. Acta*, **49**, pp. 3603–3611.
17. Liu, H., He, P., Li, Z., Liu, Y., and Li, J. (2006). A novel nickel-based mixed rare-earth oxide/activated carbon supercapacitor using room temperature ionic liquid electrolyte. *Electrochim. Acta*, **51**, pp. 1925–1931.
18. Kubo, W., Murakoshi, K., Kitamura, T., Yoshida, S., Haruki, M., Hanabusa, K., Shirai, H., Wada, Y., and Yanagida, S. (2001). Quasi-solid-state

dye-sensitized TiO_2 solar cells: Effective charge transport in mesoporous space filled with gel electrolytes containing iodide and iodine. *J. Phys. Chem. B*, **105**, pp. 12809–12815.

19. Xia, J., Masaki, N., Jiang, K., and Yanagida, S. (2006). Deposition of a thin film of TiO_x from a titanium metal target as novel blocking layers at conducting glass/TiO_2 interfaces in ionic liquid mesoscopic TiO_2 dye-sensitized solar cells. *J. Phys. Chem. B*, **110**, pp. 25222–25228.

20. Wang, P., Zakeeruddin, S. M., Moser, J. E., Humphry-Baker, R., and Gratzel, M. (2004). A solvent-free, $SeCN^-$/$(SeCN)_3^-$-based ionic liquid electrolyte for high-efficiency dye-sensitized nanocrystalline solar cells. *J. Am. Chem. Soc.*, **126**, pp. 7164–7165.

21. Coll, C., Labrador, R. H., Manez, R. M., Soto, J., Sancenon, F., Segui, M. J., and Sanchez, E. (2005). Ionic liquids promote selective responses towards the highly hydrophilic anion sulfate in PVC membrane ion-selective electrodes. *Chem. Commun.*, pp. 3033–3035.

22. Shvedene, N. V., Chernyshov, D. V., Khrenova, M. G., Formanovsky, A. A., Baulin, V. E., and Pletnev, I. V. (2006). Ionic liquids plasticize and bring ion-sensing ability to polymer membranes of selective electrodes. *Electroanal.*, **18**, pp. 1416–1421.

23. Wang, R., Okajima, T., Kitamura, F., and Ohsaka, T. (2004). A novel amperometric O_2 gas sensor based on supported room-temperature ionic liquid porous polyethylene membrane-coated electrodes. *Electroanal.*, **16**, pp. 66–72.

24. Ding, S., Xu, M., Zhao, G., and Wei, X. (2007). Direct electrochemical response of Myoglobin using a room temperature ionic liquid, 1-(2-hydroxyethyl)-3-methylimidazolium tetrafluoroborate, as supporting electrolyte. *Electrochem. Commun.*, **9**, pp. 216–220.

25. Carmichael, A. J., Hardacre, C., Holbrey, J. D., Seddon, K. R., and Niewenhuyzen, M. (2000). In Molten Salts X11: Proceedings of the International Symposium. In *Proc. Electrochem. Soc.* (Eds., Trulove, P. C., De Long, H. C., Stafford, G. R., and Deki, S.), **99–41**, pp. 209–221.

26. Hagiwara, R., and Ito, Y. (2000). Room temperature ionic liquids of alkylimidazolium cations and fluoroanions. *J. Fluorine Chem.*, **105**, pp. 221–227.

27. Matzke, M., Stolte, S., Thiele, K., Juffernholz, T., Arning, J., Ranke, J., Welz-Biermann, U., and Jastorff, B. (2007). The influence of anion species on the toxicity of 1-alkyl-3-methylimidazolium ionic liquids observed in an (eco)toxicological test battery. *Green Chem.*, **9**, pp. 1198–1207.

28. Earle, M. J., Esperanca, J. M. S. S., Gilea, A., Canongia Lopez, J. N., Rebelo, L. P. N., Magee, J. W., Seddon, K. R., and Widegren, J. A. (2006). The distillation and volatility of ionic liquids. *Nature*, **439**, pp. 831–834.

29. Lewandowski, A., and Stepniak, I. (2003). Relative molar Gibbs energies of cation transfer from a molecular liquid to ionic liquids at 298.15 K. *PCCP*, **5**, pp. 4215–4218.

30. Matsumoto, H., Yanagida, M., Tanimoto, K., Nomura, M., Kitagawa, Y., and Miyazaki, Y. (2000). Highly conductive room temperature molten salts based on small trimethylalkylammonium cations and bis(trifluoromethylsulfonyl)imide. *Chem. Lett.*, **29**, pp. 922–923.

31. Sahami, S., and Osteryoung, R.A. (1983). Voltammetric determination of water in an aluminum chloride-*N*-*n*-butyl pyridinium chloride ionic liquid. *Anal. Chem.*, **55**, pp. 1970–1973.

32. Matsumoto, H. (2005). Electrochemical windows of room-temperature ionic liquids. In *Electrochemical Aspect of Ionic Liquids* (ed. Ohno, H.), John Wiley and Sons, Inc., New Jersey, pp. 35–54.

33. Fujiwara, S., Kato, F., Watanabe, S., Inaba, M., and Tasak, A. (2009). New iodide-based molten salt systems for high temperature molten salt batteries. *J. Pow. Source.*, **194**, pp. 1180–1183.

34. Buzzeo, M. C., Evans, R. G., and Compton, R. G. (2004). Non-haloaluminate room-temperature ionic liquids in electrochemistry—a review. *Chem. Phys. Chem.*, **5**, pp. 1106–1120.

35. Hapiot, P., and Lagrost, C. (2008). Electrochemical reactivity in room-temperature ionic liquids. *Chem. Rev.*, **108**, pp. 2238–2264.

36. Carda-Broch, S., Berthod, A., and Armstrong, A. W. (2003). Solvent properties of the 1-butyl-3-methylimidazolium hexafluorophosphate ionic liquid. *Anal. Bioanal. Chem.*, **375**, pp. 191–199.

37. MacFerlane, D. R., Meakin, P., Sun, J., Amini, N., and Forsyth, M. (1999). Pyrrolidinium Imides: A new family of molten salts and conductive plastic crystal phases. *J. Phys. Chem. B*, **103**, pp. 4164–4170.

38. Barisci, J. N., Walace, G. G., MacFerlane, D. R., and Baughman, R. H. (2004). Investigation of ionic liquids as electrolytes for carbon nanotube electrodes. *Electrochem. Commun.*, **6**, pp. 22–27.

39. MacFerlane, D. R., Sun, J., Golding, J., Meakin, P., and Forsyth, M. (2000). High conductivity molten salts based on the imide ion. *Electrochim. Acta*, **45**, pp. 1271–1278.

40. McEwen, A. B., Ngo, H. L., Compte, K., and Goldman, X. L. (1999). Electrochemical properties of imidazolium salt electrolytes for

electrochemical capacitor applications. *J. Electrochem. Soc.*, **146**, pp. 1687–1695.
41. Diaz, A. F., Kanazawa, K. K., and Gardini, G. P. (1979). Electrochemical polymerization of pyrrole. *J. Chem. Soc. Chem. Commun.*, **14**, pp. 635–636.
42. Tourillon, G., and Garnier, F. (1982). New electrochemically generated organic conducting polymers. *J. Electroanal. Chem.*, **135**, pp. 173–178.
43. Diaz, A. F., and Logan, J. A. (1980). Electroactive polyaniline films. *J. Electroanal. Chem.*, **111**, pp. 111–114.
44. Kobryanskii, V. M., and Arnautov, S. A. (1992). Electrochemical synthesis of poly(p-phenylene) in an ionic liquid. *Makromol. Chem.*, **193**, pp. 455–463.
45. Lere-Porte, J. P., Radi, M., Chorro, C., Petrissans, J., Sauvajol, J. L., Gonbeau, D., Pfister-Guillouzo, G., Louarn, G., and Lefrant, S. (1993). Characterization from XPS, FT-IR and Raman spectroscopies of films of poly(*p*-phenylene) prepared by electropolymerization of benzene dissolved in ketyl pyridinium chloride-$AlCl_3$ melting salt. *Synth. Met.*, **59**, pp. 141–149.
46. Pickup, P. G., and Osteryoung, R. A. (1984). Electrochemical polymerization of pyrrole and electrochemistry of polypyrrole films in ambient temperature molten salts. *J. Am. Chem. Soc.*, **106**, pp. 2294–2299.
47. Koura, N., Ejiri, H., and Takeishi, K. (1993). Polyaniline secondary cells with ambient temperature molten salt electrolytes. *J. Electrochem. Soc.*, **140**, pp. 602–605.
48. Janiszewska, L., and Osteryoung, R. A. (1987). Electrochemistry of polythiophene and polybithiophene films in ambient temperature molten salts. *J. Electrochem. Soc.*, **134**, pp. 2787–2794.
49. Mazurkiewicz, J. H., Innis, P. C., Wallace, G. G., MacFarlane, D. R., and Forsyth, M. (2003). Conducting polymer electrochemistry in ionic liquids. *Synth. Met.*, **135**, pp. 31–32.
50. Pringle, J. M., Efthimiadis, J., Howlett, P. C., Efthimiadis, J., MacFarlane, D. R., Chaplin, A. B., Hall, S. B., Officer, D. L., Wallace, G. G., and Forsyth, M. (2004). Electrochemical synthesis of polypyrrole in ionic liquids. *Polymer*, **45**, pp. 1447–1453.
51. Sekiguchi, K., Atobe, M., and Fuchigami, T. (2002). Electropolymerization of pyrrole in 1-ethyl-3-methylimidazolium trifluoromethanesulfonate room temperature ionic liquid. *Electrochem. Commun.*, **4**, pp. 881–885.

52. Deepa, M., and Ahmad, S. (2008). Polypyrrole films electropolymerized from ionic liquids and in a traditional liquid electrolyte: A comparison of morphology and electro-optical properties. *European Polymer Journal*, **44**, pp. 3288–3299.
53. Lin, Y., and Wallace, G. G. (1994). Electropolymerisation of pyrrole under hydrodynamic conditions-effect of solution additives. *Electrochim. Acta*, **39**, pp. 1409–1413.
54. Han, D., Qiu, X., Shen, Y., Guo, H., Zhang, Y., and Niu, L. (2006). Electropolymerization of polypyrrole on PFIL–PSS-modified electrodes without added support electrolytes. *J. Electroanal. Chem.*, **596**, pp. 33–37.
55. Damlin, P., Kvarnström, C., and Ivaska, A. (2004). Electrochemical synthesis and in situ spectroelectrochemical characterization of poly(3,4-ethylenedioxythiophene) (PEDOT) in room temperature ionic liquids. *J. Electroanal. Chem.*, **570**, pp. 113–122.
56. Wagner, K., Pringle, J. M., Hall, S. B., Forsyth, M., MacFarlane, D. R., and Officer, D. C. (2005). Investigation of the electropolymerisation of EDOT in ionic liquids. *Synth. Met.*, **153**, pp. 257.
57. Randriamahazaka, H., Plesse, C., Teyssie, D., and Chevrot, C. (2004). Ions transfer mechanisms during the electrochemical oxidation of poly(3,4-ethylenedioxythiophene) in 1-ethyl-3-methylimidazolium bis((trifluoromethyl)sulfonyl)amide ionic liquid, *Electrochem. Commun.*, **6**, pp. 299–305.
58. Ispas, A., Peipmann, R., Bund, A., and Efimov, I. (2009). On the p-doping of PEDOT layers in various ionic liquids studied by EQCM and acoustic impedance. *Electrochim. Acta*, **54**, pp. 4668–4675.
59. Kim, T. Y., Lee, T. H., Kim, J. E., Kasi, R. M., Sung, C. S. P., and Suh, K. S. (2008). Organic solvent dispersion of poly(3,4-ethylenedioxythiophene) with the use of polymeric ionic liquid, *J. Poly. Sci.*, **46**, pp. 6872–6879.
60. Döbbelin, M., Pozo-Gonzalo, C., Marcilla, R., Blanco, R., Segura, J. L., Pomposo, J. A., and Mecerreyes, D. (2009). Electrochemical synthesis of PEDOT derivatives bearing imidazolium-ionic liquid moieties. *J. Poly. Sci.*, **47**, pp. 3010–3021.
61. Pringle, J. M., Forsyth, M., MacFarlane, D. R., Wagner, K., Hall S. B., and Officer, D. L. (2005). The influence of the monomer and the ionic liquid on the electrochemical preparation of polythiophene. *Polymer*, **46**, pp. 2047–2058.
62. Sekiguchi, K., Atobe, M., and Fuchigami, T. (2003). Electrooxidative polymerization of aromatic compounds in 1-ethyl-3-methylimidazolium trifluoromethanesulfonate room-temperature ionic liquid. *J. Electroanal. Chem.*, **557**, pp. 1–7.

63. Heinze, J. (2001). Electrochemistry of conducting polymers. In *Organic Electrochemistry* (ed. Hammerich, O.), Marcel Dekker, New York, pp. 1309.
64. Li, M.C., Ma, C.A., Liu, B.Y., and Jin, Z.M. (2005). A novel electrolyte 1-ethylimidazolium trifluoroacetate used for electropolymerization of aniline. *Electrochem. Commun.*, **7**, pp. 209–212.
65. Wei, D., Kvarnström, C., Lindfors, T., and Ivaska, A. (2006). Polyaniline nanotubules obtained in room-temperature ionic liquids. *Electrochem. Commun.*, **8**, pp. 1563–1566.
66. Burroughes, J. H., Bradley, D. D. C., Brown, A. R., Marks, R. N., Mackay, K. R., Friend, H., Burns, P. L., and Holmes, A. B. (1990). Light-emitting diodes based on conjugated polymers. *Nature*, **347**, pp. 539–541.
67. Sariciftci, N. S., Smilowitz, L., Heeger, A. J., and Wudl, F. (2002). Photoinduced electron transfer from a conducting polymer to buckminsterfullerene, *Science*, **258**, pp. 1474–1476.
68. Abedin, S. Z. El, Borissenko, N., and Endres, F. (2004). Electropolymerization of benzene in a room temperature ionic liquid. *Electrochem. Commun.*, **6**, pp. 422–426.
69. Schneider, O., Bund, A., Ispas, A., Borissenko, N., Abedin, S. Z. El, and Endres, F. (2005). An EQCM study of the electropolymerization of benzene in an ionic liquid and ion exchange characteristics of the resulting polymer film. *J. Phys. Chem. B*, **109**, pp. 7159.
70. Wagner, M., Kvarnström, C., and Ivaska, A. (2010). Room temperature ionic liquids in electrosynthesis and spectroelectrochemical characterization of poly(para-phenylene). *Electrochimica Acta*, **55**, pp. 2527–2535.
71. Bhattacharyya, S., Kymakis, E., and Amaratunga, G. A. J. (2004). Photovoltaic properties of dye functionalized single-Kymakis, E., and Amaratunga, G. A. J. (2002). Single-wall carbon nanotube/conjugated polymer photovoltaic devices. *Appl. Phys. Lett.*, **80**, pp. 112–114.
72. Kymakis, E., and Amaratunga, G. A. J. (2002). Single-wall carbon nanotube/conjugated polymer photovoltaic devices. *Appl. Phys. Lett.*, **80**, pp. 112–114.
73. Rahman, G. M. A., Guldi, D. M., Cagnoli, R., Mucci, A., Schenetti, L., Vaccari, L., and Prato, M. (2005). Combining single wall carbon nanotubes and photoactive polymers for photoconversion. *J. Am. Chem. Soc.*, **127**, pp. 10051–10057.
74. Zengin, H., Zhou, Z., Jin, J., Czerw, R., Smith, D. W., Echegoyen, L., Caroll, D. L., Foulger, S. H., and Ballato, J. (2002). Carbon nanotube doped polyaniline. *Adv. Mater.*, **14**, pp. 1480–1483.

75. Baibarac, M., Baltog, I., Gordon, C., Lefrant, S., and Chauvet, O. (2004). Covalent functionalization of single-walled carbon nanotubes by aniline electrochemical polymerization. *Carbon*, **42**, pp. 3143–3152.
76. Baibarac, M., Baltog, I., Lefrant, S., Mevellec, J. Y., and Chauvet, O. (2003). Polyaniline and carbon nanotubes based composites containing whole units and fragments of nanotubes. *Chem. Mater.*, **15**, pp. 4149–4156.
77. Wei, D., Kvarnström, C., Lindfors, T., and Ivaska, A. (2007). Electrochemical functionalization of single walled carbon nanotubes with polyaniline in ionic liquids. *Electrochem. Commun.*, **9**, pp. 206–210.
78. Sun, Y., Wilson, S. R., and Schuster, D. I. (2001). High dissolution and strong light emission of carbon nanotubes dissolved in aniline. *J. Am. Chem. Soc.*, **123**, pp. 5348–5349.
79. Fukushima, T., Kosaka, A., Ishimura, Y., Yamamoto, T., Takigawa, T., Ishii, N., and Aida, T. (2003). Molecular ordering of organic molten salts triggered by single-walled carbon nanotubes. *Science*, **300**, pp. 2072–2074.
80. Wei, D., Baral, J. K., Österbacka, R., and Ivaska, A. (2008). Electrochemical fabrication of a nonvolatile memory device based on polyaniline and gold particles. *J. Mater. Chem.*, **18**, pp. 1853–1857.
81. Pringle, J. M., Ngamna, O., Chen, J., Wallace, G. G., Forsyth, M., and MacFarlane, D. R. (2006). Conducting polymer nanoparticles synthesized in an ionic liquid by chemical polymerization. *Synth. Met.*, **156**, pp. 979–983.
82. Kim, J.-Y., Kim, J.-T., Song, E.-A., Min, Y.-K., and Hamaguchi, H. (2008). Polypyrrole nanostructures self-assembled in magnetic ionic liquid as a template. *Macromolecules*, **41**, pp. 2886–2889.
83. Hayashi, S., and Hamaguchi, H. (2004). Discovery of a magnetic ionic liquid [BMIM]FeCl$_4$. *Chem. Lett.*, **33**, pp. 1590–1591.
84. Pringle, J. M., Winther-Jensen, O., Lynam, C., Wallace, G. G., Forsyth, M., and MacFarlane, D. R. (2008). One-step synthesis of conducting polymer–noble metal nanoparticle composites using an ionic liquid. *Adv. Funct. Mater.*, **18**, pp. 2031–2040.
85. Wildgoose, G. G., Banks, C. E., and Compton, R. G. (2006). Metal nanoparticles and related materials supported on carbon nanotubes: methods and applications. *Small*, **2**, pp. 182.
86. Georgakilas, V., Gournis, D., Tzitzios, V., Pasquato, L., Guldi D. M., and Prato, M. (2007). Decorating carbon nanotubes with metal or semiconductor nanoparticles. *J. Mater. Chem.*, **17**, pp. 2679–2694.

87. Kim, K., Lee, S. H., Yi, W., Kim, J., Choi, J. W., Park, Y., and Jin, J. I. (2003). Efficient field emission from highly aligned, graphitic nanotubes embedded with gold nanoparticles. *Adv. Mater.*, **15**, pp. 1618–1622.
88. Chopra, N., Majumder, M., and Hinds, B. J. (2005). Bifunctional carbon nanotubes by sidewall protection. *Adv. Funct. Mater.*, **15**, 858–864.
89. Quinn, B. M., Dekker, C., and Lemay, S. G. (2005). Electrodeposition of noble metal nanoparticles on carbon nanotubes. *J. Am. Chem. Soc.*, **127**, 6146–6147.
90. Park, H. S., Choi, B. G., Yang, S. H., Shin, W. H., Kang, J. K., Jung, D., and Hong, W. H. (2009). Ionic-liquid-assisted sonochemical synthesis of carbon-nanotube-based nanohybrids: control in the structures and interfacial characteristics. *Small*, **5**, pp. 1754–1760.
91. Liu, N., Luo, F., Wu, H., Liu, Y., Zhang, C., and Chen, J. (2008). One-step ionic-liquid-assisted electrochemical synthesis of ionic-liquid-functionalized graphene sheets directly from graphite. *Adv. Funct. Mater.*, **18**, pp. 1518–1525.
92. Sun, Y. P., Zhou, B., Lin, Y., Wang, W., Fernando, K. A. S., Pathak, P., Meziani, M. J., Harruff, B. A., Wang, X., Wang, H., Luo, P. G., Yang, H., Kose, M. E., Chen, B., Veca, L. M., and Xie, S. Y. (2006). Quantum-sized carbon dots for bright and colorful photoluminescence. *J. Am. Chem. Soc.*, **128**, pp. 7756–7757.
93. Zhou, J., Booker, C., Li, R., Zhou, X., Sham, T. K., Sun, X., and Ding, Z. (2007). An electrochemical avenue to blue luminescent nanocrystals from multiwalled carbon nanotubes (MWCNTs). *J. Am. Chem. Soc.*, **129**, pp. 744–745.
94. Lotya, M., Hernandez, Y., King, P. J., Smith, R. J., Nicolosi, V., Karlsson, L. S., Blighe, F. M., De, S., Wang, Z., McGovern, I. T., Duesberg, G. S., and Coleman, J. N. (2009). Liquid phase production of graphene by exfoliation of graphite in surfactant/water solutions. *J. Am. Chem. Soc.*, **131**, pp. 3611–3620.
95. Lu, J., Yang, J. X., Wang, J., Lim, A., Wang, S., and Loh, K. P. (2009). One-pot synthesis of fluorescent carbon nanoribbons, nanoparticles, and graphene by the exfoliation of graphite in ionic liquids *ACS Nano*, **3**, pp. 2367–2375.
96. Michalet, X., Pinaud, F. F., Bentolila, L. A., Tsay, J. M., Doose, S., Li, J. J., Sundaresan, G., Wu, A. M., Gambhir, S. S., and Weiss, S. (2005). Quantum dots for live cells, in vivo imaging, and diagnostics *Science*, **307**, pp. 538–544.

Chapter 4

Imidazolium-Based Ionic Liquid Functional Materials and Their Application to Electroanalytical Chemistry

Yuanjian Zhang, Yanfei Shen, and Li Niu

State Key Laboratory of Electroanalytical Chemistry,
Changchun Institute of Applied Chemistry,
Graduate School of the Chinese Academy of Sciences,
Chinese Academy of Sciences, Changchun 130022, P. R. China
lniu@ciac.jl.cn

4.1 Introduction

Ionic liquids (ILs) are low-melting-point salts, thus forming liquids that consist only of cations and anions. They are often applied to any compounds that have a melting point less than 100°C. The first useful IL, ethylammonium nitrate, described by Walden, seems to have generated little interest; it was not until the 1980s that the physical and chemical properties of this salt were investigated [1]. This was followed by the discovery that several tetraalkylammonium salts form air- and moisture-stable ILs of interesting properties. For the overall environmental impact and economics, they were employed as solvents for electrochemistry [2–8], analytical chemistry [9], chemical synthesis [10–14], liquid/liquid separations and extractions [15–17], dissolution [18–20], catalysis [21–25], and polymerization [26]. In electrochemistry, they show relatively wide potential window and high conductivity and allow studies to be undertaken without additional supporting electrolyte [27]. Thus various applications including in electrodeposition, electropolymerization, capacitors, Li-ion batteries, and solar cell have been intensive investigated [28]. ILs have also offered many opportunities in electroanalytical chemistry [29]. Particularly, some task-specific ILs have also been designed because the structures and properties of ILs can be easily tuned by selecting proper combination of organic cations and anions. Equally importantly, these unique properties of ILs could be extended to the concept of task-specific IL-materials. No longer as simple green solvents, it would greatly expand the potential application of ILs. As excellent reviews exist describing ILs for analytical chemistry, electroanalytical chemistry or electrochemistry [9, 28–31], here we will focus on the imidazolium-based IL-materials and their applications in electroanalytical chemistry from our laboratory as well as other groups.

4.2 Electrosynthesis

ILs show a relatively wide potential window and high conductivity and allow studies to be undertaken without additional supporting electrolyte. Recently, Aida et al. have reported sing-walled carbon nanotubes (SWNTs) could be considerably untangled into much finer bundles that are physically cross-linked in ILs [32]. Thus, we were motivated to design a kind of RTIL-supported three-dimensional network SWNT electrode (as shown in Fig. 4.1a) [33]. The advantage of bucky gels of ionic liquids and SWNTs for the electrochemical functionalization of SWNTs is that the ionic liquid acts as both a dispersant of SWNTs and a supporting electrolyte. More important, it would greatly increase the effective surface area of the SWNT electrode, and the homogeneous electrochemical functionalization of the SWNTs performed well even in large quantities. This is rare for conventional electrochemical functionalization of SWNTs because the reaction occurs locally on a limited surface of bundled SWNTs deposited on metal electrodes. N-succinimidyl acrylate (NSA), as a model monomer, which bears an active ester group, was ground into a gel of ILs and SWNTs, and the mixture was placed onto gold electrode. NSA was electrografted and polymerized onto SWNTs (SWNTs-poly-NSA) by applying a reduction potential to the electrode (Fig. 4.1b). The active ester groups in the grafted poly-NSA can be utilized for further functionalization. For example, by the reaction with glucose oxidase (GOD), the modified SWNTs with an electrocatalytic activity toward

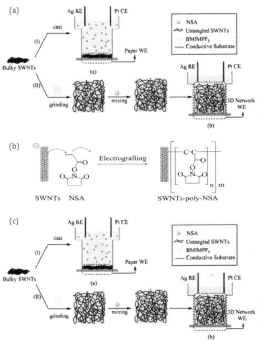

Figure 4.1 (a) Electrochemical functionalization at the SWNTs paper electrode (I) and our 3D network SWNTs electrode (II). (b) Schematic illustration of grafting NSA on SWNTs via electrochemical initiation and (c) LSV curves of electrocatalyzed oxidation of glucose in 0.05 M PBS (pH 7.4) on the SWNTs-poly-NSA-GOD electrode at different concentrations of glucose: 0, 4, 8, 12, 16, 20 mM from bottom to up. Inset is calibration curve corresponding to amperometric responses at the SWNTs-poly-NSA-GOD (solid) and SWNTs-poly-NSA (hollow) modified gold electrodes. Scan rate: 50 mV/s. Reprinted with permission from Ref. [33]. Copyright 2005 American Chemical Society.

glucose can be fabricated, which could be utilized as biosensor toward glucose (Fig. 4.1c). Similarly, Wei et al. have utilized the same method to functionalize SWNTs with polyaniline [34].

In addition, ILs could also be used as both solvent and electrolyte for the electrodeposition of copper [35, 36], aluminum [37, 38], tantalum [4], platinum [39], silver [40, 41], gold [40–42], and silicon [43]. For example, Endres et al. have reported the electrodeposition of nanocrystalline metals and alloys, such as aluminum from ILs, which previously could not be electrodeposited from aqueous or organic solutions. This method enabled the synthesis of aluminum nanocrystals with average grain sizes of about 10 nm, Al–Mn alloys, as well as Fe and Pd nanocrystals [4] (as shown in Fig. 4.2).

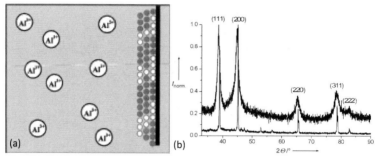

Figure 4.2 (a) Schematic representation of electrodeposition of Al in ILs and (b) Upper curve: XRD pattern of nanocrystalline Al with a grain size of 12 ± 1 nm prepared from $AlCl_3/[BMIm]^+Cl^-$ (55/45 mol%). Lower curve: Microcrystalline reference sample. From Ref. [4], Copyright Wiley-VCH Verlag GmbH & Co. KGaA. Reproduced with permission.

4.3 Functionalization of Ionic Liquids for Ease Immobilization

With the development of the synthesis of ILs, the price of ILs has been greatly decreased. However, considering the economic criteria, the ease of separation and increasing the recycle times, the immobilization of ILs on the solid supports is highly desirable. The immobilization process transferred the desired properties of ILs to the solid supports. And more important, to date, difficulties in effective immobilization of generic ILs on the electrode substrate have greatly hindered research on the electrocatalysis of ILs. It was reported that the ILs can be bound to a surface by either covalent bonds or noncovalent interactions between the ILs and the surface. Mehnert et al. have first proposed the concept of silica-supported ILs catalysis, where the ILs immobilized on the Si/SiO$_2$ surface was functionalized with alkyl siloxane appendages (Fig. 4.3) [24, 44]. Similarly, ionic species were immobilized on the surface of mesoporous nanostructured silicas by Moreau et al. (Fig. 4.4) [45]. Lee and coworkers [46] have proposed a method to immobilize ILs on gold surface, which was based on self-assembly with imidazolium ions bearing alkyl thiol (Fig. 4.5a). Interestingly, simply by anion exchange, the wettability of gold surface could be facilely controlled (Fig. 4.5b). Huck et al. developed IL based polyelectrolyte brushes, which were anchored on Au surface via thiol initiator, e.g., BrC-(CH$_3$)$_2$COO(CH$_2$)$_6$SH. Further by ion exchange with ferricyanide ions ([Fe(CN)$_6$]$^{3-}$) as redox probes, CV measurements of the modified brushes showed the typical electrochemical response corresponding to a surface-confined electroactive species and the redox counterions, as ferricyanide species form stable ion pairs with the quaternary ammonium groups of the brush (Fig. 4.6) [47]. In a noncovalent way, Mao et al. have reported that ILs could be directly immobilized on the glassy carbon electrode (GC) by casting and observed the electrocatalytic activity toward ascorbic acid (AA) and the capability to facilitate direct electron transfer of horseradish peroxidase (HRP) (Fig. 4.7) [48]. Dong et al. have reported ILs-carbon composites-pastes modified electrode and its efficient electron transfer between the electrode and the horseradish peroxidase (HRP, Fig. 4.8) [49].

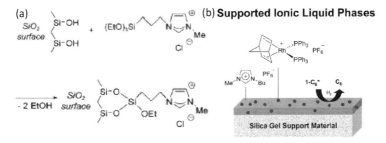

Figure 4.3 Immobilization of chloroaluminate-based ionic liquid by grafting of imidazolium chloride (a) and the scheme of the concept of supported ILs (b). From Ref. [4], Copyright Wiley-VCH Verlag GmbH & Co. KGaA. Reproduced with permission.

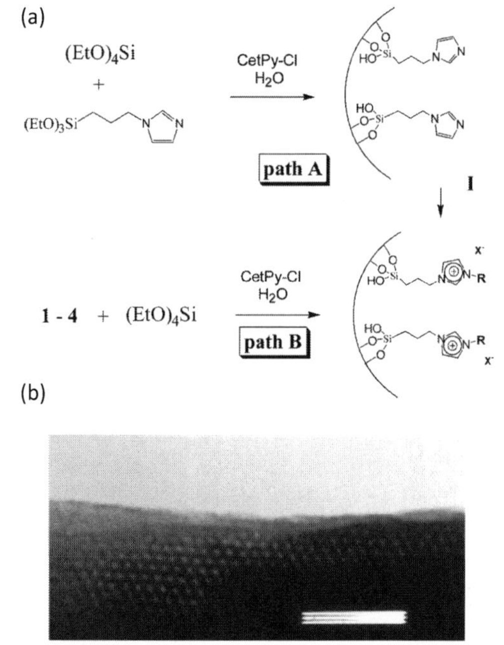

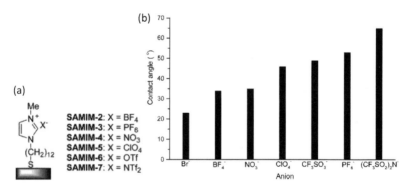

Figure 4.4 Schematic (a) and TEM image (b) of immobilization of ionic species on the surface of mesoporous nanostructured silicas [45]. Reproduced by permission of The Royal Society of Chemistry.

Figure 4.5 Self-assembled monolayers presenting imidazolium ions at the tail ends on Au having different anions (a) and the effects of counteranions on surface hydrophilicity and hydrophobicity (b). Reprinted with permission from Ref. [46]. Copyright 2004 American Chemical Society.

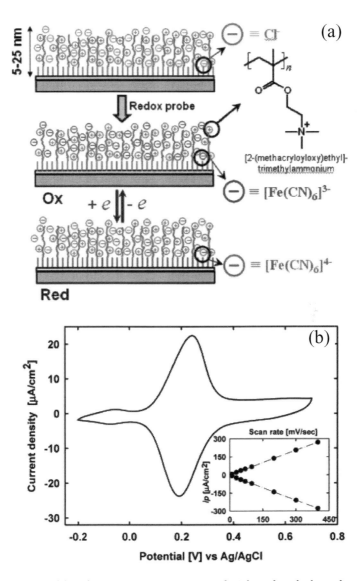

Figure 4.6 (a) Schematic representation of IL-based polyelectrolytes brushes. (b) Cyclic voltammogram corresponding to electroactive brush (14 nm). v: 30 mV/s. Supporting electrolyte: 5 mM KCl. Inset: Plot of anodic and cathodic peak currents versus v. Reprinted with permission from Ref. [47]. Copyright 2007 American Chemical Society.

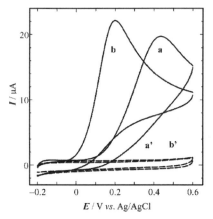

Figure 4.7 CVs for the oxidation of AA (1.0 mM) at bare (curve a) and 1-butyl-3-methyl imidazolium chloride ([BMIM][Cl])-modified (curve b) GC electrodes in 0.10 M phosphate buffer. Curves a and b represent CVs obtained at the above electrodes in the same solution containing no AA. Scan rate, 100 mV s^{-1}. Reprinted with permission from Ref. [48]. Copyright 2005 American Chemical Society.

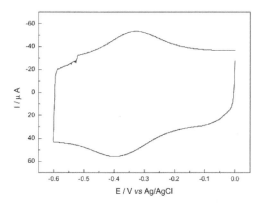

Figure 4.8 Cyclic voltammogram at the HRP/CNTs/1-butyl-3-methylimidazolium hexafluorophosphate-modified glassy carbon electrode in 0.1 M phosphate buffer solution (pH = 7.00) under nitrogen. The scan rate is 0.1 V/s. Reprinted with permission from Ref. [49]. Copyright 2004 American Chemical Society.

Our strategy for solving above challenge involves preparation of the polyelectrolyte-supported ILs (PFIL, Fig. 4.9) [50]. It was noted that polyelectrolyte could be easily immobilized onto many substrates through various methods such as electrophoresis, layer-by-layer (LbL) assembly and casting, etc. Therefore, it would be helpful for us to immobilize ILs facilely on general substrates with the aid of polyelectrolyte as carrier. Thus, the ILs could be easily immobilized on the substrate with the aid of polyelectrolyte. As illustrated, the direct electrocatalytic activity of the PFIL toward the oxidation of β-Nicotinamide adenine dinucleotide (NADH) was reported for the first time. In addition, via LbL assembly of PFIL, an electrochemically controlled tunable surface was also constructed. Such design of PFIL provided more general approaches to immobilize IL on any solid supports in spite of any size and shape, and it would be much significant for the chemical industrial processes. This practical advantage of the PFIL material was technically attractive in chemical industrial processes, and exhibited a significant future toward the application of IL wherever in surface chemistry or in catalytic chemistry.

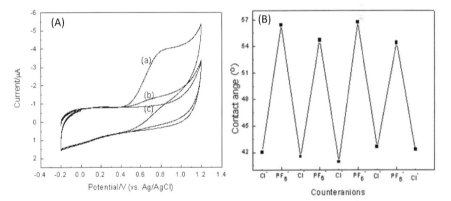

Figure 4.9 (A) Cyclic voltammograms for 0.5 mM NADH in phosphate buffer solution (0.05 M, pH = 7.4) at PFILs-Nafion (a), Nafion (b) modified GC, and only in phosphate buffer solution (0.05 M, pH = 7.4) at PFIL-Nafion modified GC (c). Inset shows scheme of PFIL. (B) Reversible change in wettability by applying an electric field onto the substrate in 10 mM NaPF$_6$ or NaCl solution. Working potential: +0.3 V; time scanning: 600 s [50]. Reproduced by permission of The Royal Society of Chemistry.

From another point of view, polyelectrolyte functionalized IL was also a novel kind of polyelectrolyte. Thus, many traditional applications of polyelectrolytes would be possible for such PFIL and some unique properties derived from ILs would also be expected. For example, ultrathin films of PFIL and Prussian blue (PB) nanoparticles were fabricated on the ITO substrate through electrostatic layer-by-layer assembly method [51]. The obtained PFIL/PB multilayer showed good reproducibility, wide linear range, and high sensitivity (even to 10^{-7} M) for the amperometric analysis of hydrogen peroxide, which was superior to a similar poly(diallyldimethyldiammonium chloride) (PDDA)/PB multilayer in the control experiments (Fig. 4.10). It indicated that the PFIL on the modified electrode played an important role in the electrocatalytic reduction of hydrogen peroxide.

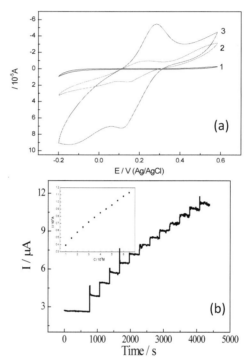

Figure 4.10 The cyclic voltammograms of the blank ITO electrode (1) ITO/(PDDA/PB) 10 multilayer films (2) and ITO/(PFIL/PB) 10 multilayer films electrodes (3) in 0.1 M KCl solution saturated with N_2 at the scan rate of 50 mV/s in the present of 0.25 mM H_2O_2 (a); Steady-state response of ITO/(PFIL/PB) 10 electrode with successive addition of 0.25 mM H_2O_2 into stirring 0.1 M KCl solution. (Applied potential: −0.1 V vs. Ag|AgCl in saturated KCl.) Inset: the calibration curve (b). Reproduced with permission from Ref. [51]. Copyright Elsevier (2008).

Similarly, polyelectrolytes functionalized IL could be incorporated into a sol–gel organic-inorganic hybrid material (PFIL/sol–gel) to improve charge-transfer efficiency and substrate diffusion in chemically modified electrodes [52]. By immobilizing GOD on a glassy carbon electrode, an enhanced current response toward glucose was obtained, relative to a control case without PFIL. In addition, chronoamperometry showed that electroactive mediators diffused at a rate 10 times higher in the apparent diffusion coefficient in PFIL-containing matrices (Fig. 4.11). These findings suggest a potential application in bioelectroanalytical chemistry.

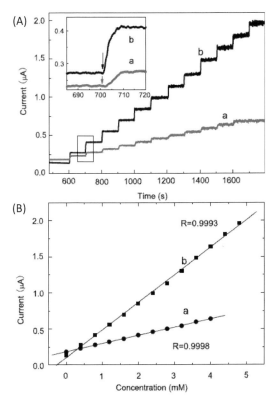

Figure 4.11 (A) Steady-state response of (a) GOD/sol–gel/GC and (b) GOD/PFIL/sol–gel/GC electrodes with successive addition of glucose concentration (0.4 mmol/L each step). Applied potential: 1.05 V vs. Ag|AgCl (in saturated KCl). Inset shows a magnification of the second addition of glucose. (B) Calibration curves of the above (a), (b) electrodes. Reproduced with permission from Ref. [52]. Copyright Elsevier (2007).

4.4 Electrolyte-Free Electrochemistry

It is common practice that electrochemical experiments should be conducted in the presence of supporting electrolytes. However, within last few years, the general necessity to add an excess of supporting electrolyte has been questioned, and it is desirable to consider even the necessity of using a supporting electrolyte in various cases. Some promising methods for electrolyte-free electrochemistry such as with ultramicroelectrodes (UMEs) or their modifications with moist ion-exchange membranes such as Nafion have been developed. We have reported an alternative pathway by a simple PFIL-modified electrode (Fig. 4.12) [53]. The studied PFIL material combines features of ionic liquids and traditional polyelectrolytes. The IL part provides high ionic conductivity and affinity to many different compounds. The polyelectrolyte part has a good stability in aqueous solution and capability to be immobilized on different substrates. The electrochemical properties of such a PFIL-modified electrode assembly in supporting electrolyte-free solution have been investigated by using an electrically neutral electroactive species, hydroquinone (HQ) as the model compound. The partition coefficient and diffusion coefficient of HQ in the PFIL film were calculated to be 0.346 and 4.74×10^{-6} cm^2 s^{-1}, respectively. Electrochemistry in PFIL is similar to electrochemistry in a solution of traditional supporting electrolytes in the solution, except that the electrochemical reaction takes place in a thin film on the surface of the electrode. PFIL is easy to be immobilized on solid substrates, inexpensive and electrochemically stable. A PFIL-modified electrode assembly is successfully used in flow analysis at HQ by amperometric detection in solution without supporting electrolyte. Similarly, polypyrrole could be electropoly-merized on PFIL-PSS modified electrodes without added support electrolytes [54].

Actually, PFIL on the electrode assembly offered a suitable electrochemical microenvironment for electrochemical reaction; in this case, it was not absolutely electrolyte free. Thus, it also hints functionalized IL could be used as solid-state electrolyte for some specific applications, such as solar cells. Figure 4.13b shows a full solid-state, flexible, dye-sensitized solar cell (DSSC) based on novel ionic liquid gel (PVA-g-VIC4Br), organic dye, ZnO nanoparticles, and CNTs thin film stamped onto a polyethylene terephthalate (PET) substrate [55]. It would pave the way for the development of a continuous roll-to-roll process for the mass production of flexible and lightweight DSSCs.

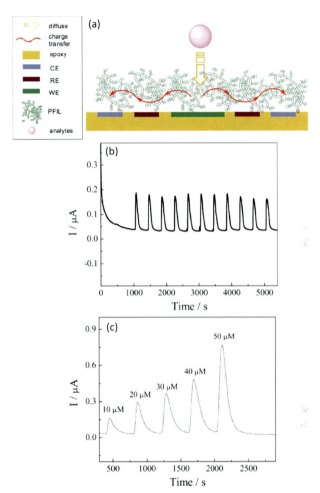

Figure 4.12 Illustration of the electrochemical process of the analytes at the PFIL-modified electrode assembly in a supporting electrolyte-free solution (a); amperometric responses of repeated injection of a 20 mM HQ solution (b); amperometric response of HQ at different concentrations. The signal was recorded at the 1.4 mm-thick PFIL-modified electrode assembly at +0.3 V. Double-distilled water was used as the carrier solution. Flow rate: 1 mL min^{-1} (c) [53]. Reproduced by permission of The Royal Society of Chemistry.

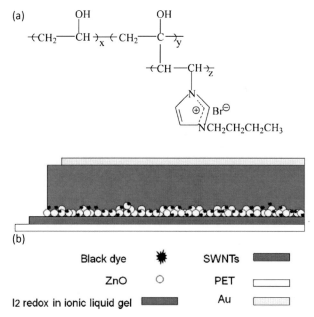

Figure 4.13 Structure of (a) PVA-g-VIC4Br and (b) solid-state DSSC. Reproduced with permission from Ref. [55].

4.5 ILs-Based Multifunctional Compounds for Electrocatalysis and Biosensors

One of the most fantastic features of ILs is that their properties can be easily and well tuned by rationally selecting proper combination of organic cations and anions. That also means that we can delicately utilize one ionic component to deliver a unique function and the other ionic component to deliver a different, completely independent function. Moreover, the components and their combination of anions and cations are various, for example, the cation of ILs can have several substituting groups (R1, R2, R3, etc., as shown in Fig. 4.14a) and these substituting groups are tunable. Therefore, it offers us a promising and facile way to combine individual functions into a target compound. In contrast, for a commonly seen compound, to achieve this multifunctional combination would encounter all sorts of synthetic challenges. We have illustrated this flexibility of ILs by selecting SWNT (R1) as a model substituting group of the imidazolium cation, and Br⁻,

PF_6^-, BF_4^-, polyoxometalates (POMs), and even glucose oxidase (GOD) as the counter-anions (Fig. 4.14a, SWNT-IL-X) [56]. The properties of SWNT and the various anions were facilely and successfully delivered into the resulting compounds. For example, the rich redox activity was also successfully transferred into SWNT-IL-POM merely by a simple and facile anions exchange. The surface-confined SWNT-IL-POM shows three couples of well-defined redox waves at scan rates up to 2 V/s (Fig. 4.14b), which was presumably attributed from electron conduction of SWNT, ionic conduction of IL, and redox conduction of POM. It was unusual for an often-seen compound in electrochemical systems.

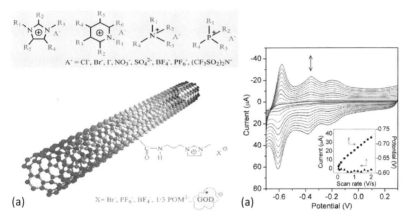

Figure 4.14 (a) Examples of cations and anions commonly used for the formation of ionic liquids and SWNT-IL-X. (b) Cyclic voltammograms (CVs) of SWNT-IL-POM modified GC electrode (d = 3 mm) at different scan rate in 0.5 M H_2SO_4. Scan rate: 0.05, 0.1, 0.2, 0.4, 0.6, 0.8, 1.0, 1.2, 1.4, 1.6, 1.8 and 2.0 V/s from inner to outer. The inset shows the peak current (square) and peak potential (triangle) of the third reduction wave as a function of scan rate. From Ref. [56], Copyright Wiley-VCH Verlag GmbH & Co. KGaA. Reproduced with permission.

Moreover, due to the controllable wettability by counter-anion exchange [56, 57], SWNT-IL-Br, which was both hydrophobic and hydrophilic, could assemble at the water/oil interface in a controllable manner, i.e., from monolayer to multilayers [58]. Thus, it enabled a novel SWNT-sandwiched water/oil interface for the electron transfer, one of the most fundamental chemical processes across the interfaces. It will promote both fundamental electron transfer research such as in biological and artificial membrane and homo/heterogeneous catalysis. For example, it indicated SWNTs accelerated the electron transfer at water/oil interface by using scanning electrochemical microscopy (Fig. 4.15).

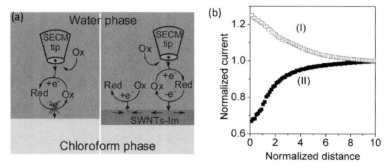

Figure 4.15 (a) SECM arrangement (not to scale) for feedback measurement at pure (left) and SWNTs-Im-sandwiched (right) water/chloroform interface; (b) Experimental approach curves (CHI 900) for a tip in aqueous solution approaching SWNTs-Im-sandwiched (I) and pure (II) water/chloroform interface. Currents are normalized to the steady-state diffusion limiting current, $i_{T,\infty}$ and distance to tip radius. The aqueous solution contained 0.5 mM $Ru(NH_3)_6^{3+}$ and 100 mM KCl. The tip (Pt, 10 mm radius, RG = 10) was held at −0.35 V vs. Ag|AgCl (saturated KCl) for $Ru(NH_3)_6^{3+}$ reduction and approached at 1 μm/s. The counter electrode was Pt wire [58]. Reproduced by permission of The Royal Society of Chemistry.

For its potential bioelectroanalytical and biomedical application, integration of carbon nanostructures such as CNTs and graphene nanosheets with biomolecules such as redox enzyme is highly anticipated. Therein, carbon nanostructures are expected to act not only as an electron transfer promoter or mediator, but also as substrates for biomolecules immobilization. Based on ionic liquid-backbone, a novel immobilization method for biomolecules on carbon nanostructures could also be achieved via ionic interaction, which is of universality and widespread use in biological system. As illustrated, glucose oxidase (GOD) and SWNTs and graphene nanosheets were integrated into one bionanocomposite with the aid of ionic liquid-like unit on functionalized SWNTs [59] or PFIL wrapped polyvinylpyrrolidone (PPV)-graphene [60].

Results indicated that the ionic liquid-like unit on SWNTs, which did not depend on pH, would provide an extra and stronger ionic affinity between SWNTs and GOD so as to enhance the loading efficiency of GOD on SWNTs (Fig. 4.16). The direct electrochemistry of GOD was also achieved on PFIL wrapped polyvinylpyrrolidone (PVP)-graphene (Fig. 4.16a). Moreover, in these novel graphene-bionanocomposites, the specific substrate-enzyme bioactivity of GOD in the bionanocomposites was reserved, which could be utilized to build the electrochemical biosensor toward detection of glucose (Fig. 4.16b).

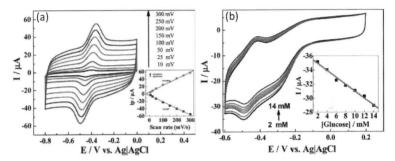

Figure 4.16 Cyclic voltammetric measurements at GCE/graphene-IL-GOD electrode in 0.05 M N_2 saturated PBS solution at different scan rates. Scan rate: 0.02, 0.05, 0.1, 0.15, 0.2 V s^{-1} from inner to outer. Inserted is the calibrated plot of peak currents versus scan rates (a) (reprinted from Ref. [59], copyright 2007, with permission from Elsevier). Cyclic voltammetric curves at the GCE/graphene-IL-GOD electrode in various concentrations (2, 4, 6, 8, 10, 12, and 14 mM) of glucose PBS solution saturated with O_2. The inset is the calibration curve ($R = 0.994$) corresponding to amperometric responses at −0.49 V. Scan rate: 0.05 V s^{-1} (b) (reprinted with permission from Ref. [60], copyright 2009 American Chemical Society).

Similarly, nanoparticles (NPs) could also be integrated in to nanocomposites based on the ILs-backbone. The approach involved the functionalization of CNTs with amine-terminated ILs and in situ reduction of $HAuCl_4$ without any additional step to introduce other reducing agents (Fig. 4.17) [61]. This composite material showed good electrocatalytic activity toward reduction of oxygen. In addition, the results of control experiments confirmed that the presence of ILs and the small size of gold nanoparticles increase the electrocatalytic activity of CNTs greatly.

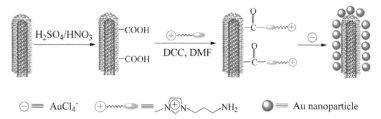

Figure 4.17 Illustration of the preparation procedures of CNT-IL-Au Reprinted from Ref. [61], copyright 2008, with permission from Elsevier.

Besides, ILs unit could be attached to the sidewall of CNTs by radical grafting, in which acid-oxidation pretreatment of CNTs could be avoided. Chen et al. reported that thermal-initiation free radical polymerization of the IL monomer 3-ethyl-1-vinylimidazolium tetrafluoroborate ([VEIM]BF$_4$) on the CNT surface (Fig. 4.18a) [62]. Then under similar method, the Pt and PtRu nanoparticles with narrow size distribution (average diameter: (1.3 ± 0.4) nm for PtRu, (1.9 ± 0.5) nm for Pt) are dispersed uniformly on the CNTs and show better performance in methanol electrooxidation than that without ILs units (Fig. 4.18b).

4.6 ILs-Protected Nanostructures as Electrocatalysts for Some Key Reactions

It was noted that without CNTs, very small Au NPs with an average diameter 1.7 nm could be obtained, where the Au NPs were presumably stabilized by amine-terminated ILs (Fig. 4.19a) [63]. They could be kept stable without any special protection and showed much better electrocatalysis toward reduction of dioxygen (ORR) comparing with those stabilized by thiol or citrate (Fig. 4.19b,c). Moreover, by further immobilizing IL-protected Au NPs on carbon nanostructures, e.g., carbon nanotubes or graphene nanosheets, by static interactions, the ORR current increased up to 6 times, in which the unique electronic property of carbon nanostructures as well as their large active surface areas were believed to play key roles [64, 65]. Here, not only Au NPs but also other noble metal NPs, such as Pt [66], and even graphene nanosheets themselves (Fig. 4.20) [67] could be stabilized by amine-terminated IL, and high electrocatalytic activity toward ORR and oxidation of methanol was also observed.

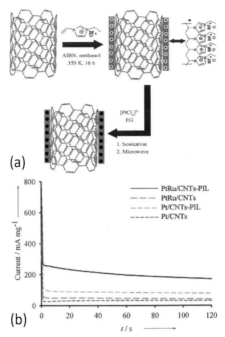

Figure 4.18 (a) Schematic diagram of the modification of CNTs with PIL and the preparation of Pt/CNTs-PIL nanohybrids. EG: ethylene glycol, AIBN: 2,2′-azobisisobutyronitrile. (b) Transient current of PtRu/CNTs-PIL, PtRu/CNTs, Pt/CNTs-PIL, and Pt/CNTs catalysts for methanol electrooxidation at 0.50 V in nitrogen-saturated 0.5 M H_2SO_4 + 1.0 M CH_3OH aqueous solution. From Ref. [62], Copyright Wiley-VCH Verlag GmbH & Co. KGaA. Reproduced with permission.

Besides, polyelectrolyte-functionalized IL (PFIL) also showed great potential in stabilization of Au NPs [68]. It was found that the size distribution of resulting Au NPs (PFIL-AuNPs) was narrow and could be tuned by the concentration of $HAuCl_4$. Moreover, they showed a high stability in water at room temperature for at least one month in solutions of pH 7–13 (even in the present of high concentration of NaCl). In addition, the PFIL-AuNPs exhibited obvious electrocatalytical activity toward NADH oxidation, suggesting a potential application for bioelectroanalysis (Fig. 4.21). By further cooperating with electric conductive CNTs, the resulting nanocomposites (CNT-AuNPs-PFIL) showed obvious electrocatalysis toward reduction of H_2O_2 and O_2. In addition, when glucose oxidase was immobilized into CNT-AuNPs-PFIL thin films, it demonstrated favorable linear catalytic response to glucose [69].

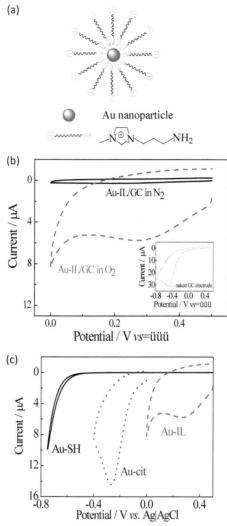

Figure 4.19 (a) Structure illustration of the resulting Au-IL nanoparticles. (b) Cyclic voltammograms of Au-IL nanoparticles modified GC electrode in 0.5 M H_2SO_4 solution saturated with O_2 (dashed) and N_2 (solid), respectively. Inset is bare GC electrode in 0.5 M H_2SO_4 solution saturated with O_2. (c) Cyclic voltammograms of Au-IL nanoparticles (dashed), Au-cit nanoparticles (dotted), and Au-SH nanoparticles (solid) modified GC electrodes in 0.5 M H_2SO_4 solution, respectively. Scan rate: 0.05 Vs^{-1} [63]. Reproduced by permission of The Royal Society of Chemistry.

Figure 4.20 Illustration of the preparation of polydisperse chemically converted graphene (p-CCG) from graphite oxide (GO) by amine-terminated IL [67]. Reproduced by permission of The Royal Society of Chemistry.

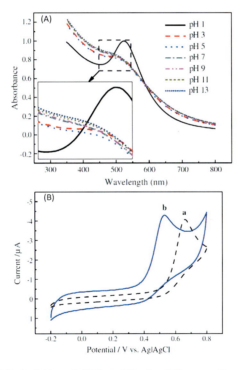

Figure 4.21 (A) Stability of PFIL-AuNPs in different pH solutions after 2 days storage. The inset shows an enlarged view of the rectangular region. (B) Cyclic voltammograms for 0.5 mM NADH in phosphate buffer solution (0.05 M, pH 7.4) at (a) PFIL, (b) PFIL-AuNP-modified GCE at a scan rate of 0.05 V/s. Reproduced with permission from Ref. [68].

Huck et al. prepared Au NPs inside IL-based polyelectrolytes [70]. The nanocomposite synthesis relies on loading the macromolecular film with $AuCl^{4-}$ precursor ions followed by their in situ reduction to Au nanoparticles. It was observed that the nanoparticles are uniform in size and are fully stabilized by the surrounding polyelectrolyte chains. Moreover, XRR analysis revealed that the Au NPs are formed within the polymer-brush layer. Interestingly, AFM experiments confirmed that the swelling behavior of the brush layer is not perturbed by the presence of the loaded NPs (Fig. 4.22). The Au NP-poly-METAC nanocomposite is remarkably stable to aqueous environments, suggesting the feasibility of using this kind of nanocomposite systems as robust and reliable stimuli-responsive platforms.

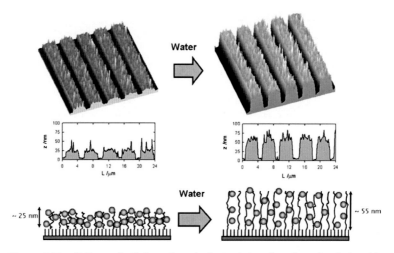

Figure 4.22 Patterned IL-based polyelectrolytes brushes loaded with AuNPs, in compressed and swollen states (left and right, respectively). Top, middle: AFM images and cross-sectional analyses of patterned brush layers, obtained in air and water; bottom: schematic corresponding to AFM data [70]. Reproduced by permission of The Royal Society of Chemistry.

4.7 Others

Counterions of ILs could be facilely exchanged in solution, thus based on a surfactant with an imidazolium unit, the ion-exchange reaction from bromine ion to hexafluorophosphate ion resulted in the amphiphilic-to-hydrophobic transition of the imidazolium salt, leading to the destruction of the micelles (Fig. 4.23), i.e., an ion-responsive micelle system was proposed [71]. This simple process was rapid, taking only one or two seconds from addition of the hexafluorophosphate ions to precipitation being formed. Furthermore, this procedure was much quicker than dissolution of micelles in response to other external stimuli such as light, heat or pH. Potential application for transportation of oil was highly anticipated, which is easier to pump through pipelines in an emulsified state that is less viscous than oil alone. However, the oil must be easily recoverable from this emulsified state after transportation. Here the ILs surfactant approach satisfied these criteria. Moreover, these smart surfactants would find other applications in the fields of nanoscience and polymer science, which would greatly benefit both fundamental research and industry.

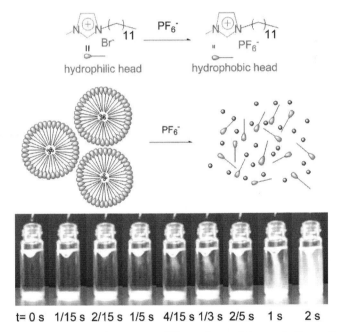

Figure 4.23 Anion exchange to PF_6^- and resulting micelle collapse (above panel), and high-speed photographs showing precipitation after 12 μL of 3.75 M $NaPF_6$ was pipetted into 15 mL of 10 mM aqueous $C_{12}MIMBr$ solution (0.3 equiv. relative to $C_{12}MIMBr$) with mild agitation (bottom panel). From Ref. [71], Copyright Wiley-VCH Verlag GmbH & Co. KGaA. Reproduced with permission.

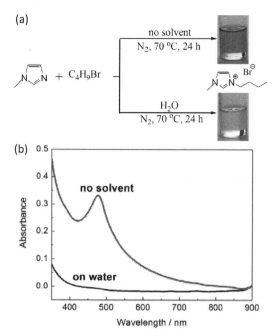

Figure 4.24 Photographs (a) and UV-vis absorption spectra (b) of BMIMBr in liquid prepared on water (lower) and without any solvent (upper). Reprinted from Ref. [72], copyright 2009, with permission from Elsevier.

Although colorless ILs are most desirable, as prepared ILs frequently bear color, despite appearing pure by most analytical techniques, except for UV-vis spectrum. Thus, it leads to some uncertainties and limits for the fundamental research and applications of ILs, which has been a longstanding challenge in the field. Using 1-butyl-3-methylimidazolium bromide (BMIMBr), 1-butyl-3-methylimidazolium tetrafluoroborate (BMIMBF$_4$) and 1-hexyl-3-methylimidazolium bromide (HMIMBr) as examples, it was demonstrated that following traditional synthesizing method except that the water was added as solvent, colorless ILs could be facilely obtained (Fig. 4.24) [72]. In this case, neither critical pretreatment of starting materials and precautions during the reaction nor time-consuming and costly post-decolor-purification was needed. It was believed that in the presence of water, the reactants were biphasic; thus local hot spots due to the higher concentration could be effectively eliminated, which directly produce colorless ILs. It will not only pave the way for economical synthesis of a broad variety of colorless ILs for spectroscopy so as to eliminate the uncertainties and limits but also motivate theoretical and experimental studies to deepen fundamental understanding of the source(s) of color in ILs and enrich the applications of ILs in spectroscopic analytic researches and applications.

4.8 Conclusions

Owing to high ionic conductivity and wide electrochemical window, ILs have the intrinsic advantages for potential applications in electroanalytical chemistry. IL functional materials extend their scope of applications in traditional electroanalytical chemistry merely as green solvents and enable them to be modified on the electrode more conveniently and effectively. Moreover, many other advanced functional compounds could be integrated into multifunctional materials in which ILs act as the backbone. The unique properties of ILs have been well transferred into these ILs materials, and due to synergistic effect, some new electrocatalytic activities are observed. These features enable ILs materials to be promising materials for electroanalytical chemistry. Besides, the concept illustrated here can also be applied to ILs applications in other fields.

References

1. Wasserscheid, P., and Welton, T. (2002). *Ionic Liquids in Synthesis*; Wiley-VCH: Weinheim.
2. Li, J. H., Shen, Y. F., Zhang, Y. J., and Liu, Y. (2005). Room-temperature ionic liquids as media to enhance the electrochemical stability of self-assembled monolayers of alkanethiols on gold electrodes. *Chem. Commun.*, pp. 360–362.
3. Choi, D. S., Kim, D. H., Shin, U. S., Deshmukh, R. R., Lee, S. G., and Song, C. E. (2007). The dramatic acceleration effect of imidazolium ionic liquids on electron transfer reactions. *Chem. Commun.*, pp. 3467–3469.
4. Endres, F., Bukowski, M., Hempelmann, R., and Natter, H. (2003). Electrodeposition of nanocrystalline metals and alloys from ionic liquids. *Angew. Chem.*, **42**, pp. 3428–3430.

5. Dobbs, W., Suisse, J.-M., Douce, L., and Welter, R. (2006). Electrodepositionof silver particles and gold nanoparticles from ionic liquid-crystal precursors. *Angew. Chem. Int. Ed.*, **45**, pp. 4179–4182.
6. Ohno, H. (2005). *Electrochemical Aspects of Ionic Liquids*; John Wiley & Sons: Hoboken, New Jersey.
7. Fukushima, T., and Aida, T. (2007). Ionic liquids for soft functional materials with carbon nanotubes. *Chem. Eur. J.*, **13**, pp. 5048–5058.
8. Abbott, A. P., Capper, G., McKenzie, K. J., and Ryder, K. S. (2007). Electrodeposition of zinc-tin alloys from deep eutectic solvents based on choline chloride. *J. Electroanal. Chem.*, **599**, pp. 288–294.
9. Anderson, J. L., Armstrong, D. W., and Wei, G. T. (2006). Ionic liquids in analytical chemistry. *Anal. Chem.*, **78**, pp. 2892–2902.
10. Buehler, G., and Feldmann, C. (2006). Microwave-assisted synthesis of luminescent $LaPO_4$: Ce,Tb nanocrystals in ionic liquids. *Angew. Chem. Int. Edit.*, **45**, pp. 4864–4867.
11. Zhou, Y., and Antonietti, M. (2003). Preparation of highly ordered monolithic super-microporous lamellar silica with a room-temperature ionic liquid as template via the nanocasting technique. *Adv. Mater.*, **15**, pp. 1452–1455.
12. Antonietti, M., Kuang, D. B., Smarsly, B., and Zhou, Y. (2004). Ionic liquids for the convenient synthesis of functional nanoparticles and other inorganic nanostructures. *Angew. Chem. Int. Ed.*, **43**, pp. 4988–4992.
13. Welton, T. (1999). Room-temperature ionic liquids: solvents for synthesis and catalysis. *Chem. Rev.*, **99**, pp. 2071–2084.
14. El Seoud, O. A., Koschella, A., Fidale, L. C., Dorn, S., and Heinze, T. (2007). Applications of ionic liquids in carbohydrate chemistry: A window of opportunities. *Biomacromolecules*, **8**, pp. 2629–2647.
15. Abbott, A. P., Cullis, P. M., Gibson, M. J., Harris, R. C., and Raven, E. (2007). Extraction of glycerol from biodiesel into a eutectic based ionic liquid. *Green Chem.*, **9**, pp. 868–872.
16. Blanchard, L. A., Hancu, D., Beckman, E. J., and Brennecke, J. F. (1999). Green processing using ionic liquids and CO_2. *Nature*, **399**, pp. 28–29.
17. Swatloski, R. P., Visser, A. E., Reichert, W. M., Broker, G. A., Farina, L. M., Holbrey, J. D., and Rogers, R. D. (2002). On the solubilization of water with ethanol in hydrophobic hexafluorophosphate ionic liquids. *Green Chem.*, **4**, pp. 81–87.
18. Swatloski, R. P., Spear, S. K., Holbrey, J. D., and Rogers, R. D. (2002). Dissolution of cellose with ionic liquids. *J. Am. Chem. Soc.*, **124**, pp. 4974–4975.

19. Winterton, N. (2006). Solubilization of polymers by ionic liquids. *J. Mater. Chem.*, **16**, pp. 4281–4293.
20. Liu, J., Cheng, S., Zhang, J., Feng, X., Fu, X., and Han, B. (2007). Reverse micelles in carbon dioxide with ionic-liquid domains. *Angew. Chem. Int. Ed.*, **46**, pp. 3313–3315.
21. Miller, A. L., and Bowden, N. B. (2007). Room temperature ionic liquids: new solvents for Schrock's catalyst and removal using polydimethylsiloxane membranes. *Chem. Commun.*, pp. 2051–2053.
22. Choi, D. S., Kim, J. H., Shin, U. S., Deshmukh, R. R., and Song, C. E. (2007). Thermodynamically- and kinetically-controlled Friedel–Crafts alkenylation of arenes with alkynes using an acidic fluoroantimonate (V) ionic liquid as catalyst. *Chem. Commun.*, pp. 3482–3484.
23. Mehnert, C. P. (2005). Supported ionic liquid catalysis. *Chem. Eur. J.*, **11**, pp. 50–56.
24. Parvulescu, V. I., and Hardacre, C. (2007). Catalysis in ionic liquids. *Chem. Rev.*, **107**, pp. 2615–2665.
25. Gu, Y. L., Karam, A., Jerome, F., and Barrault, J. (2007). Selectivity enhancement of silica-supported sulfonic acid catalysts in water by coating of ionic liquid. *Org. Lett.*, **9**, pp. 3145–3148.
26. Pringle, J. M., Ngamna, O., Lynam, C., Wallace, G. G., Forsyth, M., and MacFarlane, D. R. (2007). Conducting polymers with fibrillar morphology synthesized in a biphasic ionic liquid/water system. *Macromolecules*, **40**, pp. 2702–2711.
27. Hapiot, P., and Lagrost, C. (2008). Electrochemical reactivity in room-temperature ionic liquids. *Chem. Rev.*, **108**, pp. 2238–2264.
28. Armand, M., Endres, F., MacFarlane, D. R., Ohno, H., and Scrosati, B. (2009). Ionic-liquid materials for the electrochemical challenges of the future. *Nat Mater*, **8**, pp. 621–629.
29. Wei, D., and Ivaska, A. (2008). Applications of ionic liquids in electrochemical sensors. *Anal. Chim. Acta*, **607**, pp. 126–135.
30. Soukup-Hein, R. J., Warnke, M. M., and Armstrong, D. W. (2009). Ionic liquids in analytical chemistry. *Annu. Rev. Anal. Chem.*, **2**, pp. 145–168.
31. Liu, J. F., Jonsson, J. A., and Jiang, G. B. (2005). Application of ionic liquids in analytical chemistry. *TrAC-Trend. Anal. Chem.*, **24**, pp. 20–27.
32. Fukushima, T., Kosaka, A., Ishimura, Y., Yamamoto, T., Takigawa, T., Ishii, N., and Aida, T. (2003). Molecular ordering of organic molten salts triggered by single-walled carbon nanotubes. *Science*, **300**, pp. 2072–2074.
33. Zhang, Y. J., Shen, Y. F., Li, J. H., Niu, L., Dong, S. J., and Ivaska, A. (2005). Electrochemical functionalization of single-walled carbon nanotubes

in large quantities at a room-temperature ionic liquid supported three-dimensional network electrode. *Langmuir*, **21**, pp. 4797–4800.

34. Wei, D., Kvarnstrom, C., Lindfors, T., and Ivaska, A. (2007). Electrochemical functionalization of single walled carbon nanotubes with polyaniline in ionic liquids. *Electrochem. Commun.*, **9**, pp. 206–210.

35. El Abedin, S. Z., Polleth, M., Meiss, S. A., Janek, J., and Endres, F. (2007). Ionic liquids as green electrolytes for the electrodeposition of nanomaterials. *Green Chem.*, **9**, pp. 549–553.

36. Lin, Y. W., Tai, C. C., and Sun, I. W. (2007). Electrochemical preparation of porous copper surfaces in zinc chloride-1-ethyl-3-methyl imidazolium chloride ionic liquid. *J. Electrochem. Soc.*, **154**, pp. D316–D321.

37. Moustafa, E. M., El Abedin, S. Z., Shkurankov, A., Zschippang, E., Saad, A. Y., Bund, A., and Endres, F. (2007). Electrodeposition of al in 1-butyl-1-methylpyrrolidinium bis(trifluoromethylsulfonyl)amide and 1-ethyl-3-methylimidazolium bis(trifluoromethylsulfonyl)amide ionic liquids: In situ STM and EQCM studies. *J. Phys. Chem. B*, **111**, pp. 4693–4704.

38. Mukhopadhyay, I., and Freyland, W. (2003). Electrodeposition of Ti nanowires on highly oriented pyrolytic graphite from an ionic liquid at room temperature. *Langmuir*, **19**, pp. 1951–1953.

39. Yu, P., Yan, J., Zhang, J., and Mao, L. Q. (2007). Cost-effective electrodeposition of platinum nanoparticles with ionic liquid droplet confined onto electrode surface as micro-media. *Electrochem. Commun.*, **9**, pp. 1139–1144.

40. Bhatt, A. I., Mechler, A., Martin, L. L., and Bond, A. M. (2007). Synthesis of Ag and Au nanostructures in an ionic liquid: Thermodynamic and kinetic effects underlying nanoparticle, cluster and nanowire formation. *J. Mater. Chem.*, **17**, pp. 2241–2250.

41. Jia, F. L., Yu, C. F., Deng, K. J., and Zhang, L. H. (2007). Nanoporous metal (Cu, Ag, Au) films with high surface area: General fabrication and preliminary electrochemical performance. *J. Phys. Chem. C*, **111**, pp. 8424–8431.

42. Oyama, T., Okajima, T., and Ohsaka, T. (2007). Electrodeposition of gold at glassy carbon electrodes in room-temperature ionic liquids. *J. Electrochem. Soc.*, **154**, pp. D322–D327.

43. Borisenko, N., Zein El Abedin, S., and Endres, F. (2006). In situ STM investigation of gold reconstruction and of silicon electrodeposition on Au(111) in the room temperature ionic liquid 1-butyl-1-

methylpyrrolidinium bis(trifluoromethylsulfonyl)imide. *J. Phys. Chem. B*, **110**, pp. 6250–6256.

44. Mehnert, C. P., Cook, R. A., Dispenziere, N. C., and Afeworki, M. (2002). Supported ionic liquid catalysis—a new concept for homogeneous hydroformylation catalysis. *J. Am. Chem. Soc.*, **124**, pp. 12932–12933.

45. Gadenne, B., Hesemann, P., and Moreau, J. J. E. (2004). Supported ionic liquids: Ordered mesoporous silicas containing covalently linked ionic species. *Chem. Commun.*, pp. 1768–1769.

46. Lee, B. S., Chi, Y. S., Lee, J. K., Choi, I. S., Song, C. E., Namgoong, S. K., and Lee, S.-G. (2004). Imidazolium ion-terminated self-assembled monolayers on Au: Effects of counteranions on surface wettability. *J. Am. Chem. Soc.*, **126**, pp. 480–481.

47. Choi, E. Y., Azzaroni, O., Cheng, N., Zhou, F., Kelby, T., and Huck, W. T. S. (2007). Electrochemical characteristics of polyelectrolyte brushes with electroactive counterions. *Langmuir*, **23**, pp. 10389–10394.

48. Yu, P., Lin, Y., Xiang, L., Su, L., Zhang, J., and Mao, L. (2005). Molecular films of water-miscible ionic liquids formed on glassy carbon electrodes: Characterization and electrochemical applications. *Langmuir*, **21**, pp. 9000–9006.

49. Zhao, F., Wu, X., Wang, M. K., Liu, Y., Gao, L. X., and Dong, S. J. (2004). Electrochemical and bioelectrochemistry properties of room-temperature ionic liquids and carbon composite materials. *Anal. Chem.*, **76**, pp. 4960–4967.

50. Shen, Y. F., Zhang, Y. J., Zhang, Q. X., Niu, L., You, T. Y., and Ivaska, A. (2005). Immobilization of ionic liquid with polyelectrolyte as carrier. *Chem. Commun.*, pp. 4193–4195.

51. Li, F., Shan, C., Bu, X., Shen, Y., Yang, G., and Niu, L. (2008). Fabrication and electrochemical characterization of electrostatic assembly of polyelectrolyte-functionalized ionic liquid and Prussian blue ultrathin films. *J. Electroanal. Chem.*, **616**, pp. 1–6.

52. Yang, F., Jiao, L., Shen, Y., Xu, X., Zhang, Y., and Niu, L. (2007). Enhanced response induced by polyelectrolyte-functionalized ionic liquid in glucose biosensor based on sol–gel organic-inorganic hybrid material. *J. Electroanal. Chem.*, **608**, pp. 78–83.

53. Shen, Y. F., Zhang, Y. J., Qiu, X. P., Guo, H. Q., Niu, L., and A., I. (2007). Polyelectrolyte-functionalized ionic liquid for electrochemistry in supporting electrolyte-free aqueous solutions and application in amperometric flow injection analysis. *Green Chem.*, **9**, pp. 746–753.

54. Han, D., Qiu, X., Shen, Y., Guo, H., Zhang, Y., and Niu, L. (2006). lectropolymerization of polypyrrole on PFIL-PSS-modified electrodes without added support electrolytes. *J. Electroanal. Chem.*, **596**, pp. 33–37.

55. Wei, D., Unalan, H. E., Han, D. X., Zhang, Q. X., Niu, L., Amaratunga, G., and Ryhanen, T. (2008). A solid-state dye-sensitized solar cell based on a novel ionic liquid gel and ZnO nanoparticles on a flexible polymer substrate. *Nanotechnology*, **19**, pp. 424006–424010.

56. Zhang, Y. J., Shen, Y. F., Yuan, J. H., Han, D. X., Wang, Z. J., Zhang, Q. X., and Niu, L. (2006). Design and synthesis of multifunctional materials based on an ionic-liquid backbone. *Angew. Chem. Int. Edit.*, **45**, pp. 5867–5870.

57. Yu, B., Zhou, F., Liu, G., Liang, Y., Huck, W. T. S., and Liu, W. M. (2006). The electrolyte switchable solubility of multi-walled carbon nanotube/ionic liquid (MWCNT/IL) hybrids. *Chem. Commun.*, pp. 2356–2358.

58. Zhang, Y. J., Shen, Y. F., Kuehner, D., Wu, S. X., Su, Z. M., Ye, S., and Niu, L. (2008). Directing single-walled carbon nanotubes to self-assemble at water/oil interfaces and facilitate electron transfer. *Chem. Commun.*, pp. 4273–4275.

59. Zhang, Y., Shen, Y., Han, D., Wang, Z., Song, J., Li, F., and Niu, L. (2007). Carbon nanotubes and glucose oxidase bionanocomposite bridged by ionic liquid-like unit: Preparation and electrochemical properties. *Biosens. Bioelectron.*, **23**, pp. 438–443.

60. Shan, C., Yang, H., Song, J., Han, D., Ivaska, A., and Niu, L. (2009). Direct electrochemistry of glucose oxidase and biosensing for glucose based on graphene. *Anal Chem*, **81**, pp. 2378–2382.

61. Wang, Z., Zhang, Q., Kuehner, D., Xu, X., Ivaska, A., and Niu, L. (2008). The synthesis of ionic-liquid-functionalized multiwalled carbon nanotubes decorated with highly dispersed Au nanoparticles and their use in oxygen reduction by electrocatalysis. *Carbon*, **46**, pp. 1687–1692.

62. Wu, B. H., Hu, D., Kuang, Y. J., Liu, B., Zhang, X. H., and Chen, J. H. (2009). Functionalization of carbon nanotubes by an ionic-liquid polymer: dispersion of Pt and PtRu nanoparticles on carbon nanotubes and their electrocatalytic oxidation of methanol. *Angew. Chem. Int. Edit.*, **48**, pp. 4751–4754.

63. Wang, Z. J., Zhang, Q. X., Kuehner, D., Ivaska, A., and Niu, L. (2008). Green synthesis of 1–2 nm gold nanoparticles stabilized by amine-terminated ionic liquid and their electrocatalytic activity in oxygen reduction. *Green Chem.*, **10**, pp. 907–909.

64. Li, F. H., Yang, H. F., Shan, C. S., Zhang, Q. X., Han, D. X., Ivaska, A., and Niu, L. (2009). The synthesis of perylene-coated graphene sheets decorated with Au nanoparticles and its electrocatalysis toward oxygen reduction. *J. Mater. Chem.*, **19**, pp. 4022–4025.

65. Li, F. H., Wang, Z. H., Shan, C. S., Song, J. F., Han, D. X., and Niu, L. (2009). Preparation of gold nanoparticles/functionalized multiwalled carbon nanotube nanocomposites and its glucose biosensing application. *Biosens. Bioelectron.*, **24**, pp. 1765–1770.

66. Li, F. H., Li, F., Song, J. X., Song, J. F., Han, D. X., and Niu, L. (2009). Green synthesis of highly stable platinum nanoparticles stabilized by amino-terminated ionic liquid and its electrocatalysts for dioxygen reduction and methanol oxidation. *Electrochem. Commun.*, **11**, pp. 351–354.

67. Yang, H. F., Shan, C. S., Li, F. H., Han, D. X., Zhang, Q. X., and Niu, L. (2009). Covalent functionalization of polydisperse chemically-converted graphene sheets with amine-terminated ionic liquid. *Chem. Commun.*, pp. 3880–3882.

68. Shan, C. S., Li, F. H., Yuan, F. Y., Yang, G. F., Niu, L., and Zhang, Q. (2008). Size-controlled synthesis of monodispersed gold nanoparticles stabilized by polyelectrolyte-functionalized ionic liquid. *Nanotechnology*, **19**, pp. 285601–285606.

69. Jia, F., Shan, C., Li, F., and Niu, L. (2008). Carbon nanotube/gold nanoparticles/polyethylenimine-functionalized ionic liquid thin film composites for glucose biosensing. *Biosens. Bioelectron.*, **24**, pp. 951–956.

70. Azzaroni, O., Brown, A. A., Cheng, N., Wei, A., Jonas, A. M., and Huck, W. T. S. (2007). Synthesis of gold nanoparticles inside polyelectrolyte brushes. *J. Mater. Chem.*, **17**, pp. 3433–3439.

71. Shen, Y. F., Zhang, Y. J., Kuehner, D., Yang, G. F., Yuan, F. Y., and Niu, L. (2008). Ion-responsive behavior of ionic-liquid surfactant aggregates with applications in controlled release and emulsification. *Chem. Phys. Chem.*, **9**, pp. 2198–2202.

72. Shen, Y., Zhang, Y., Han, D., Wang, Z., Kuehner, D., and Niu, L. (2009). Preparation of colorless ionic liquids "on water" for spectroscopy. *Talanta*, **78**, pp. 805–808.

Chapter 5

Nanostructured TiO$_2$ Materials for New-Generation Li-Ion Batteries

Gregorio F. Ortiz,[a] Pedro Lavela,[a] José L. Tirado,[a] Ilie Hanzu,[b] Thierry Djenizian,[b] and Philippe Knauth[b]

[a]*Inorganic Chemistry Department,*
University of Córdoba Campus Universitario Rabanales,
Marie Curie Building, 14071 Córdoba, Spain
[b]*Electrochemistry of Materials Research Group,*
Laboratoire Chimie Provence (UMR 6264),
University of Aix-Marseille I, II, III–CNRS, Marseille F-13397, France

iq1ticoj@uco.es

5.1 Introduction

Few topics raise as much interest as energy-related issues do today. As fuel costs have been continuously spiraling for the last two decades, alternative solutions have been developed. For instance, a significant progress has been made in the development of renewable energy technologies such as solar cells, fuel cells, biofuels, and wind energy. Some of them are mature enough to be exploited commercially; for example, wind or solar energy. However, in spite of the ease of electricity generation by these new technologies, the absence of a cheap and reliable energy storage option constitutes a major inconvenience and an obstacle for the further development of renewable energy sources. In the past, this type of energy sources have been discriminated, but new technology makes alternative energy more attractive, practical, and price competitive than fossil fuels. It is expected that future generations will usher in a long-expected transition away from oil and gasoline as our primary fuel. High performance of such technologies is mainly achieved by designing sophisticated device structures with multiple materials; for example, tandem cells in photovoltaic devices. For example, inorganic/polymer semiconductors with poor charge carrier mobility limit the energy conversion efficiency of organic photovoltaic cells to less than 6%. Thermoelectric devices typically possess an energy conversion efficiency of less than 3%. Portable electric power sources have lower energy and power density largely resulting from poor charge and mass transport properties.

The demand for advanced energy storage devices is continuously increasing. It comes from many different directions having in common the potentially high social and technological benefits of each target application. The lithium-ion battery is a representative system for such electrochemical energy storage and conversion. Nowadays, the effectiveness of lithium-ion batteries is palpable as—having the advantage of light weight and a reduced size—it is used for consumer electronics such as cellular phones, digital cameras, and laptop computers [1]. Other potential uses of renewable energy are in (i) miniaturized portable electronics (battery on a chip, thin-film microbatteries), (ii) rural electricity, (iii) aerospace applications, and (iv) electric and hybrid vehicles (EVs, HEVs), among others. Application (iv) requires high energy and power densities. However, it is a challenge for scientific community to develop a high power density because of a large polarization at high discharge– charge rates of the battery. This polarization is caused by slow lithium diffusion in the active material and increases in the resistance of the electrolyte when the discharging–charging rate is increased. The characteristics of the energy storage device may vary notably depending of the potential application. Safety issues, environmental protection, and manufacturing costs of the commercial batteries still require further improvement.

A potential solution to overcome these problems is to fabricate, design, and shape nanostructured electrode materials that provide high surface area and short diffusion paths for ionic transport and electronic conduction. One-dimensional (1D) nanowires, nanorods or nanotubes, two-dimensional (2D) layered nanostructures, and tree-dimensional (3D) nanomateriales have attracted lot of attention because of their distinctive properties and potential application [2, 3]. The main objective of nanoscale science and technology has been to synthesize nanomaterials with controlled size and shape, as well as search for new properties that are not realized in microscale morphologies [4]. Research is needed to develop nanoscale materials by design, self-assembly for functionality; nanoscale materials characterization and metrology; and properties of materials at nanoscale. New research directions need to be addressed in addition to current research efforts going on in many industries, universities and government research centers and laboratories. First, heterogeneous processes need integration, such as the combination of hierarchical directed assembly techniques with other processing techniques. The second is nanoscale metrology tools, such as in-line or in situ monitoring and feedback. The third is high-throughput hierarchical directed assembly; the fourth is nanoscale components and interconnect reliability. The fifth is nanoscale defect mitigation and removal and defect-tolerant materials, structures, and processes, e.g., self-healing. The sixth is probabilistic design for manufacturing that addresses variability and noise at the atomic scale.

5.2 Economic and Scientific Context of Battery Technology over the World

The world battery market is estimated at over $70 billion with a continuous growth at 4.8% annually through 2012 (www.freedoniagroup.com). Non-lead-acid secondary battery demand will surpass primary and lead-acid secondary batteries. The demand will be headed by Asian countries, Poland, Brazil, and Russia, while markets in the industrialized countries will grow slowly. China is estimated to record the largest growth, with a demand for primary and secondary batteries expecting to annually rise 13% through 2010 to $16 billion.

The major driving force for the wide commercialization of rechargeable batteries was the extended use of portable devices.

Especially, cell phones must be able to operate continuously in standby mode. It involves significant low power consumption from a small battery. Considering that primary batteries only return around 20% of the energy needed for their manufacturing, rechargeable batteries are the lowest cost solution. Other driving forces for the extended use of rechargeable batteries in portable devices are their high sensitivity to minor power losses or fluctuations; the increased demand for wireless game products; lower prices of cell phones, PDAs, and cameras; and the progressive substitution of desktop computers by laptops and netbooks. In the near future, there will be many other application fields demanding very small power sources. Thus, nanobatteries can be very useful as small, high-power, active RFID (radio frequency identification) batteries. MEMS (micro-electro-mechanical systems) batteries will be useful not only in the medical and pharmaceutical industry as power suppliers in implantable medical devices (several microns in size) but also for automotive industry as parts for projectors, optical switching, printers, and various sensors. These batteries could also be suitable for Smartdust. Smartdust is a concept for wireless MEMS sensors that can detect anything from light and temperature to vibrations. Professor Christopher Pister from UC Berkeley suggested in 2001 creating a new type of micro sensor that could theoretically be as small as a grain of sand. Research into this idea is under way and DARPA (the Defense Advanced Research Projects Agency) is funding the work. If these sensors are developed successfully, they will require tiny, powerful batteries to operate and transmit their data. The nanobatteries developed by the group of Peled [5, 6] might be able to supply the necessary power while keeping the size of the sensor as small as possible.

Several rechargeable electrochemical systems have demonstrated to be useful accomplishing the needs of electronic devices in last decades. Mainly, nickel–cadmium and nickel–metal hydride have successfully contributed to the implementation of portable communication and consumer electronic devices. Nevertheless, notorious disadvantages such as element toxicity, memory effects affecting the charge/discharge behavior, and limited energy densities have influenced the progressive shifting to lithium batteries. Rechargeable lithium batteries are light and compact. Their high energy densities make them suitable for portable applications. Moreover, the possibilities in further developments of high-performance lithium batteries are not still limited. Their charge/discharge lifespan and rates can be improved by research in new materials and designs. This will allow the expansion of their use to new applications deserving great expectations. In fact, Li-ion batteries share 80% of the battery industry market. The global sales of Li-ion batteries were 2.5 billion units in 2006 and continue to grow. In 2008, 2.71 billion units of Li-ion batteries were sold across the world, with a sales value of $8.03 billion. Cellular phone market accounts for 60% of the total consumption. The sales are expected to reach 3 billion units in 2010 in the world, with power and polymer batteries to be the new sources of growth of the battery market. Figure 5.1 illustrates a market that is far from exhausting its growing potential.

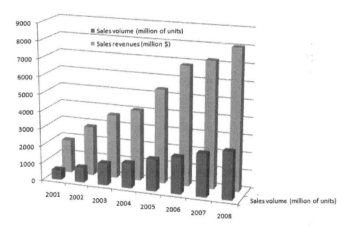

Figure 5.1 Global Li-ion Battery Market Scale, 2001–2008. *Source*: Market Avenue.

The role of energy storage means is a key to face the future energy challenges. An optimal energy management requires an increasing efficiency in electricity production and distribution. The present transmission and distribution systems are uniquely operated from large power plants to consumers. Thus, electricity needs to be used when produced. This situation is not generally achieved and a misfit between demand and consumption is daily and seasonally observed. It is commonly resolved by building overdesigned and expensive plants that currently work under their maximum capacity. Load leveling and peak shaving operations, using large electricity reservoirs, would undoubtedly reduce the costs of energy production. Power plant would be built to supply only sufficient generating capacity to meet average electrical demand rather than peak demands.

Perspectives for the world consumption of energy in its various forms over the next 20 years forecast an overall growth of around 40% [7]. To overcome oil scarcity and environmental degradation, energy sustainability urges for the substitution of fossil fuels by renewable sources and promotion of electric transport. Wind and solar photovoltaic sources will increase their contributions to electricity in the coming decades. The noncontinuous and even erratic energy supply of natural forces makes necessary the availability of suitable ways to store electricity. Facilities such as pumped-hydroelectric and compressed air energy storages are more suitable to store large amounts of energy. Nevertheless, the dependence on a geographical location, delayed response, and long manufacturing time are the nonnegligible drawbacks that must be considered.

The potential significance of lithium batteries can be inferred from the fact that a 1000 m^3 lithium battery system can store 400 MWh. This performance is comparable to a wind farm [8]. There are some attempts to produce large-scale Li-ion batteries. Enterprises in Japan have produced 3 kW demonstration modules. However, the high cost (>$600/kWh) precludes an extended commercialization for grid management purposes [9].

Spreading the use of electric vehicles (EV) and/or hybrid electric vehicles (HEV) is a long-pursued alternative to public and private transport. Recent success in the market introduction of several car models has prompted other manufacturers to increase the offers for the next years. Ni–Cd, Ni–MH, and Pb-acid-type batteries can supply excellent pulsed power, but their large size and heavy weight hinders an optimal application in EV. On the contrary, lithium-ion batteries are well-suited for their use in high-performance electric vehicles. Lithium-ion provides very high specific energy and a large number of charge–discharge cycles [10]. In turn, several drawbacks remain unsolved when it comes to ensure the reliability of these systems when applied to transport vehicles. Thus, safety issues and higher charge/discharge power rates require fundamental research in materials science to provide more efficient electrodes and electrolytes.

5.2.1 Li-Ion Battery Technology: Past and Present

In order to improve the performances of a Li-ion battery, one must understand the processes involved as well as their limitations. Generally speaking, any kind of battery is formed by a negative and a positive electrode, also known as anode and cathode (during the discharge phase), respectively, separated by an electrolyte. As the processes that occur at the two electrodes are generally different, the study of negative and positive electrodes is performed separately. In this chapter, we shall focus on the negative electrodes.

Scrosati and coworkers [11, 12] introduced the term "rocking-chair" batteries in 1980. In this pioneering concept, known as the first generation of rocking-chair batteries, both electrodes intercalate reversibly lithium and show a back-and-forth motion of the lithium ions during cell charge and discharge. The anodic material in these systems was a lithium insertion compound, such as $Li_xFe_2O_3$. The basic requirement for the rocking-chair cell was the existence of two intercalation compounds stable within different potential ranges. The intercalation/insertion reactions must be highly reversible and are desired to provide sufficiently large cell voltages. Figure 5.2 shows the approximate voltage expected in the first-generation cells.

During the late 1980s, research at Sony Energytech [13] developed the first patents and commercial products that can be considered the advent of a second generation of rocking-chair cells. Simultaneously, the term "lithium-ion" was used to describe the batteries using a carbon-based material as the anode that inserts lithium at low voltage during the charge of the cell, and $Li_{1-x}CoO_2$ as cathode material. Larger specific capacities and higher cell voltages were obtained for the second-generation batteries, compared with the first-generation batteries (Fig. 5.2). During the past 10 years, remarkable efforts have been made by many research groups in order to improve the performances of the electrodes and the electrolyte of the lithium-ion battery [14–16]. Two lines of research can be distinguished: (i) improvement of $LiCoO_2$ and carbon-based materials and (ii) replacement of the electrode materials with others with different composition and structure. Concerning the positive electrode, the replacement of lithium cobaltate has been shown to be a difficult task. In this direction, Tarascon et al. gave new impetus to the development of rocking-chair batteries by using $LiMn_2O_4$ as cathode [17]. Lithium manganese spinel oxide and the olivine LiFePO$_4$ are the most promising candidates up to now [18]. These materials have their useful electrochemical reaction in the 3–4 V region, which is appropriate when combined with a negative electrode having a potential sufficiently close to lithium. More recently, 5 V materials have emerged as a different option, e.g., LiNi$_{0.5}$Mn$_{0.5}$O$_4$ and LiCoPO$_4$ [19, 20]. The high working potential of this material makes possible its combination with a high-voltage anode material such as TiO$_2$. A cell built according to this concept belongs to the category of third-generation Li-ion batteries.

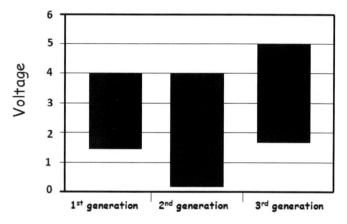

Figure 5.2 Schematic comparison between the different potential windows in the three successive generations of rocking-chair systems. Reprinted from Ref. [41], with permission from Elsevier.

For commercial Li-ion cells, the most used anode remains graphite, which has the advantage of a very good cycle life and reasonably low cost. A graphite electrode works by the intercalation/deintercalation of the Li ions between the graphene layers that correspond to charging/discharging reaction of the Li-ion cell. Concomitantly, electrons are injected or removed from the conduction band of the graphite electrode. However, the graphite and other carbonaceous materials have some major shortcomings. It has to be noted from the beginning that, after a series of improvements in the last decade, the graphite negative electrodes operate very close to their theoretical capacity leaving little place for further improvement. The theoretical specific capacity of graphite is 372 mAh g^{-1} corresponding to a composition of LiC$_6$. The very large majority of commercial systems already operate close to the theoretical value of the capacity. Thus, if significant further capacity improvement of a Li-ion battery is desired, a new negative active material is required to replace graphite.

Another disadvantage of the graphite electrode is the very low intercalation potential, which is only several tens of mV above the potential of the Li/Li$^+$ redox couple. As a consequence, the negative electrode is at a strongly reducing potential. Up to date, there is no electrolyte and solvent combination that can fully withstand the strong reducing conditions at the graphite electrode. Nevertheless, the graphite electrode is stable as a compact passive layer, usually referred to as the solid electrolyte interface (SEI), formed on its surface by the reduction of the electrolyte and the solvent. The large irreversible capacity of a Li-ion battery in the first cycle is a direct consequence of this process. Although stable at room temperature, the SEI passivating abilities degrade rather fast when the temperature increases, hence making the Li-ion batteries quite sensitive to temperatures as low as 50°C.

An intercalation potential very close to the reduction of Li$^+$ ions may also lead to metallic Li electrodeposition on the graphite electrode. The metallic lithium formed in this way is a finely divided powder that, unsurprisingly, is highly reactive, making the battery very unstable. This situation may occur in the event of an accidental overcharge of the battery. Practically, today all the commercial battery packs include electronics that monitor the battery and prevent overcharging.

Needless to say, that an insufficient thermal design of the battery assembly or an accidental slight overcharge may lead to the catastrophic destruction of the battery by the phenomenon of thermal runaway. The losses suffered recently by some battery and laptop computers manufacturers that needed to recall a large number of faulty batteries constitute an eloquent illustration, in economic terms, of the importance that this phenomenon may acquire.

From an economical point of view, graphite will probably never be surpassed by other materials, but the fabrication of electrodes showing better properties is still required. Therefore, other active materials that can serve as anodes have to be investigated. For instance, titanium dioxide has been known for a relatively long time to be able to insert Li ions. Among the known polymorphs of this oxide, it has been reported that the anatase structure is potential candidate for the fabrication of anode in lithium-ion batteries due to the very good lithium intercalation behavior [21].

A clear advantage of the TiO_2 on graphite negative electrodes emerges from the relatively high insertion potential of TiO_2 (around 1.75 V vs. Li/Li$^+$) with respect to the intercalation potential of Li$^+$ into graphite (less than 0.1 V vs. Li/Li$^+$). Such a high potential eliminates the risk of overcharging that may lead to growth of metallic lithium dendrites. Hence, the irreversible capacity due to the formation of the passive layer can be avoided as the passive layer (SEI) would be very thin or inexistent. Without the passive layer, the overall kinetics of the electrode should increase as there are no diffusion penalties introduced by a thick solid SEI.

It can be noted that the use of a high-potential negative electrode may also have an impact on the design of positive electrode active materials. Due to the generalized use of graphite, the large majority of commercial positive electrodes belong to the so-called "4 Volts" (4 V) category, among which $LiCoO_2$ is the most used. A cell formed between a graphite electrode and $LiCoO_2$ has a nominal voltage of 3.7 V. Although positive electrode active materials reaching a potential of 5 V vs. Li/Li$^+$ exist, they cannot be used in functional batteries together with graphite negative electrodes. The resulting 5 V potential difference across such a cell exceeds by more than 1 V the electrochemical stability window of the electrolyte and solvent making the cell unstable, prone to self-discharge and to a poor cycle life. With a high intercalation potential, TiO_2 is also a promising candidate for the "5 Volts" (5 V) category electrodes.

Very recently, improvement in rate capability, capacity, and cycling behavior has been observed in the case of nanostructured titania, strengthening the possibility to use this oxide as a negative material for Li-ion batteries. Bruce and Lindsay have reported that the crystallite size and shape play key roles for improving the performances of the electrodes [22, 23]. It also has to be noted that their investigation concerned either hydrothermal synthesized TiO_2 nanotubes or anodic oxide thin films. None of these two materials is actually organized, but they rather consist of a collection of objects with random orientations.

In all cases, the fabrication of nanostructured electrodes seems to be one of the most promising tracks for improving the performances of power sources because of several advantages. Nanomaterial-based electrodes mitigate the rate-limiting effects of sluggish electron-kinetics and mass-transport. The large surface area of nanomaterials serves to distribute the current density improving electron-kinetics, while the small size ensures that intercalation sites reside close to the surface. Furthermore, when nanomaterials are used to assemble electrodes, new reactions mechanisms are possible. It is proven that conversion reactions of iron oxides work only when the oxide is nanostructured. One of the most spectacular examples of taking the advantages of nanostructured electrodes has been reported by Simon et al. [24]. In this approach, electrodeposition of active negative material has been performed onto current collectors constituted by copper nanopillars. They have clearly demonstrated that kinetic limitations encountered in the case of conventional electrodes can be tackled by designing such highly nanostructured electrodes.

The properties of TiO_2 electrodes may improve dramatically when the material is porous because of a large surface area. Due to exciting properties offered by self-organized TiO_2 nanotube ($ntTiO_2$) layers such as porosity and many potential applications, the preparation of this kind of nanostructured titania by electrochemical techniques has attracted the interest of researchers [25–31]. An early report [32] demonstrated that the anodization of Ti foils in fluoride-containing medium leads to porous nanostructured titania. Since Macklin and Neat [33] reported the high capacity and reversibility of lithium insertion into titanium oxide electrodes in lithium batteries, a considerable research effort has been aimed at optimizing and understanding the titanium oxide anode [23].

The electrochemical performances of self-organized TiO_2 nanotube obtained from Ti foils (references [26, 28] provide examples of the fabrication process) and Ti thin films deposited onto Si substrates have been examined according to a procedure described in reference [34]. Layers of nanotubular titania ($ntTiO_2$) are directly produced by the anodization of Ti in fluoride-containing medium and without the use of any template. We have demonstrated that this kind of nanotubular morphology can be beneficial for an extra lithium accommodation into the structure apart from the interstitial and octahedral sites and allows the manufacture of electrodes without the use of additives and binders [35] in agreement with the study by Liu et al. [36]. The use of TiO_2 reduces the overall cell voltage, but it is reported to provide cells with enhanced safety, low self-discharge, good capacity retention on cycling, and high power due to the possibility to be used in the nanoscale form. In addition, TiO_2 is chemically stable, economically competitive, nontoxic, and an environmental "White Knight."

5.3 Fabrication of Self-Assembled TiO_2 Nanotubes

5.3.1 Anodization of Titanium: Experimental Aspects

The fabrication of self-organized titania nanotube layers ($ntTiO_2$) onto two different substrates is one part of the aim of this chapter and is presented here. In a typical experiment, Ti thin films with a thickness of about 2 μm were deposited by cathodic sputtering using a D.C. triode system onto Si substrates engraved from p-type Si(1 0 0) wafers with a resistivity of 1–10 Ω cm (Wafer World, Inc.), following a similar procedure given in reference [34]. The second Ti substrate consisted of commercial Ti foils from Sigma Aldrich with a thickness of 250 and 127 μm and 99.7% purity. Figure 5.3 shows the schematic representation of the fabrication of self-organized TiO_2 nanotube layers onto different substrates; the principle of fabrication will be discussed in subsequent sections.

Anodization experiments were carried out by applying a constant voltage of 20 V during a time that oscillates between 20 min and 4 h in a conventional two-electrode cell by using an EG&G PARSTAT 2273 potentiostat/galvanostat. The working electrode consisted of titanium (Ti films sputtered onto Si and/or alternatively commercial Ti foil), while a platinum grid served as a counterelectrode. Both electrodes were kept in a distance of 3 cm and the electrolyte was a solution of 1 M H_3PO_4, 1 M NaOH and 0.5 wt% HF. Prior to anodization, Ti was cleaned by sonicating in acetone, isopropanol, and methanol for 30 min, and afterward, the substrates were rinsed with distilled water and dried in Ar flow. To ensure good electronic contact of the electrodes, a fast-drying silver paint [Ted Pella, Inc.] was used. Optional thermal treatment at 450°C in air was performed for 3 h.

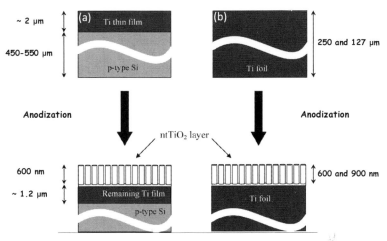

Figure 5.3 Schematic representation of ntTiO$_2$ grown onto two different substrates: (a) silicon and (b) Ti foil. Reprinted from Ref. [86], with permission from Elsevier.

The formation of the TiO$_2$ nanotubes by anodization follows several distinct phases that can be observed during the anodization experiment directly by following the change of colors of the titanium layer or by following the current response of the anodization cell. For the latter, the current response of the cell as function of the time is shown in Fig. 5.4.

After applying the high potential, the current drops quickly in the first seconds and a compact oxide layer on Ti surface is formed (step I, inset, Fig. 5.4). The color of the titanium layer turns violet within this time interval. When the experiment is performed in fluoride-free electrolytes (1 M H$_3$PO$_4$ + 1 M NaOH), the anodization should stop here. The compact oxide layer that is 40–60 nm thick now acts as a very efficient barrier layer and no longer increases in thickness. The violet color seems to be given by a phenomenon of light dispersion as it does not change when observed under different angles. The dispersion of light may occur on the small defects in the oxide layer. When the current density reaches the first minimum, the titanium surface turns blue for several seconds. The blue color is a marker for the beginning of the second region of anodization (step II, inset, Fig. 5.4). At this stage the defect sites present in the compact oxide layer are attacked and dissolved by the fluoride (F⁻) ions present in solution. This process represents the pores nucleation. Then, the current density increases (2.5 min), because the pores are soaking through the barrier oxide layer exposing the titanium to the anodization solution. The beginning of the third region of anodization (step III, inset, Fig. 5.4) is characterized by the maximum of the current density and by a change in color turning to intense green. This corresponds to the maximum pore formation rate. As the viable pores are selected (by a mechanism that will be discussed in a following section), TiO$_2$ nanotubes begin to form. The length of the tubes increases throughout the whole third region. The diffusion length of the fluoride ions and of the anodization by-products increase as well [37]. This is probably the reason for the slow decay of the current density value. Stage IV corresponds to the quasi-equilibrium state in which the growth rate of the tubes equals their dissolution rate.

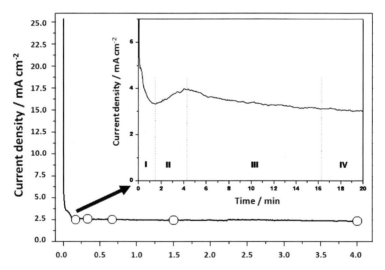

Figure 5.4 Current–time curves for the 4 h and 20 min (inset) of anodization under a potentiostatic regime at 20 V in 1 M H$_3$PO$_4$ + 1 M NaOH + 0.5 wt% HF solution. The four distinct phases of anodization are marked with roman numerals in the inset.

In fact, the growth and dissolution of the oxide layer continues but is much slower at this stage. During this stage, the initial compact layer (that has become porous meanwhile) is completely etched off the surface of the TiO$_2$ nanotubes. SEM pictures of the resulting titanium dioxide nanotubes are shown in Section 5.4.

5.3.2 The Principle of Fabrication

Several studies on the formation of self-organized TiO$_2$ nanotube layers [26, 38] have shown that the growth of oxide nanotubes results from a competition between two reactions. The first reaction is the electrochemical formation of titanium dioxide and may be written in two forms:

$$Ti + 4H_2O - 4e^- \rightarrow Ti(OH)_4 + 4H^+ \quad (5.1a)$$

$$Ti + 2H_2O - 4e^- \rightarrow TiO_2 + 4H^+ \quad (5.1b)$$

In the second reaction, the dissolution of the oxide takes place through the complexation of titanium (IV) by fluoride ions. The reaction can be written as follows:

$$TiO_2 + 6F^- + 4H^+ \rightarrow TiF_6^{2-} + 2H_2O \quad (5.2)$$

The successive stages leading to the final product can be summarized as follows:

(A) Growth of a compact surface film by the electrochemical oxidation of titanium to give oxide products with different hydroxyl content (Fig. 5.5a): Concerning the exact nature of this layer, the following stoichiometry TiO$_{2-x}$(OH)$_{2x}$ can be found in the literature. However, XPS results have been reported that suggest that the surface layer of the film is composed of Ti(OH)$_4$ with traces of fluorine, while the composition of the inner layers is closer to TiO$_2$ [38]. It might be possible that the composition of the initial barrier layer has a significant influence on the subsequent development of the oxide layer. Further studies are required to elucidate this aspect. The Ti(OH)$_4$ layer may also be responsible for the hyperhydrophilic properties of the nanotubular TiO$_2$ layer.

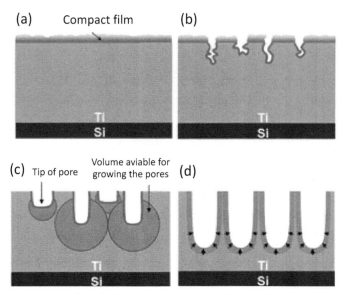

Figure 5.5 Schematic representation of self-organized TiO$_2$ nanotubes fabrication on titanium metal deposited into silicon wafer to improve planarity: (a) compact oxide film formation, (b) breakdown of the compact film and initial growth of wormlike pores, (c) natural selection of deeper pores at the expense of shorter pores, due to their smaller available volume for growth, and (d) conversion of pores into nanotubes by formation of cracks as a result of expansion along the directions of oxidation (arrows). Adapted from Refs. [38] and [39].

(B) Breakdown of the surface film in selected sites by the dissolution reaction (Eq. 5.2): The presence of F$^-$ ions induces the localized breakdown of the barrier layer, probably starting on the existent defect sites. SEM investigations show that the breakdown sites are randomly distributed and there is no sign of self-organization at this stage. These sites act as nucleation points of disordered wormlike porous structures (Fig. 5.5b).

(C) Growth of the wormlike pores: This is a result of the differential dissolution of the oxide film. In the deeper parts of the pores the electrochemical processes (Eq. 5.1) are taking place, and the protons generated favor reaction (Eq. 5.2) by an autocatalytic process. Thus, the oxide dissolution is favored in the lower end of the pore that deepens progressively in the titanium layer. The maximum length achieved by these pores is conditioned by the fluoride ion concentration.

(D) Dissolution of surface film: Simultaneous to pore growth, the initial compact surface film is by the same chemical process (Eq. 5.2). However, the process is slow as it is not enhanced by the autocatalytic effect that leads to local acidification at the bottom of the pores. However, by the end of the anodization, it is commonly found that the film is etched away by chemical dissolution.

(E) Conversion to straight pores: The growth direction of the strong pore is also affected by the neighboring pore. If the oxidation area is larger on one side than on the opposite, due to possible overlapping areas with neighbor pores, the oxidation and H^+ production are larger on one side, leading to the bending of the pore (Fig. 5.5c). This bent growth produces a wormlike structure. When the competition for oxidizable area is equilibrated, the formation of the wormlike structure finishes and all the remaining pores grow straight until the end of anodization, thus producing a tubular morphology.

(F) Self-organization: The mechanism of self-organization was proposed by Yasuda et al. [39], and it is based on a natural selection process. Deeper pores grow at the expense of shorter ones. There is a competence between pores for the volume available for oxidation. When shorter pores are surrounded by deeper pores, the oxide layer grows too thick to allow the short pore to continue growing (Fig. 5.5c).

(G) Conversion to nanotubes: Self-organization of the porous structures could lead to hexagonal arrangements of pores similar to those found in anodic alumina oxide (AAO) membranes commonly used as templates in the synthesis of nanotubes and nanobars. On the contrary, the separation of nanotubes takes place in anodized titanium, which converts the self-organized porous system to a TiO_2 nanotube arrangement. The formation of cracks between the pores that leads to the nanotube morphology results from the mechanical stress in the oxide tubes. In turn, the mechanical stress results from the volume expansion when the metal phase is converted to the oxide phase during anodization [40]. The stress is classified into compressive stress and tensile stress depending on the transport number of the anion and cation in the oxides. Taking into account that the density of Ti (A = 47.87) is 4.506 g·cm^{-3}; while the density of anatase (M = 79.88) is 3.9 g·cm^{-3}, an expansion factor of 1.93 can be calculated. Such high expansion leads to crack formation in the highly stressed regions between deep pores, due to the opposite directions of oxidation-expansion in neighbor pores (Fig. 5.5d).

5.4 Characterization of Titania Nanotubes Layers

5.4.1 Scanning Electron Micrograph (SEM) and X-Ray Diffraction (XRD)

SEM pictures of the resulting titanium dioxide nanotubes are shown in this section. Depending on the electrochemical conditions during anodization experiments, the obtained nanotube length (or thickness) differs. From the top-view image, it is possible to observe the nanotube morphology at different times of anodization. In the beginning, after 5 min, the presence of F$^-$ ions induces the localized breakdown of the compact oxide layer, probably starting on defect sites (Fig. 5.6).

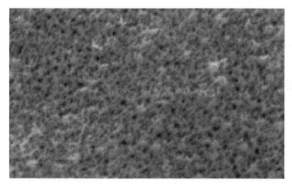

Figure 5.6 SEM micrograph of the Ti foils anodized for 5 min at 20 V in 1 M H$_3$PO$_4$ + 1 M NaOH + 0.3%wt. HF (*Note*: showing the nucleation points).

SEM investigations show that the breakdown sites are randomly distributed and there is no sign of self-organization at this stage, as we mentioned in Section 5.3.2 (Fig. 5.5b). These sites act as nucleation points of a disordered wormlike structure (Fig. 5.5c). Highly ordered array of TiO_2 nanotubes is observable after 20 min and 4 h of anodization, as discussed in Section 5.3.2. $ntTiO_2$ presents a tube diameter that varies from about 50 nm to about 200 nm and wall thickness of about 20 nm whatever the time employed during the experiment (Fig. 5.7a,b). From cross-sectional images, it is possible to observe that the nanotube length is increased during anodization for longer times. Thus, the samples obtained after 20 min and 4 h exhibit about 600 nm and 900 nm length, respectively (Fig. 5.7c,d).

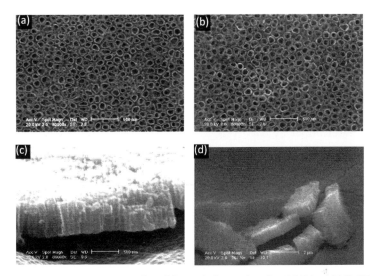

Figure 5.7 SEM micrographs of the Ti foils anodized at 20 V in 1 M H_3PO_4 + 1 M NaOH + 0.3%wt. HF for (a) 20 min and (b) 4 h (top view). Idem from cross section for (c) 20min and (d) 4 h.

On the other hand, when using Ti thin film onto silicon substrates (Fig. 5.8), self-organized TiO_2 nanotubes layers with thickness of about 600 nm are obtained after anodization under the same conditions. Moreover, the nanotube diameter and wall thickness are in the range 50–200 nm and about 20 nm, respectively. Hence, a layer of titania nanotubes is successfully fabricated on Ti foils and Ti thin films. Moreover, the titania nanotubes were heated at 450°C in air for 3 h and the nanotube morphology remained unmodified, as we will discuss later.

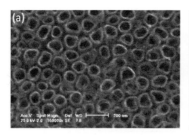 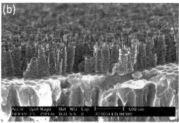

Figure 5.8 SEM images of TiO_2 nanotube layers after anodization (20 V, 20 min) in a fluoride-containing electrolyte: (a) top view, and (b) cross section of the amorphous TiO_2 nanotube layers onto Si Substrate.

XRD characterization can be helpful in knowing which TiO_2 phase or which kind of amorphization is obtained just after anodization and thermal treatment. For that purpose, X-ray-diffraction (XRD) patterns were recorded at room temperature using a Siemens D5000 diffractometer with Cu Kα radiation (1.5406 Å) with a recording scan 0.133°/min.

As-formed ntTiO_2 nanotube layers onto Ti foils are amorphous as determined by XRD (Fig. 5.9a,b) irrespective of the thickness (600 or 900 nm) obtained because the only peaks that can be indexed are for Ti (JCPDS file no. 5-682). The annealing of ntTiO_2 layers at 450°C in air for 3 h (Fig. 5.10) leads to a partial crystallization into anatase structure. Two new peaks at about 25.3° and 47.9° are detected, corresponding to the (101) and (200) reflections of the anatase phase (JCPDS file 21-1272), respectively. On the other hand, the Ti layer deposed on Si substrate was identified as hexagonally close-packed α-Ti (JCPDS file no.: 44-1294) and the ntTiO_2 nanotubes obtained at room temperature are X-ray amorphous. On heating we observe again the partial crystallization of titania to anatase (Fig. 5.9c). Peaks of Ti are still present after growing the ntTiO_2 not only onto Ti foils but also onto Si because a layer of Ti (1.25 μm in thickness) remains after anodization (see SEM image in reference [34]). In the particular case of the silicon substrate, 35% of the initial Ti layer is converted into ntTiO_2. This remaining Ti layer is not electroactive versus lithium, as it is discussed later.

The robustness of the two substrates should be addressed considering its integrity under these annealing conditions since no differences were observed by XRD and SEM. Moreover, the microstructure of the nanotubes remains basically unmodified as it can be seen from the SEM image given in Fig. 5.9. It is clear that no morphological changes appear in the porous structure after this thermal treatment, which is in agreement with previous data reported by Schmuki's group [42].

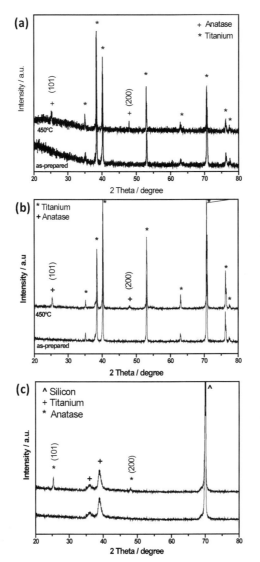

Figure 5.9 XRD patterns of (a) as-prepared and annealed ntTiO$_2$ nanotube layers on Ti foils with 600 nm length. (b) As-prepared and annealed ntTiO$_2$ nanotube layers on Ti foils with 900 nm length. (c) As-prepared and after heat treatment at 450°C of TiO$_2$ nanotube layers onto Si substrate. Note that reflections of anatase are written in the figure.

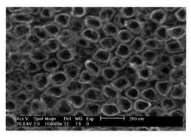 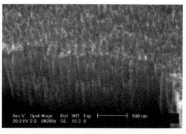

Figure 5.10 Top-view (left) and cross-section (right) SEM images of ntTiO$_2$ after annealing at 450°C in air for 3 h.

The thickness of nanotubes can be estimated from Eq. 5.1 and using Faraday's law:

$$L = \frac{QM}{Fn\delta}, \qquad (5.3)$$

where Q is the circulated charge (C cm^{-2}), M the molecular weight of the oxide (79.9 g mol^{-1}), F the Faraday constant (96500 C equiv^{-1}), n the number of electrons involved in the reaction, and δ the density of TiO$_2$ (3.8–4.1 g cm^{-3}). Therefore, the theoretical thickness of the nanotube layers should be 3500 nm for 20 min of anodization, leading to an anodization efficiency of about 20%. This relative low value can be explained by the competition between Eqs. 1.1 and 1.2 shown in Section 5.3.2, i.e., between growth and dissolution of the nanotube layer [26].

5.4.2 TiO$_2$ Polymorph: The Interest of Anatase and Amorphous Titania

Several TiO$_2$ polymorphs have been reported and named as rutile, anatase, brookite, TiO$_2$-B (bronze), TiO$_2$-R (ramsdellite), TiO$_2$-H (hollandite), TiO$_2$-II (columbite), and TiO$_2$-III (baddeleyite). Since a long time, the practical use of TiO$_2$ had been as white pigment in paint substituting the toxic lead oxides. One of the reasons for this is its whiteness attributed to the relatively high refractive index, which, in combination with small particle sizes, results in strong light scattering in a broad range of wavelengths. Other commercial applications are in products such as sun lotion and toothpaste due to its remarkable optical properties and chemical stability and as a pigment when the crystalline form has a high refractive index (rutile). Concerning energy conversion and storage applications, titanium dioxide (TiO$_2$) is one of the most important wide-gap semiconductors and is widely envisaged for use in photoelectrochemical solar cells, supercapacitors, and Li-ion batteries [41, 42], among others.

Photoelectrochemical solar cells: The use of TiO$_2$ in solar energy conversion is directly related with the early development of photoelectrocemical cells to either water cleavage and H$_2$ production or electricity generation. In the early 1970s, Fujishima and Honda proposed a design of solar cell benefiting from the semiconductor properties of TiO$_2$. Its bandgap energy is suitable for adsorbing ultraviolet-light-promoting electrons to conduction bands. The potential difference created in the electrode is appropriated to oxidize oxide anions while H$_2$ is produced in the Pt counterelectrode [43]. However, the absorption efficiency is rather low and its development was further abandoned for commercial purposes. In the 1990s, dye-sensitized photoelectrochemical cells proposed combining the better absorption properties of dyes with TiO$_2$ semiconductivity [44, 45], even though efficiency was limited by the deposition of thick dye films deposited on titania compact layers. In this way, photon absorption was enhanced, but recombination was also promoted.

A notorious improvement was achieved by using nanostructured TiO$_2$ providing large surface area for the deposition of dye thin films. Thus, an enhanced absorption was allowed without being penalized by recombination. Parellel TiO$_2$ nanotube arrays prepared by anodization of metallic titanium layer not only provide a large surface area but also ordered pore geometry. The nanotubes disposal seems to be more suitable for pore geometry of the nanotubes for solid-state cells manufacturing.

An improved solar harvesting efficiency compared with conventional sol–gel-derived TiO$_2$ has been demonstrated [46]. In fact, the electron diffusion length is consistent with a collection efficiency for electrons close to 100%, even for thick nanotube films [47].

The efficiency of dye-sensitized solar cell has reached as much as 11% [48], but values higher than 15% are required to make them competitive with Si-based solar cells. Electrons transport time is delayed in the nanotube by the presence of grain boundaries, defects, and trap sites. This is an important limitation in manufacturing long TiO$_2$ nanotube for efficient dye-sensitized solar cells. For this reason, several solutions have been proposed, including ZnO thin film coating [49], substitutional doping (C, N, etc.) [50, 51], and combining TiO_2 with quantum dots (QDs) [52].

Charge storage application: charge can be directly store in capacitors and batteries. In capacitors, charge is electrostatically attracted and retain at the electrode surface, the capacitance being related to the amount of stored charge and proportional to area. Capacitors are able to deliver high power pulses with a very long cycle life, but the energy density is somewhat poor. In turn, electrochemical batteries rely in a reversible redox reaction to accumulate charge. The limited reaction rate is responsible for lower charge and discharge rates though allowing higher energy densities. In both devices, TiO_2 has demonstrated to be a suitable material to provide high performances.

In order to enhance the charge storage, supercapacitor devices propose the use of high surface electrodes. Two main inorganic materials have been successfully essayed. In activated carbons, an electrochemical double-layer is formed at the electrode–electrolyte interfacial regions. However, the low capacitance values (10–40 F cm^{-2}) exhibited are an important drawback for their extended use of nonfaradaic supercapacitors. Nanostructured metal oxides supercapacitors are based on a pseudocapacitive effect involving a dependency between voltage and a faradaic reaction. It has been demonstrated that higher energy densities can be achieved without significantly penalizing power rate and long life cycling of double layer capacitors. These asymmetric supercapacitors are assembled with a high surface carbon and the metal oxides using an aqueous solution as electrolyte. The faradaic reaction here involved is expected to provide charge devices to fill the gap between capacitors and batteries. Hydrous ruthenium oxide provides high capacitance values (720 F g^{-1}) but is very expensive. Hence, less alternative less costly materials were searched including manganese oxide, iron oxide, vanadium oxide, etc. [53].

TiO_2 is a low cost material, however, TiO_2 is a not electronically conductive material which is usually used in the dielectrically capacitors but should not be suitable for supercapacitors based on pseudocapacitive effects. In turn, titania has demonstrated to be an excellent matrix to favor the dispersion of other active materials. Moreover, an ordered multiporous morphology should provide more real surface areas for the faradaic reaction through an accessible interface for mass and charge transfer reactions. Dunn et al. have reported that the pseudocapacitive contribution to charge storage in anatase TiO_2 nanocrystals is significantly enhanced for particle sizes lower than 10 nm. Thus, total stored charge and charge/discharge kinetics were improved [54]. Thus, several examples have been reported including RuO_2 on TiO_2 nanotubes [55] $Co(OH)_2$ on TiO_2 nanotubes [56]. However, the powdered TiO_2 nanotubes are usually disorderly arranged and features very small tube size (<10 nm) hindering the entering of the transition metal particle into the inner surfaces of nanotubes during the compositing process. Self organized TiO_2 nanotube arrays with a controlled pore size and tube length are feasible by an electrochemical anodization method. The superior energy-storage performance of NiO on TiO_2 nanotubes has been mostly attributed to its highly accessible redox reaction sites [57]. Brezesinski et al. have revealed the importance of the combination of nanocrystallinity and mesoporosity to attain greater power densities and energy densities in noncomposited TiO_2 films. According to these authors polymer-directed assembly of preformed nanocrystals into three-dimensional frameworks is an alternative route to overcome the loss of periodicity and porosity upon crystallization [58].

Recently, Telcordia Technologies has proposed nonaqueous asymmetric hybrid supercapacitor as a new concept to provide high energy densities to pseudocapacitors [US Patent no. 6,252,762]. This device utilizes a nonaqueous electrolyte coupled with a Faradaic lithium-intercalation anode. The stored charge usually comes from the reversible adsorption/desorption reaction of anions on the surface of the carbonaceous cathode material and the reversible lithium insertion/extraction reaction in the anode material [59]. The first asymmetric supercapacitors were based on $Li_4Ti_5O_{12}$ anodes though others titanium oxides have been further researched as TiO_2-B nanowires [60] and TiO_2 nanotubes [61].

Anode in Li-ion batteries: the interest on titanium oxides based materials comes from the early nineties when several reports revealed the lithium insertion properties of spinel related structures as $LiTi_2O_4$ and $Li_4Ti_5O_{12}$ [62]. However, the specific capacity is limited to 175 mAh g^{-1}. TiO_2 is recently regaining interest as anode for Li-ion batteries due to its higher theoretical capacity (335 mAh g^{-1}) when 1 Li is inserted per TiO$_2$ unit. Additional advantageous features are its high mechanical stability upon continuous battery cycling, low cost and lack of toxicity. Its potential versus the Li$^+$/Li redox couple is about 1.75 V. It limits undesirable side reaction with the electrolyte and avoids Li electroplating, inherent of graphite anodes and origin of safety concerns. Nevertheless, the poor ionic and electronic of TiO$_2$ polymorphs limit the electrochemical insertion of lithium. Previous electrochemical studies have shown that the different modifications of TiO$_2$ accommodate lithium in different ratios [63, 64]. Thus, rutile, being the most thermodynamically stable polymorph of TiO$_2$, insert less than 0.1 Li ions per formula unit in its bulk crystalline [65] Li diffusion in rutile is highly anisotropic. While lithium diffuses rapidly through c-axis channels (ca. 10–6 cm^2 s^{-1}), a very slow diffusion proceed through the ab plane (ca. 10–15 cm^2 s^{-1}). Consequently, the repulsive Li-Li interaction in c channels and the lithium blocking in ab plane significantly restrict the insertion reaction in crystalline bulk materials [66]. Similarly, brookite inserts only small amounts of lithium [67]. However, it has been recently reported that lithium can be efficiently inserted in both phases when produced as nanometric particles. It can be ascribed to the stability of the morphology and to better accommodation of the structural changes at nanometer size [68, 69]. Contrarily, bulk anatase [70] and the synthetic polymorph TiO$_2$ (B) [71] react widely with stoichiometries up to 0.8–1 Li per titanium.

Natural mineral anatase has a three-dimensional network that consists of strongly distorted edge sharing TiO$_6$ octahedra as shown in Fig. 5.11. It can be considered a stacking of one-dimensional zigzag, equivalent in [100] and [010] directions, leading to empty zigzag channels in the anatase framework. When reacting with lithium, the ions are accommodated on highly distorted octahedral interstices and consequently undergo an orthorhombic distortion which is better described as a five-fold coordination of lithium with oxygen. The overall distortion of the atomic positions in the change from TiO$_2$ (anatase) to Li$_x$TiO$_2$ (Li-titanate) is small and leads to a more regular TiO$_6$ octahedra in the inserted compound. A volume increase of the unit cell is supposed to be about 4% as Nuspl et al. reported. Thus, a decrease of unit cell along the *c*-axis and an increase along *b*-axis takes place. The structural change can be explained as occupation of Ti–Ti bonding atomic orbitals by the electron that enters the TiO_2 lattice with each Li-ion to maintain charge neutrality. As a result of lithium intercalation, the symmetry of anatase is lowered from I41/amd to Imma in $Li_{0.5}TiO_2$, where lithium is statistically distributed over one four-fold site in Imma since only one half of the crystallographic sites are occupied. Consequently, the lithium intercalation can be viewed as a filling of titanium *d* levels by neglecting the influence of lithium levels. The partially filled titanium levels are the t_{2g} set in an ideal octahedral coordination. Such behavior is expected for the compounds that are capable of incorporating lithium simply and reversibly.

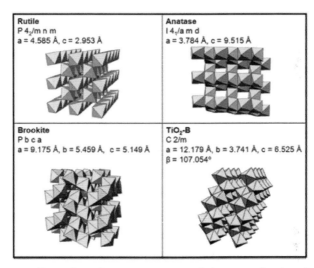

Figure 5.11 Examples of some structures of the natural mineral: rutile, anatase, brookite, and TiO_2–B.

In an ideal octahedral symmetry, the t_{2g} position of the titanium "d" orbitals is not degenerated, but this symmetry is split in the case of anatase and $Li_{0.5}TiO_2$ because the local TiO_6 geometry is lowered from O_h in an idealized framework to D_{2d} and C_{2v}, respectively. Figure 5.12 shows in short the splitting of titanium *d* orbitals. This splitting of the t_{2g} set can be easily derived by using arguments of ligand-field theory [72]. This phase transition is reflected in the electrochemical charge and discharge curves by the appearance of a large plateau. The continuous structural changes on charge/discharge cycling are responsible for the capacity fade [73]. Instead, lithium insertion in nanostructured anatase revealed a continuous voltage decrease on discharge, which is related to a solid solution process, overcoming the detrimental effect of the phase transition [74].

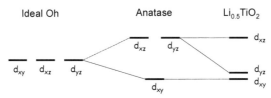

Figure 5.12 Splitting of titanium d orbitals when lithium intercalation takes place into TiO_2. Adapted from Ref. [72].

Size is not the only factor that matters in preparing new nanostructured anodes for Li-ion batteries. The control of shape may provide interesting advantages. The improved electrochemical behavior of TiO_2–B nanowires with diameters similar to that of anatase and TiO_2–B nanoparticles has been recently reported [75].

A number of reports have been published related to the improved electrochemical properties of TiO_2 nanotubes. Zhou et al. used a hydrothermal process to obtain uniform anatase nanotubes with a rate capacity of 182 mAh g^{-1} and a minimal electrochemical resistance although the discharge current density was low (ca. 80 mA g^{-1}) [76]. Li et al. prepared anatase nanotubes by calcining protonic titanate nanotubes precursors. The lithium test cell assembled with this electrode exhibited high discharge capacity with 314.4 mAh g^{-1} and good capacity retention [77]. The hydrothermal process has been also used to prepare TiO_2 nanorods with 40–80 nm in diameters and 400–1500 nm lengths. This active material was able to deliver 280 mAh g^{-1} after 40 cycles [78].

Nevertheless, the hydrothermal method provides nanotubes with different sizes and random orientations, thus hindering the electronic and ionic transport. Hence, it makes necessary the presence of a conductive second phase mix uniformly even if it is present as a nanostructured material. The electron-conductive additive must be intimately mixed or even better, as nanostructured as the active material. Otherwise, the electron transport may remain limited. For this purpose, nanocomposited materials in which the electron-conductive material is included at the nanoscale are being developed. Particularly, it has been reported the beneficial effect of the addition of small amounts of silica or RuO_2 for improving the rate performance anatase TiO_2 nanotubes [79]. Also, nanosized TiO_2 and carbon nanotubes were composited to prepared electrodes delivering a reversible capacity of 168 mAh g^{-1} at the first cycle which remained almost constant during long-term cycling [80].

An alternative approach is constituted by electrochemical induced self-organization promoted by a controlled anodic oxidation of a titanium substrate. Synthesis of these titania nanotubes and its application to lithium ion batteries are described in this work, which constitute the main part. The well-aligned tubes are expected to facilitate both electrode/electrolyte contact, and electron and lithium-ion transport. Moreover, the morphological parameters such as length, diameter, and wall thickness are easily tailored by tuning the synthesis conditions. Besides, the high packing density and long tube length provide a high storage capacity, which is described in following section.

5.5 Battery Applications

5.5.1 Electrochemical Behavior of Samples in Lithium Cells

Because of its relevance to batteries, the mechanism of lithium insertion into anatase TiO_2 has been extensively studied. The electrochemical insertion/extraction of Li is believed to be driven by the accumulation of electrons in TiO_2 electrodes in contact with Li$^+$-containing electrolytes, and the overall cell reaction can be written as

$$TiO_2 + xLi^+ + xe^- \leftrightarrow Li_xTiO_2 \tag{5.4}$$

The crystalline structure of anatase is tetragonal (s.g. I4$_1$/amd) and contains distorted TiO_6 octahedra, which define a series of octahedral and tetrahedral vacant sites. These sites allow lithium uptake of 0.5 Li per formula unit, corresponding to a theoretical capacity of 168 mAh g^{-1} [81]. A two-phase mechanism has been suggested to describe the electrochemical insertion of lithium into anatase, involving equilibrium between Li-poor (tetragonal) and Li-rich (orthorhombic) phases [82] as we mentioned above. The latter phase results from a structural distortion caused by a cooperative Jahn–Teller effect, as the incoming electrons increase the d electron density in localized Ti-d orbitals above a critical intercalation concentration. The maximum theoretical capacity for $x = 1$ should be 336 mAh g^{-1}, in which lithium transfers its valence electron to the host lattice and enters the one-dimensional channels as Li$^+$ ion denoted by $(Li_x)^{x+}(TiO_2)^{x-}$. The average net charge of lithium (+1.071) determined by a Mulliken population analysis supports a complete ionic formulation [83].

All the electrochemical performance of the ntTiO$_2$ layers was studied by experiments carried out in Li/LiPF$_6$ (EC:DEC)/ntTiO$_2$ cells. The electrolyte supplied by Merck was embedded in a Whatman glass microfiber acting as a separator and the current collector for the ntTiO$_2$ was a copper foil (99.99% purity). For these experiments, no additives such as poly(vinyl difluoride), which acts as binder agent, and carbon black (conductive agent) were used. Assembling of the cells was performed in a glove box filled with purified argon in which moisture content and oxygen level were less than 2 ppm. Lithium cells were galvanostatically cycled using an Arbin potentiostat/galvanostat multichannel system. For the discharge/charge reaction, a constant current density of 100, 20, and 5 µA cm^{-2} was applied to the assembled cells in the range between 2.6 and 1.0 V.

In Table 5.1, all capacity values for ntTiO$_2$ with about 600 nm length are expressed per electrode area to allow better comparison with literature data for thin-film microbatteries. Figures 5.13 and 5.14 show the galvanostatic discharge/charge curves versus composition of the electrodes grown on Si and Ti foils (Si-free) substrates using a current of 5 µA cm^{-2} (C/8), respectively. Galvanostatic curves are helpful to identify the behavior for lithium intercalation into the electroactive materials. As we have materials with different morphology and crystalline structure, we will study the different behavior in curves. For instance, the crystallized electrode exhibits a pseudo-plateau at 1.75 V during the discharge, and during the charge, a potential plateau at 1.95 V is visible. These two potential plateaus (Fig. 5.13b) correspond to Li$^+$ insertion and deinsertion from interstitial and octahedral sites of crystalline anatase TiO$_2$ nanotubes.

Table 5.1 Summary of capacity values for ntTiO$_2$ with about 600 nm length, which are expressed per electrode area to allow better comparison with literature data for thin-film microbatteries. Reversible and irreversible areal capacities and efficiency on cycling of the ntTiO$_2$ layers as electrodes in experimental Li test cells (reprinted from Ref. [86], with permission from Elsevier.)

Electrode	Current density (μA cm^{-2})	First discharge capacity (μAh cm^{-2})	Irreversible capacity (μAh cm^{-2})	1st reversible capacity (μAh cm^{-2})	50th reversible capacity (μAh cm^{-2})	Efficiency (%)
Amorphous ntTiO$_2$ on Si	5	196	107	89	56	63
	100	45	19	26	17	65
Crystalline ntTiO$_2$ on Si	5	165	89	76	40	53
	100	34	10	24	23	96
Amorphous ntTiO$_2$ on Ti foil	5	129	76	53	37	70
	100	41	18	23	15	65
Crystalline ntTiO$_2$ on Ti foil	5	83	34	49	29	60
	100	39	13	26	20	77

Note: Values for layers of 600 nm length onto Si substrates and Ti foils.

A total capacity of about 165 μAh cm^{-2} was obtained in the first discharge that includes reversible and irreversible reactions. The first reversible capacity obtained in the second discharge was 76 μAh cm^{-2}. For amorphous nanotubes, no plateau is observed during the discharge/charge profile of the curve. Only one pseudo-plateau at about 1.15 V contributes to a large irreversible capacity. The total capacity in the first discharge (196 μAh cm^{-2}) and the first reversible capacity (89 μAhcm^{-2}) are higher than for crystalline structure. From these results, one can admit that the irreversible capacity is strongly dependent on the structure of ntTiO$_2$. The relatively high irreversible capacity can be attributed to different phenomena: first, the irreversible reaction of Li$^+$ with adsorbed water molecules onto the ntTiO$_2$ electrode [84] and, second, the formation of a very thin, disordered layer at the electrode surface that may appear on both amorphous and crystalline electrodes [35, 85–86]. Water is certainly still present in some proportion in annealed samples, because the annealing temperature is not high enough to completely remove strongly chemisorbed water [35] or bound water [40]. The difference in the irreversible capacity can be explained by the fact that an annealing treatment removes structural and chemical defects in the amorphous phase that act as Li$^+$ ion traps.

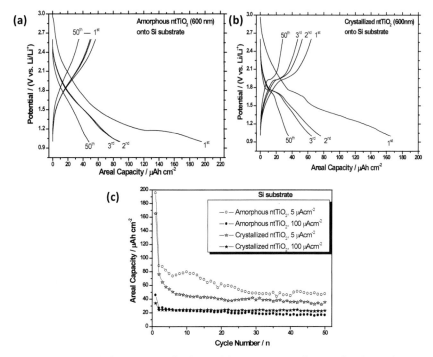

Figure 5.13 Galvanostatic discharge/charge curves during the 50 cycles of ntTiO$_2$ nanotube layers obtained onto Si substrates: (a) as-prepared, and after thermal treatment at 450°C (b). (c) Evolution of its areal capacity as a function of cycle number using a kinetic of 5 (C/8) and 100 (2.5C) μA cm^{-2}.

A cycling life study at two different kinetics was carried out (Fig. 5.13c). The lithium intercalation can be viewed as a filling of titanium d levels by neglecting the influence of Li levels. The partially filled titanium levels are the t_{2g} set in an ideal octahedral coordination as mention above. Such behavior is expected for compounds that are able to incorporate lithium ions easily and reversibly [83]. At 5 μA cm^{-2}, the capacity for amorphous ntTiO$_2$ onto Si is about 56 μAh cm^{-2} after 50 cycles leading to an efficiency of about 63%. The ntTiO$_2$ annealed at 450°C presented a maximum capacity of 76 μAh cm^{-2} and after 50 cycles 40 μAh cm^{-2} resulting in an efficiency of about 53%. These results suggest that amorphous nanotubes can accommodate extra Li into its structure compared with heated sample. Part of this enhancement is due to the amorphous structure of titania nanotubes, which facilitates an extra lithium insertion.

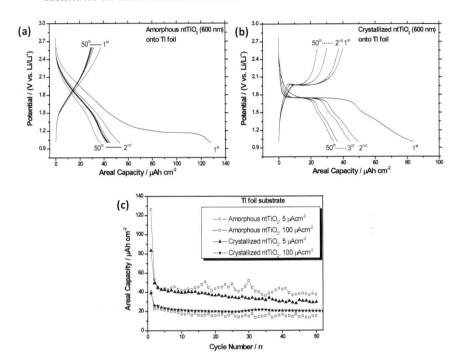

Figure 5.14 Galvanostatic discharge/charge curves during the 50 cycles of ntTiO$_2$ nanotube layers obtained onto titanium foils (Si-free substrates): (a) as-prepared, and after thermal treatment at 450°C (b). (c) Evolution of its areal capacity as a function of cycle number using a kinetic of 5 (C/8) and 100 (2.5C) μA cm^{-2}.

Other possible reasons for the high experimental capacity values are the high surface area and highly organized 1D structure of titania nanotube layers. When a rate of 100 μA cm^{-2} (2.5C) is used, the reversible and irreversible capacities are lower than at a rate of 5 μA cm^{-2}, but the capacity retention for crystalline is around 96% (close and star-shaped symbols in Fig. 5.13c; Table 5.1).

A similar behavior was found in the galvanostatic curves of ntTiO$_2$ samples with 600 nm length manufactured on commercial Ti foil (Fig. 5.14). The main difference between amorphous and crystalline nanotubes resides in the observation of a voltage plateau during the discharge (1.75 V) and charge (1.95 V) of the cell during the 50 cycles. Under low kinetics, the amount of lithium inserted into amorphous (53 µAh cm^{-2}) is higher than in crystalline ntTiO$_2$ (49 µAh cm^{-2}), and they exhibit after 50 cycles an efficiency of 70% and 60%, respectively. A cycling life study of the electrodes under fast kinetics (100 µA cm^{-2}, 2.5 C) show that capacity retention is better in crystalline (77%) than amorphous ntTiO$_2$ (65%), but the capacity value is about 20 µAh cm^{-2} (Fig. 5.14 and Table 5.1).

The better electrochemical performances exhibited by the ntTiO$_2$ layers grown onto Si substrate are attributed solely to the presence of the nanotubes. According to literature, silicon and titanium cannot react with Li$^+$ between 2.6 and 1.0 V. Indeed, Si is only electroactive in the region that ranges between 0.45 and 0.01 V versus lithium [87, 88] forming LiSi and Li$_{3.6}$Si alloys during the discharge of the cell. Ti metal is not electroactive versus Li$^+$ and do not form alloys as corroborated also by other authors who have also used Ti foils for the production of TiO$_2$ nanotubes [36, 85].

Therefore, the better capacities obtained from ntTiO$_2$ layers formed onto Si are attributed to the nature of Ti thin film. Compared with ntTiO$_2$ layers grown from mechanically polished Ti foils, anodization of titanium thin film obtained by physical vapor deposition (PVD) process leads to the formation of highly flat ntTiO$_2$ layer. Thus, we assume that the contact between the ntTiO$_2$ layers and the electrolyte is drastically improved.

Now we describe the electrochemical behavior of ntTiO$_2$ grown on Ti foils for 4 h at 20 V, which resulted in about 900 nm of length. For the purpose of comparing its electrochemical performance with other thin films described in literature, we show the data in Table 5.2. Table 5.2 presents a comparison of nanotubes versus the compact oxide layer to really compare the nanotube morphology exhibit improved electrochemical behavior in lithium cell. Figure 5.15a,b shows the galvanostatic discharge/charge curves versus the composition of the as-formed ntTiO$_2$-based electrode cycled between 2.6 and 1.0 V at a rate of 20 µA cm^{-2} (C/2). During the first discharge (lithium insertion), the voltage plateau at 1.72 V corresponding to lithium insertion into crystalline TiO$_2$ is

Table 5.2 Values of the reversible and irreversible areal capacities and efficiency on cycling of the ntTiO$_2$ layers used as electrodes in experimental test cells at different kinetics (reprinted with permission from Ref. [35], Copyright 2009 American Chemical Society)

Electrode	Current density (μA cm^{-2})	First reversible capacity (1st cycle) (μAh cm^{-2})	Irreversible capacity (μAh cm^{-2})	Reversible capacity in 50th cycle (μAh cm^{-2})	Efficiency (%)
Amorphous ntTiO$_2$	5	77	60	55	71
	20	50	55	43	86
	100	44	36	33	75
Crystalline ntTiO$_2$	5	68	27	48	71
	20	38	12	31	82
	100	30	16	27	90
TiO$_2$ C.L.	5	9	8	6	67

Data corresponding to titania nanotube layers with 900 nm length fabricated onto Ti foils.

TiO$_2$ C.L: titania compact layer that is obtained by anodization in fluorine-free electrolyte.

not present. Only one pseudo plateau at about 1.15 V contributes to a large irreversible reaction of Li. A total capacity of 105 μAh cm^{-2} is measured at the end of the first discharge with a reversible capacity of 50 μAh cm^{-2}, leading to an irreversible capacity of 55 μAh cm^{-2}. After 50 cycles, the as-prepared ntTiO$_2$ layer shows a reversible capacity of 43 μAh cm^{-2}. Nevertheless, the overlapping of the quasi-plateaus in galvanostatic curves often hinders a clear resolution of these processes. For this purpose, the derivative curves were calculated and plotted (Fig. 5.15c,d). Now the horizontal quasi-plateaus become resolved bands. Derivative curves of the as-formed ntTiO$_2$ demonstrate clearly that the process at 1.15 V is irreversible (Fig. 5.15c,d). Moreover, a very broad band in the range of 2.4–1.4 V during discharge and charge is still visible on further cycling, suggesting that the insertion/deinsertion process is reversible. Compared with usual large anatase particles, the surface area that the nanotube layers expose to the electrolyte is higher, and this leads to a more facile insertion of lithium. Furthermore, the existence of organized 1D nanotubular structures might contribute to a homogeneous insertion process.

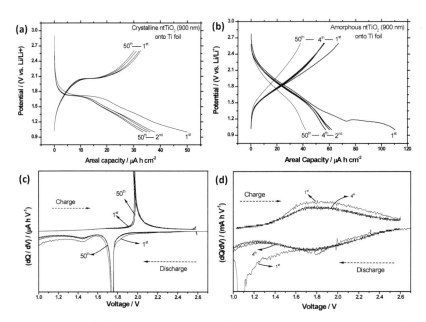

Figure 5.15 Galvanostatic discharge/charge curves vs. composition of TiO$_2$ nanotube layers: (a) as-formed and (b) annealed using 20 µA cm^{-2}. Voltage limits: 2.6–1.0 V. Derivative curves during the 1st, 2nd, 3rd, and 4th discharge/charge of (c) crystalline ntTiO$_2$ and (d) amorphous ntTiO$_2$ electrodes. Note that the bottom part indicates the lithiation (discharge) process and the upper part shows the delithiation (charge) process.

When sample of ntTiO$_2$ with 900 nm length heated at 450°C is tested in lithium cell, the voltage plateaus are observed during the discharge and charge (Fig. 5.15a) at potential of 1.72 and 2.0 V, respectively. The plateaus are attributed to insertion and deinsertion of Li$^+$ from tetrahedral and octahedral sites of crystalline anatase TiO$_2$ nanotubes. The deinsertion potential is slightly higher than in crystalline anatase [89]. After the first discharge, the capacity is 50 µAh cm^{-2} and the reversible capacity is 38 µAh cm^{-2} leading to an irreversible capacity of 12 µAh cm^{-2}. In its derivative curves (Fig. 5.15c), one can see clearly the peaks, which are very intense and narrow, and correspond to reduction/oxidation steps of anatase when cycling. It is worth noting that the intensity of the peaks is maintained on further cycling and, thus, indicates that crystalline anatase is accommodating lithium reversibly in its structure on further cycles (Fig. 5.15a–d shows 50 cycles).

The electrode material made of titania nanotubes shows a high rate of charge–discharge, which is probably due to the 3D network structure of porous titania nanotubes. It worth noting that we are considering the nanotubes as a perfect cylinder, as is represented in Fig. 5.16 with the dimension of wall thickness, nanotube diameter, and length. Moreover, Fig. 5.16 shows the structure of a porous nanotube configuration and transport path of both lithium ions and electrons. The lithium ions and electrolyte are easily transported in the uniform channels of the pores and electrons transport quickly through the arrangement nanotubular structure.

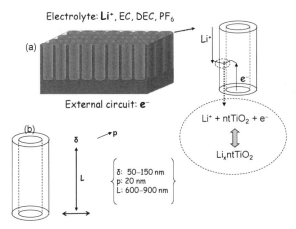

Figure 5.16 (a) Schematic representation of the self-organized titania nanotubes used for the assessment of the specific area. (b) Transport path of lithium ions and electrons in mesoporous titania nanotube.

From these results, the irreversible capacity is strongly dependent on the structure of $ntTiO_2$. To explain the relatively high irreversible capacity, the influence of the reaction between adsorbed water molecules onto the $ntTiO_2$ electrode and Li^+ has been considered [84]. From the SEM images (Figs. 5.7 and 5.8), and assuming that the nanotube is a perfect cylinder (Fig. 5.16), it is apparent that the nanotubes are not connected along the vertical axis, leading to a specific area of about 40 cm². Assuming that 1×10^{15} H_2O molecules/cm² can be adsorbed onto the nanotubes, the number of H_2O molecules present onto the $ntTiO_2$ electrode should be around 4×10^{16}, which can react with Li^+ ions according to the following reaction:

$$H_2O + Li^+ + xe^- \rightleftharpoons LiOH + \frac{1}{2}H_2 \tag{5.5}$$

However, the number of Li^+/cm^2 corresponding to the irreversible capacity of 55 and 12 µAh cm^{-2} for both amorphous and crystalline materials is significantly higher, i.e., 1.26×10^{18} and 2.7×10^{17} Li^+/cm^2, respectively. Therefore, the large irreversible capacity obtained for both samples can be attributed not only to the presence of adsorbed water but also to the formation of a very thin disordered layer at the electrode surface. Considering the numbers given above for irreversibly inserted Li ions and available sites per unit area, this would correspond to insertion in about 30 atomic planes for amorphous tubes and 7 atomic planes for crystalline nanotubes.

It can also be noted that the ratio between the irreversible capacity of the amorphous and the annealed materials is around 5. This difference can be explained by the fact that annealing treatment of the amorphous electrode removes structural and chemical defects that act as Li^+ ion traps, which are responsible for the irreversible insertion of Li^+. This explanation is for titania nanotubes with 900 nm, but it can be also applied for the shorter nanotubes show above.

The cycle performance of as-formed and annealed $ntTiO_2$-based electrodes at different kinetics (5, 20 and 100 µA cm^{-2}) is illustrated in Fig. 5.17 for 50 cycles. First, these results suggest that $ntTiO_2$ can be used as an alternative electrode for rechargeable Li-ion microbatteries. The specific capacity obtained for the as-prepared samples is higher than for the annealed materials confirming that the amount of lithium ions inserted into amorphous $ntTiO_2$ is higher than in the crystalline structure. The highest specific capacity is obtained with as-prepared $ntTiO_2$ layers using a relative slow kinetic (5 µA cm^{-2} or C/8), which delivered a maximum reversible capacity of 77 µAh cm^{-2} after the first cycle and 55 µAh cm^{-2} after 50 cycles (see Fig. 5.17a). The average capacity loss of 0.45 µAh cm^{-2} per cycle leads to a cycling efficiency of 71%. Although lower values of specific capacity are observed using faster kinetics, the efficiency on cycling is improved and can reach 90% in the case of crystalline $ntTiO_2$ at a rate of 100 µA cm^{-2} or 2.5 C (Fig. 5.17c). The higher efficiency on cycling obtained with crystalline materials compared with amorphous structures is due to the lower amount of structural defects and Li-ion trap sites.

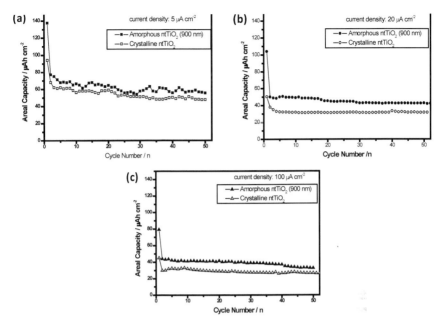

Figure 5.17 Specific reversible capacity vs. cycle number in lithium cells for the as-formed or "amorphous" (closed and colored symbols) and annealed TiO$_2$ (open symbols) nanotube layer with 900 nm length cycled at (a) 5 µA cm^{-2} (C/8), (b) 20 µA cm^{-2} (C/2), and (C) 100 µA cm^{-2} (2.5C). *Note:* C/n rate means that the total capacities of the cell correspond after 1 h in discharge.

To demonstrate that ntTiO$_2$ layers obtained by anodization in fluorine-containing electrolyte and whose lengths are about 600–900 nm show enhanced electrochemical properties, we have also fabricated two sorts of new TiO$_2$ samples, and the variation in the areal capacity vs. cycling has been studied in

(i) TiO$_2$ compact layer electrochemically grown at 20 V for 60 min in 1 M H$_3$PO$_4$ + 1 M NaOH electrolyte. The principles of fabrication are shown in Section 5.3.2, in which the compact oxide layer or barrier layer is explained in details (see also Section 5.3.1).

(ii) TiO$_2$ synthesized by a sol–gel method: Particularly, the formation of microparticles prepared by a microemulsion technique. As a precursor's titanium isopropoxide, anionic surfactant sodium *bis*(2-ethylhexyl) sulfosuccinate (AOT from Sigma), hexane, and water were used as received. Titanium isopropoxide was added to an AOT/hexano/water microemulsion to initiate the synthesis. First, 3 mL of titanium isoproposide was mixed with 15 mL of hexane followed by stirring at room temperature for 30 min. Then, the microemulsion was formed by adding 3 mL of water to a mixture made of 15 mL of AOT and hexane followed by sonication and stirring for 1 h. Finally, we added drop by drop the solution containing AOT + H_2O to the solution composed of titanium isopropoxide. The resultant solution was stirred and sonicated for 1 h. To separate the formed solid from the aqueous solution, the mixture was washed with distilled water by spinning it repeatedly. The powder was then dried and cured at room temperature for 24 h. Afterward, the obtained precursor was successively heated at 100°C, 200°C, 300°C, and 450°C.

From Fig. 5.18, it is apparent that the presence of nanotubes leads to a significant improvement of the areal capacity compared with compact layer and micro-TiO_2 elaborated by a reverse micelle synthesis (R.M.S.). Amorphous titania compact layer exhibits areal capacity about 20 µAh cm^{-2} in first discharge. Meanwhile, the first reversible capacity is about 10 µAh cm^{-2}, and this value is almost kept after 50 cycles. On the other hand, the areal capacities values of micro-TiO_2 at 100°C, 300°C, and 450°C are 30, 40, and 37 µAh cm^{-2}, respectively. On cycling, the capacity retention is very poor (less than 20 µAh cm^{-2} after 10 cycles), except for sample heated at 450°C, which retains the capacity slightly better and after 35 cycles exhibits about 15 µAh cm^{-2}. We have fabricated an amorphous TiO_2 nanotube layer that presents high areal capacity of 77 µAh cm^{-2} (in first reversible cycle) and after 50 cycles is 55 µAh cm^{-2}. Compared with other thin-layer-based electrodes that have been reported in literature—(i) MoS_2-based planar Li-ion battery [6] and (ii) TiO_2 deposited on an Al nanorod current collector [90]—the ntTiO_2 thin films shown in this work present an areal capacity of 55 µAh cm^{-2} after 50 cycles, suggesting that this electrode can be a potential candidate for the fabrication of 2D microbatteries. Moreover, it worthy to note that ntTiO_2 layers can be used to fabricate nanocomposite electrodes, e.g., ntTiO_2/SnO_x [91]. It can be noted that the morphology of tin crystallites depends on the electrochemical parameters. For instance, when electrodeposition is performed at a lower current density (–5 mA cm^{-2}), the shape of tin oxide is sponge-like structure into ntTiO$_2$ [91]. The different crystallites morphology [92], as well as the filling of nanotubes [93], is governed solely by the electrodepositing parameters. All these nano-architectured electrodes are of special interest as possible alternative for negative electrodes in Li-ion batteries.

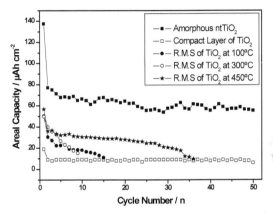

Figure 5.18 Capacity evolution vs. cycle number for amorphous ntTiO$_2$ thin film, compact layer of TiO$_2$ electrochemical grown at 20 V for 60 min in 1 M H$_3$PO$_4$ + 1 M NaOH electrolyte. Micro-TiO$_2$ electrode prepared by reverse micelle synthesis (R.M.S.) at different temperatures. Current density of 5 µA cm^{-2} between 2.6 and 1 V.

5.5.2 SEM Study of Cycled Electrodes

The structural rigidity of nanostructured materials is a positive property in order to achieve extended cycle life in advanced electrodes. To assess the stability of the self-assembled titania nanotubes, ex situ SEM images of cycled electrodes after 50 cycles at 2.6 V charge voltage are collected in Fig. 5.19. For the sample supported onto a Si substrate, the characteristic nanotube geometry—about 600 nm nanotube length (cross-sectional view), and between 50 and 200 nm diameter (top view)—is preserved. These analyses demonstrate the robustness of the nanotube morphology for those samples grown onto Si substrate, which survive after repeated cycling.

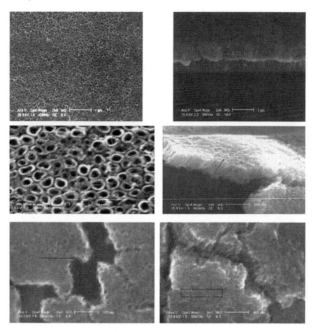

Figure 5.19 Titania nanotube layer after reaction with lithium.

Moreover, for nanotubes with 900 nm length, it is observed that the characteristic nanotube geometry is unchanged after cycling: a tube diameter of about 80 nm, wall thickness of 20 nm (top view), and nanotube length of 920 nm (cross-sectional view). The chemical analysis carried out using energy dispersive X-ray (EDX) spectroscopy (not shown) reveals that there are phosphorus and fluorine impurities. These analyses demonstrate the robustness of the nanotube morphology, which survives after repeated cycling. However, doing a more extended study of the whole area of the electrodes, we found parts where there are some partially opened nanotubes; it is also possible to see some closed nanotubes (Fig. 5.19).

5.6 Conclusions

Self-organized TiO_2 nanotubes are key materials for the negative electrode of a new generation of Li-ion Batteries. Working voltages well above graphite avoid lithium electrodeposition and extended SEI layer formation. Short lithium diffusion and electron conducting distances improve rate performance and an efficient capacity usage. The materials are apt for miniaturization and use in advanced NEMS. No toxicity and environmental friendliness are additional positive features of these materials.

Eventually, self-organized TiO_2 nanotubes can be easily obtained by a powder-free fabrication method based on the electrochemical anodization of titanium foil or titanium sputtered onto a silicon substrate. The TiO_2 nanotubes prepared onto Si substrate had an inner diameter between 50 and 200 nm and about 600 nm length. This method ensures good electrical contact between titania nanotube layers and the current collector. Also, the resulting array shows a special mechanical stability and buffers possible volume changes arising from the structural tetragonal–orthorhombic conversion of anatase, or possible expansion on insertion in amorphous titania. Using Ti foils, it is possible to obtain nanotubes with 600 and 900 nm of thickness (or length) exhibiting the same diameter and wall thickness. The powder-free fabrication method used to prepare the active material directly on the current collector tackles the use of additives such as polymer binders and conductive agents (e.g., carbon black).

Lithium test cells using self-organized TiO_2 nanotubes as active electrode material onto Si deliver reversible areal capacity after 50 cycles of about 56 $\mu Ah\ cm^{-2}$ leading to an efficiency of 63%. It can be noticed that crystalline $ntTiO_2$ onto Si shows the best efficiency (96%) at 100 $\mu A\ cm^{-2}$ although areal capacity is relatively low. The analysis of cycled electrodes suggests that the nanotube morphology persists on cycling. These results suggest that the highly flat $ntTiO_2$ layer supported onto Si substrate can be used as an alternative electrode for rechargeable on-chip 2D Li-ion microbatteries.

Moreover, we also studied the possibility of using self-organized TiO_2 nanotubes onto Ti foils as an alternative electrode for lithium-ion batteries. A specific reversible capacity up to 77 $\mu Ah\ cm^{-2}$ is achieved in lithium test cells for as-formed $ntTiO_2$ layers that were ~80 nm in diameter and 900 nm in length. We showed that using a relatively fast kinetic (100 $\mu A\ cm^{-2}$), the efficiency after 50 cycles is 75%. The possible reasons for the high experimental capacity values are the high surface area and highly organized 1D structure of titania nanotube layers. After thermal treatment, the efficiency on cycling is 90% using faster kinetic (100 μA cm^{-2}) for crystalline nanotubes, suggesting that ntTiO$_2$ can be used an alternative electrode for rechargeable batteries. Postmortem analyses have shown that the ntTiO$_2$ morphology survives at least 50 cycles.

Acknowledgment

Gregorio F. Ortiz acknowledges the financial support from the Spanish Ministry of Science and Innovation within the Program "Ramón y Cajal" (Ref. RYC-2010-05596).

References

1. Van Schalkwijk, W., and Scrosati, B. (2002), *Advances in Lithium-Ion Batteries* (Eds: Van Schalkwijk, W., and Scrosati, B.), Springer, Berlin.
2. Service, R. F. (2005). How far can we push chemical self-assembly. *Science*, **309**, pp. 95–95.
3. Peng, X. G., Manna, L., Yang, W., Wickham, J., and Scher, E. (2000). Shape control of CdSe nanocrystals, *Nature*, **404**, pp. 59–61.
4. Sun, Y. G., Mayers, B., and Xia, Y. N. (2003). Metal nanostructures with hollow interiors. *Adv. Mater.* **15**, pp. 641–646.
5. Golodnitsky, D., Nathan, M., Yufit, V., Strauss, E., Freedman, K., Burstein, L., Gladkich, A., and Peled, E. (2006). Progress in three-dimensional (3D) Li-ion microbatteries. *Solid State Ionics*, **177**, pp. 2811–2819.
6. Golodnitsky, D., Yufit, V., Nathan, M., Shechtman, I., Ripenbein, T., Strauss, E., Menkin, S., and Peled, E. (2006). Advanced materials for the 3D microbattery. *J. Power Sources*, **153**, pp. 281–287.
7. Dell, R. M., and Rand, D. A. (2001). Energy storage — a key technology for global energy sustainability. *J. Power Sources*, **100**, pp. 2–17.
8. Hall, P. J., and Bain, E. J. (2008). Energy-storage technologies and electricity generation. *Energy Policy*, **36**, pp. 4352–4355.
9. Chen, H., Ngoc Cong, T., Yang, W., Tan, C., Li, Y., and Ding, Y. (2009). Progress in electrical energy storage system: A critical review. *Prog. Nat. Sci.*, **19**, pp 291–312.
10. Kennedy, B., Patterson, D., and Camilleri, S. (2000). Use of lithium-ion batteries in electric vehicles. *J. Power Sources*, **90**, pp. 156–162.

11. Lazzari, M., and Scrosati, B. (1980). Cyclable lithium organic electrolyte cell based on 2 intercalation electrodes. *J. Electrochem. Soc.*, **127**, pp. 773–774.
12. DiPietro, B., Patriarca, M., and Scrosati, B. (1982). On the use of rocking chair configurations for cyclable lithium organic electrolyte batteries. *J. Power Sources*, **8**, pp. 289–299.
13. Nagaura, T., and Tozawa, K. (1990). Lithium ion rechargeable battery. *Prog. Batt. Solar Cells*, **9**, pp. 209–217.
14. Winter, M., Besenhard, J. O., Spahr, M. E., and Novak, P. (1998). Insertion electrode materials for rechargeable lithium batteries. *Adv. Mater.*, **10**, pp. 725–763.
15. Tarascon, J. M., and Armand, M. (2001). Issues and challenges facing rechargeable lithium batteries. *Nature*, **414**, pp. 359–367.
16. Pistoia, G. (Ed.) (1994), *Lithium Batteries*, Elsevier, New York.
17. Guyomard, D., and Tarascon, J. M. (1992). Li metal-free rechargeable $LiMn_2O_4$/carbon cells — their understanding and optimization. *J. Electrochem. Soc.*, **139**, pp. 937–948.
18. Padhi, A. K., Nanjundaswamy, K. S., and Goodenough, J. B. (1997). Phospho-olivines as positive-electrode materials for rechargeable lithium batteries. *J. Electrochem. Soc.*, **144**, pp. 1188–1194.
19. Alcántara, R., Jaraba, M., Lavela, P., and Tirado, J. L. (2002). Optimizing preparation conditions for 5 V electrode performance, and structural changes in $Li_{1-x}Ni_{0.5}Mn_{1.5}O_4$ spinel. *Electrochim. Acta*, **47**, pp. 1829–1835.
20. Lloris, J. M., Pérez-Vicente, C., and Tirado, J. L. (2002). Improvement of the electrochemical performance of $LiCoPO_4$ 5 V material using a novel synthesis procedure. *Electrochem. Solid State Lett.*, **5**, pp. A234–A237.
21. Huang, S., Kavan, L., Exnar, I., and Grätzel, M. (1995). Rocking chair lithium battery based on nanocrystalline TiO_2 (Anatase). *J. Electrochem. Soc.*, **142**, pp. L142–L144.
22. Armstrong, G., Armstrong, A. R., Canales, J., and Bruce, P. G. (2006). $TiO_2(B)$ nanotubes as negative electrodes for rechargeable lithium batteries. *Electrochem. Solid-State Lett.*, **9**, pp. A139–A143.
23. Lindsay, M. J., Backford, M. G., Attard, D. J., Luca, V., Skyllas-Kazacos, M., and Griffith, C. S. (2007). Anodic titania films as anode materials for lithium ion batteries. *Electrochim. Acta*, **52**, pp. 6401–6411.

24. Taberna, P. L., Mitra, S., Poizot, P., Simon, P., and Tarascon, J. M. (2006). High rate capabilities Fe_3O_4-based Cu nano-architectured electrodes for lithium-ion battery applications. *Nature Materials*, **5**, pp. 567–573.
25. Zwilling, V., Aucouturier, M., and Darque-Ceretti, E. (1999). Anodic oxidation of titanium and TA 6 V alloy in chromic media: An electrochemical approach. *Electrochim. Acta*. **45**, pp. 921–929.
26. Macak, J. M., Tsuchiya, H., and Schmuki, P. (2005). High-aspect-ratio TiO_2 nanotubes by anodization of titanium. *Angew. Chem., Int. Ed.,* **44**, pp. 2100–2102.
27. Albu, S. P., Ghicov, A., Macak, J. M., Hahn, R., and Schmuki, P. (2007). Self-organized, free-standing TiO_2 nanotube membrane for flow-through photocatalytic applications. *Nano Lett.,* **7**, pp. 1286–1289.
28. Mor, G. K., Shankar, K., Paulose, M., Varghese, O. K., and Grimes, C. A. (2006). Use of highly-ordered TiO_2 nanotube arrays in dye-sensitized solar cells. *Nano Lett.,* **6**, pp. 215–218.
29. Macak, J. M., Taveira, L. V., Tsuchiya, H., Sirotna, K., Macak, J., and Schmuki, P. (2006). Influence of different fluoride containing electrolytes on the formation of self-organized titania nanotubes by Ti anodization *J. Electroceram.,* **16**, pp. 29–34.
30. Tian, M., Wu, G. S., Adams, B., Wen, J. L., and Chen, A. C. (2008). Kinetics of photoelectrocatalytic degradation of nitrophenols on nanostructured TiO_2 electrodes. *J. Phys. Chem. C.,* **112**, pp. 825–831.
31. Djenizian, T., Hanzu, I., Premchand, Y. D., Vacandio, F., and Knauth, P. (2008). Electrochemical fabrication of Sn nanowires on titania nanotube guide layers. *Nanotechnology*, **19**, pp. 205601.
32. Kelly, J. J. (1979). Influence of fluoride ions on the passive dissolution of titanium. *Electrochim. Acta*, **24**, pp. 1273–1282.
33. Macklin, W. J., and Neat, R. J. (1992). Performance of titanium dioxide-based cathodes in a lithium polymer electrolyte cell. *Solid State Ionics*, **53**, pp. 694–700.
34. Premchand, Y. D., Djenizian, T., Vacandio, F., and Knauth, P. (2006). Fabrication of self-organized TiO_2 nanotubes from columnar titanium thin films sputtered on semiconductor surfaces. *Electrochem. Commun.* **8**, pp. 1840–1844.
35. Ortiz, G. F., Hanzu, I., Djenizian, T., Lavela, P., Tirado, J. L., and Knauth, P. (2009). Alternative Li-ion battery electrode based on self-organized titania nanotubes. *Chem. Mater.,* **21**, pp. 63–67.
36. Liu, D., Xiao, P., Zhang, Y., Garcia, B. B., Zhang, Q., Guo, Q., Champion, R., and Cao, G. (2008). TiO_2 nanotube arrays annealed in

N-2 for efficient lithium-ion intercalation. *J. Phys. Chem. C.*, **112**, pp. 11175–11180.

37. Yasuda, K., and Schmuki, P. (2007). Control of morphology and composition of self-organized zirconium titanate nanotubes formed in $(NH_4)_2SO_4/NH_4F$ electrolytes. *Electrochim. Acta.*, **52**, pp. 4053–4061.

38. Taveira, L. V., Macak, J. M., Tsuchiya, H., Dick, L. F. P., and Schmuki, P. (2005). Initiation and growth of self-organized TiO_2 nanotubes anodically formed in $NH_4F/(NH_4)_2SO_4$ electrolytes. *J. Electrochem. Soc.*, **152**, pp. B405–B410.

39. Yasuda, K., Macak, J. M., Berger, S., Ghicov, A., and Schmuki, P. (2007). Mechanistic aspects of the self-organization process for oxide nanotube formation on valve metals. *J. Electrochem. Soc.*, **154**, pp. C472–C478.

40. Shibata, T., and Zhu, Y. C. (1995). Effect of film formation conditions on the structure and composition of anodic oxide-films on titanium. *Corrosion Sci.*, **37**, pp. 253–270.

41. Tirado, J. L. (2003). Inorganic materials for the negative electrode of lithium-ion batteries: state-of-the-art and future prospects. *Mater. Sci. Eng. R.*, **40**, pp. 103–136.

42. Hahn, R., Ghicov, A., Tsuchiya, H., Macak, J. M., Muñoz, A. G., and Schmuki, P. (2007). Lithium-ion insertion in anodic TiO_2 nanotubes resulting in high electrochromic contrast. 5th *International Conference on Porous Semiconductors — Science and Technology. Physica Status Solidi A — Applications and Materials Science.*, **204**, pp. 1281–1285.

43. Fujishima, A., and Honda, K. (1972). Electrochemical photolysis of water at a semiconductor electrode. *Nature*, **238**, pp. 37–38.

44. O'Regan, B., and Grätzel, M. (1991). A low-cost, high-efficiency solar cell based on dye-sensitized colloidal TiO_2 films. *Nature*, **353**, pp. 737–740.

45. Jayaweera, P. V. V., Perera, A. G. U., and Tennakone, K. (2008). Why Gratzel's cell works so well. *Inorg. Chim. Acta*, **361**, pp. 707–711.

46. Zhu, K., Neale, N. R., Miedaner, A., and Frank, A. J. (2007). Enhanced charge-collection efficiencies and light scattering in dye-sensitized solar cells using oriented TiO_2 nanotubes. Arrays, *Nano Lett.*, **7**, pp. 69–74.

47. Jennings, J. R., Ghicov, A., Peter, L. M., Schmuki, P., and Walker, A. B. (2008). Dye-sensitized solar cells based on oriented TiO_2 nanotube

arrays: transport, trapping, and transfer of electrons. *J. Am. Chem. Soc.*, **130**, pp. 13364–13372.

48. Gratzel, M. (2004). Conversion of sunlight to electric power by nanocrystalline dye-sensitized solar cells. *J. Photochem. Photobiol., A: Chem.*, **164**, pp. 3–14.

49. Kang, S. H., Kim, J. Y., Kim, Y., Kim, H. S., and Sung, Y. E. (2007). Surface modification of stretched TiO_2 nanotubes for solid-state dye-sensitized solar cells. *J. Phys. Chem. C*, **111**, pp. 9614–9623.

50. Shaban, Y. A., and Khan, S. U. M. (2009). Carbon modified (CM)-n-TiO_2 thin films for efficient water splitting to H_2 and O_2 under xenon lamp light and natural sunlight illuminations. *J. Solid State Electrochem.* **13**, pp. 1025–1036.

51. Ghicov, A., Macak, J. M., Tsuchiya, H., Kunze, J., Haeublein, V., Frey, L., and Schmuki, P. (2006). Ion implantation and annealing for an efficient n-doping of TiO_2 nanotubes. *Nano Lett.*, **6**, pp. 1080–1082.

52. Sun, W. T., Yu, Y., Pan, H. Y., Gao, X. F., Chen, Q., and Peng, L. M. (2008). Cds quantum dots sensitized TiO_2 nanotube-array photoelectrodes. *J. Am. Chem. Soc.*, **130**, pp. 1124–1125.

53. Cottineau, T., Toupin, M., Delahaye, T., Brousse, T., and Bélanger, D. (2006). Nanostructured transition metal oxides for aqueous hybrid electrochemical supercapacitors. *Appl. Phys. A*, **82**, pp. 599–606.

54. Wang, J., Polleux, J., Lim, J., and Dunn, B. (2007). Pseudocapacitive contributions to electrochemical energy storage in TiO_2 (anatase) nanoparticles. *J. Phys. Chem. C*, **111**, pp. 14925–14931.

55. Wong, Y. G., and Zhang, X. G. (2004). Preparation and electrochemical capacitance of RuO_2/TiO_2 nanotubes composites. *Electrochim. Acta*, **49**, pp. 1957–1962.

56. Tao, F., Shen, Y., Liang, Y., and Li, H. (2007). Synthesis and characterization of $Co(OH)_2$/TiO_2 nanotube composites as supercapacitor materials. *J. Solid State Electrochem.*, **11**, 853–858.

57. Xie, Y., Huang, C., Zhou, L., Liu, Y., and Huang, H. (2009). Supercapacitor application of nickel oxide–titania nanocomposites. *Comp. Sci. Tech.*, **69**, pp. 2108–2114.

58. Brezesinski, T., Wang, J., Polleux, J., Dunn, B., and Tolbert, S. H. (2009). Templated nanocrystal-based porous TiO_2 films for next-generation electrochemical capacitors. *J. Am. Chem. Soc.*, **131**, pp. 1802–1809.

59. Amatucci, G. G., Badway, F., Pasquier, A. D., and Zheng, T. (2001). An asymmetric hybrid nonaqueous energy storage cell. *J. Electrochem. Soc.*, **148**, pp. A930–A939.

60. Wang, Q., Wen, Z., and Li, J., (2006). A hybrid supercapacitor fabricated with a carbon nanotube cathode and a TiO_2-B nanowire anode. *Adv. Funct. Mater.*, **16**, p. 2141–2146.
61. Qiang, W., Zhenhai, W., and Jinghong, L. (2007). Carbon nanotubes/TiO_2 nanotubes hybrid supercapacitor. *J. Nanosci. Nanotechnol.*, **7**, pp. 3328–3331.
62. Colbow, K. K., Dahn, J. R., and Haering, R. R. J. (1989). Structure and electrochemistry of the spinel oxides $LiTi_2O_4$ and $Li_4/3Ti_5/3O_4$. *J. Power Sources,* **26**, pp. 397–402.
63. Bonino, F., Busani, L., Manstretta, M., Rivolta, B., and Scrosati, B. (1981). Anatase as a cathode material in lithium–organic electrolyte rechargeable batteries. *J. Power Sources,* **6**, pp. 261–270.
64. Murphy, D. W., Di Salvo, F. J., Carides, J. N., and Waszczak, J. V. (1978). Topochemical reactions of rutile related structures with lithium. *Mater. Res. Bull.,* **13**, pp. 1395–1402.
65. Restori, R., Schwarzenbach, D., and Schneider, J. R. (1987). Charge-density in rutile, TiO_2. *Acta Crystallogr, Sect. B: Struct. Sci.*, **43**, pp. 251–257.
66. Yang, Z., Choi, D., Kerisit, S., Rosso, K. M., Wang, D., Zhang, J., Graff, G., and Liu, J. (2009). Nanostructures and lithium electrochemical reactivity of lithium titanites and titanium oxides: A review. *J. Power Sources*, **192**, pp. 588–598.
67. Murphy, D. W., Cava, R. J., Zahurak, S. M., and Santoro, A. (1983). Ternary Li_xTiO_2 phases from insertion reactions. *Solid State Ionics*, **9–10**, pp. 413–418.
68. Hu, Y.-S., Kienle, L., Guo, Y.-G., and Maier, J. (2006). High lithium electroactivity of nanometer-sized rutile TiO_2. *Adv. Mater.*, **18**, pp.1421–1426.
69. Reddy, M. A., Kishore, M. S., Pralong, V., Varadaraju, U. V., and Raveau, B. (2007). Lithium intercalation into nanocrystalline brookite TiO_2. *Electrochem. Solid-State Lett.* **10**, pp. A29–A31.
70. Cromer, D. T., and Herrington, K. (1955). The structures of anatase and rutile, *J. Am. Chem. Soc.*, **77**, pp. 4708–4709.
71. Marchand, R., Brohan, L., and Tournoux, M. (1980). TiO_2(B) a new form of titanium dioxide and the potassium octatitanate $K_2Ti_8O_{17}$. *Mater. Res. Bull.*, **15**, pp. 1129–1133.
72. Nuspl, G., Yoshizawa, K., and Yamabe, T. (1997). Lithium intercalation in TiO_2 modifications. *J. Mater. Chem.*, **7**, pp. 2529–2536.

73. Sudant, G., Baudrin, E., Larcher, D., and Tarascon, J. M. (2005). Electrochemical lithium reactivity with nanotextured anatase-type TiO_2. *J. Mater. Chem.*, **15**, pp. 1263–1269.
74. Wagemaker, M., Borghols, W. J. H., and Mulder, F. M. (2007). Large impact of particle size on insertion reactions. a case for anatase Li_xTiO_2. *J. Am. Chem. Soc.*, **129**, pp. 4323–4327.
75. Armstrong, A. R., Armstrong, G., Canales-Garcia, J. R., and Bruce, P. G. (2005). Lithium-ion intercalation into TiO_2-B nanowires. *Adv. Mat.*, **17**, pp. 862–865.
76. Zhou, Y.-K., Cao, L., Zhang, F.-B., He, B.-L., and Li, H.-L. (2003). Lithium insertion into TiO_2 nanotube prepared by the hydrothermal process. *J. Electrochem. Soc.*, **150**, pp. A1246–A1249.
77. Li, J., Tang, Z., and Zhang, Z. (2005). Preparation and novel lithium intercalation properties of titanium oxide nanotubes. *Electrochem. Solid-State Lett.*, **8**, pp. A316–A319.
78. Wang, Y., Wu, M., and Zhang W. F. (2008). Preparation and electrochemical characterization of TiO_2 nanowires as an electrode material for lithium-ion batteries. *Electrochim. Acta*, **53**, pp. 7863–7868.
79. Erjavec, B., Dominko, R., Umek, P., Sturm, S., Pintar, A., and Gaberscek, M. (2009). Tailoring nanostructured TiO_2 for high power Li-ion batteries. *J. Power Sources*, **189**, pp. 869–874.
80. Huang, H., Zhang, W. K., Gan, X. P., Wang, C., and Zhang, L. (2007). Electrochemical investigation of TiO_2/carbon nanotubes nanocomposite as anode materials for lithium-ion batteries. *Mater. Lett.*, **61**, pp. 296–299.
81. Kavan, L., Grätzel, M., Rathousky, J., and Zukal, A. (1996). Nanocrystalline TiO_2 (anatase) electrodes: Surface morphology, adsorption, and electrochemical properties. *J. Electrochem. Soc.*, **143**, pp. 394–400.
82. Koudriachova, M. V., Harrison, N. M., and de Leeuw, S. W. (2002). Density-functional simulations of lithium intercalation in rutile. *Phys. Rev. B: Condens. Matter*, **65**, pp. 235423.
83. Nuspl, G., Yoshizawa, K., and Yamabe, T. (1997). Lithium intercalation in TiO_2 modifications. *J. Mater. Chem.*, **7**, pp. 2529–2536.
84. Kim, J., and Cho, J. (2007). Rate characteristics of anatase TiO_2 nanotubes and nanorods for lithium battery anode materials at room temperature. *J. Electrochem. Soc.*, **154**, pp. A542–A546.
85. Lindsay, M. J., Skyllas-Kazacos, M., and Luca, V. (2009). Anodically synthesized titania films for lithium batteries: Effect of titanium substrate and surface treatment. *Electrochim. Acta*, **54**, pp. 3501–3509.

86. Ortiz, G. F., Hanzu, I., Knauth, P., Lavela, P., Tirado, J. L., and Djenizian, T. (2009). TiO_2 nanotubes manufactured by anodization of Ti thin films for on-chip Li-ion 2D microbatteries. *Electrochim. Acta*, **54**, pp. 4262–4268.
87. Guo, H., Zhao, H., Yin, C., and Qiu, W. (2006). A nanosized silicon thin film as high capacity anode material for Li-ion rechargeable batteries. *Mater. Sci. Eng. B*, **131**, pp. 173–176.
88. Park, M. S., Wang, G. X., Liu, H. K., and Dou, S. X. (2006). Electrochemical properties of Si thin film prepared by pulsed laser deposition for lithium ion micro-batteries. *Electrochim. Acta*, **51**, pp. 5246–5249.
89. Xu, J., Jia, C., Cao, B., and Zhang, W. F. (2007). Electrochemical properties of anatase TiO_2 nanotubes as an anode material for lithium-ion batteries. *Electrochim. Acta*, **52**, pp. 8044–8047.
90. Cheah, S. K., Perre, E., Rooth, M., Fondell, M., Hårsta, A., Nyholm, L., Boman, M., Gustafsson, T., Lu, J., Simon, P., and Edström, K. (2009). Self-supported three-dimensional nanoelectrodes for microbattery applications. *Nano Lett.*, **9**, pp. 3230–3233.
91. Ortiz, G. F., Hanzu, I., Knauth, P., Lavela, P., Tirado, J. L., and Djenizian, T. (2009). Nanocomposite electrode for Li-Ion microbatteries based on SnO on nanotubular titania matrix. *Electrochem. Solid-State Lett.*, **12**, pp. A186–A189.
92. Hanzu, I., Djenizian, T., Ortiz, G. F., and Knauth, P. (2009). Mechanistic Study of Sn Electrodeposition on TiO_2 nanotube layers: thermodynamics, kinetics, nucleation, and growth modes. *J. Phys. Chem. C*, in press.
93. Macak, J. M., Gong, B. G., Hueppe, M., and Schmuki, P. (2007). Filling of TiO_2 nanotubes by self-doping and electrodeposition. *Adv. Mater.*, **19**, pp. 3027–3031.

Chapter 6

Hierarchically Nanostructured Electrode Materials for Lithium-Ion Batteries

Yu-Guo Guo, Sen Xin, and Li-Jun Wan

Key Laboratory of Molecular Nanostructure and Nanotechnology, Institute of Chemistry, Chinese Academy of Sciences (CAS), Zhongguancun North First Street No. 2, Beijing, 100190, P. R. China

ygguo@iccas.ac.cn, wanlijun@iccas.ac.cn

This chapter introduces readers to a new kind of hierarchically nanostructured electrode materials for lithium-ion batteries. In this chapter, readers will learn the basic working principle of a typical lithium-ion battery as well as each component in the battery, and the materials used for fabricating the battery electrodes. In addition, it discusses both advantages and disadvantages in the use of nanomaterials and makes a comparison between nanomaterials and their micron-scale counterparts. Thus, readers understand the necessities to employ new materials, such as hierarchically nanostructured materials for the purpose of eliminating the flaws of nanomaterials to the greatest extent while retaining their inherent merits. Various examples of such structures, including sphere-like and flower-like nano/micro hierarchical structures as well as hierarchical three-dimensional (3D) mixed conducting networks, are taken to show their promising use as electrode materials in high-performance lithium-ion batteries.

Electrochemical Nanofabrication: Principles and Applications (Second Edition)
Edited by Di Wei
Copyright © 2016 Pan Stanford Publishing Pte. Ltd.
ISBN 978-981-4613-86-6 (Hardcover), 978-981-4613-87-3 (eBook)
www.panstanford.com

6.1 Brief Introduction to Lithium-Ion Batteries

Lithium-ion batteries are important types of secondary batteries in which lithium ions (Li^+) keep traveling back and forth between the anode and the cathode through the electrolyte in charging/discharging cycles, thus gaining the name "rocking chair batteries" (see Fig. 6.1) [1–3].

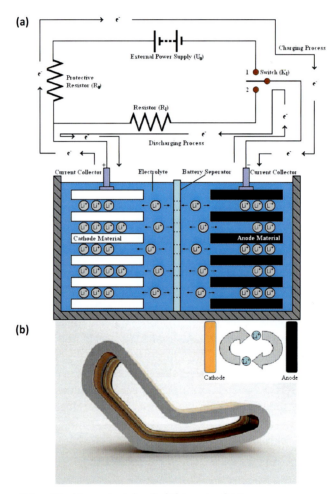

Figure 6.1 Working principle of a lithium-ion battery.

M. Armand proposed the first rechargeable lithium battery based on the intercalation concept in 1972 (and then mentioned in his book in 1980) [4]. The prototype in laboratory finally became commercially available at SONY in 1991. Since then, with its various advantages (e.g., high energy density, larger capacity, high energy-to-weight ratios, lack of memory effect, longer lifespan, and portable design) compared with other traditional rechargeable batteries such as lead-acid and nickel–cadmium batteries, lithium-ion battery soon became the key enabling technology in the energy industry and made a great contribution to the sustainable economy. It has stepped into people's daily lives in many aspects, ranging from consumer electronics and household appliances to the quickly developing hybrid electric vehicles (HEVs) and electric vehicles (EVs) [5–9]. In addition, lithium-ion batteries are now gradually extending their applications to defense and aerospace [10]. Moreover, it is widely believed that lithium-ion batteries, along with other new and renewable sources of energy, may finally replace traditional fossil energy in the near future.

Generally, a typical lithium-ion battery, as shown in Fig. 6.1a, consists of three fundamental components: an anode, a cathode, and an electrolyte. The anode, which is mainly made of carbon-based compounds and lithium alloys, is the receptor for the lithium ions, and the cathode, typically consisting of oxides of transition metals, is responsible for the source of lithium ions. Since both the anode and the cathode are the materials into which and from which Li ions can migrate, they can be compared to some kinds of "lithium storage containers." On the other hand, the electrolyte, commonly organic solvents (e.g., $LiPF_6$, which has been widely adopted in the manufacture of commercial lithium-ion batteries [however, solid electrolytes, including polymers and other inorganic materials, may also be used]), provides paths for the migration of Li ions as well as the separation of ionic transport and electronic transport in a lithium-ion battery. When the circuit is closed (i.e., switch K_1 in Fig. 6.1a is turned down to the contact 2), Li ions migrate between the anode and the cathode through the electrolyte.

It can be seen from Fig. 6.1a that during the charging process, lithium ions are extracted from the cathode and flow through the electrolyte (usually they need to penetrate the battery separator) to reach and move into the anode; meanwhile, electrons flow in the same direction from the cathode to the anode in the external circuit. As a result, external work can be applied on the battery. When discharging, the reverse process occurs, and the battery does electrical work to the external circuit. Figure 6.1b shows the comparison between the schematic diagram of a lithium-ion battery and a rocking chair. As can be seen, the back-and-forth movement builds the common ground for the two seemingly unrelated objects.

For a typical lithium-ion battery made from the cathode of $LiCoO_2$ and the anode of graphite, the cathode half reaction can be expressed as while the anode half reaction can be expressed as

$$LiCoO_2 \rightleftharpoons Li_{1-x}CoO_2 + xLi^+ + xe^-, \quad (6.1)$$

while the anode half reaction can be expressed as

$$xLi^+ + xe^- + 6C \rightleftharpoons Li_xC_6 \quad (6.2)$$

Meanwhile, the electrons flow through the external circuit; in this way, devices can be powered. The movements of lithium ions in both electrodes of the battery are generally called "lithium insertion" (Li^+ moving in) and "lithium extraction" (Li^+ moving out), respectively [11, 12].

Table 6.1 Summary of cathode and anode materials used for different types of lithium-ion batteries (all reproduced with permission from [14], copyright 2008, Wiley-VCH Verlag GmbH & Co. KGaA)

Types	Cathodes	Anodes
High energy	$LiNi_xCo_yM_{1-x-y}O_2$ (layered)[a] $LiMn_{2-x}M_xO_4$ (spinel)[b] MF_x[c]	Si, Sn, Sb MO_x[d] graphite
High power	$LiMn_{2-x}Al_xO_{4+\delta}$ (spinel) $LiNi_xCo_{1-2x}Mn_xO_2$ $LiFePO_4$ (olivine)	hard carbon graphite $Li_4Ti_5O_{12}$
Long cycle life	$LiFePO_4$ (olivine) $LiMn_{2-x}Al_xO_{4+\delta}$	$Li_4Ti_5O_{12}$ graphite

[a]M = Mn, Al, and Cr; [b]5 V systems, M = Ni, Cu, and Cr; [c]M = Cu, Ni, and Fe; [d]M = Fe, Co, Ni, Cr, Mn, Cu, and Sn.

Although they have many merits, lithium-ion batteries have suffered from some limitations. For example, their power density is relatively low at high charge–discharge rates due to a large polarization caused by slow lithium diffusion in the active material [13]. Also, when developing and using Li-ion batteries, safety issues should be taken into consideration since the introduction of excessive active materials may lead to a serious consequence such as explosion. So far, extensive work has been done on further improving the performance of lithium-ion batteries, most of which are related to improvements on the cathode, the anode, and the electrolyte through various methods, including the introduction of new materials, the chemical modification on the known compounds currently being used, and the use of nanomaterials. Table 6.1 summarizes different cathode and anode materials used in lithium-ion batteries.

6.2 Anode Materials

In the lithium batteries proposed by Whittingham [15], the cathode was made of titanium(II) sulfide (TiS), whereas lithium metal was taken as the corresponding anode. However, the use of metallic lithium as anode suffered from severe safety issues; therefore, in the design of lithium-ion batteries, the anode has been successively made of alloys, oxides, chalcogenides and carbonaceous materials [16]. In 1981, researchers at Bell Labs found graphite, which has a theoretical capacity of 372 mAh g^{-1}, could be used as the anode material in lithium-ion batteries [17], thus laying the foundation for the large-scale industrialized production and application of lithium-ion batteries. After that, graphite and graphitized materials (such as graphitized mesophase microbeads) were widely used in the manufacture of lithium-ion batteries and remained ever since [18]. Currently, the research on new anode materials focuses mainly on the compounds of the following three categories: nongraphitized carbon materials, metal oxides, and alloys [19].

6.2.1 Graphite

Graphite is the first kind of anode material for commercial lithium-ion batteries, and it is still widely used. However, the co-intercalation of solvents such as propylene carbonate during lithium insertion may lead to the decomposition and intensive exfoliation of graphene sheets [20]. Besides, graphite is very sensitive to the electrolyte, and if not properly dealt with, it may even hinder lithium ions from reversibly intercalating into graphite thus causing great capacity loss in the first several cycles [21]. Therefore, to improve the safety and cycling performance of lithium-ion battery, solvent co-intercalation and thermal stability of graphite anode must be taken into consideration, and modifications should be made on the material; yet much work has been done on this front.

6.2.2 Nongraphitized Carbon Materials

Generally, nongraphitized carbon materials consist of two categories: soft carbon and hard carbon. Although the soft carbon material possesses a very high reversible Li-storage capacity, it suffers from a serious voltage hysteresis during delithiation; the hard carbon material, on the other hand, with its relatively high capacity (200–600 mAh g^{-1}) over a wide voltage range, shows

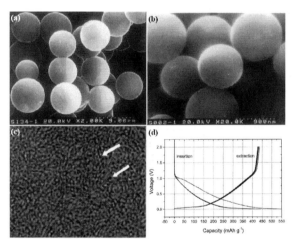

Figure 6.2 Results of electronic microscopy characterization and electrochemical test of as-prepared HCSs. (a) Scanning electron microscopy image of monodispersed HCSs with particle size of 5 μm. (b) Scanning electron microscopy image of monodispersed HCSs with particle size of 1 μm. (c) High-resolution transmission electron microscopy (HRTEM) image of the interior structure of an HCS, in which micropores can be clearly seen. (d) Typical plots of voltage vs. capacity of the HCSs as negative electrode in an Li/HCS cell during the first 10 charge/discharge cycles (all reproduced with permission from [24], copyright 2001, Elsevier Science Ltd.).

deficiencies in its initial coulombic efficiency and tap density [22, 23]. Among all the carbon materials, spherical hard carbon materials seem to be promising. However, they were difficult to prepare until 2001, when a hydrothermal method to prepare hard carbon spheres (HCSs) from sugar solution was introduced [24]. With sugar as the starting material, a series of HCS materials with a perfect spherical morphology and a controllable size ranging from 100 nm to 5 μm can be obtained by this method [24–26]. In the transmission electron microscopy (TEM) image shown in Fig. 6.2, a large quantity of uniform nanopores (ca. 0.4 nm) can be observed distributing in monodispersed hard carbon spherules, which make the material has a BET specific surface area of 400 $m^2 g^{-1}$, and can reversibly store lithium up to 430 mAh g^{-1}, which might be partially attributed to the tiny nanopores inside of HCSs. It has been reported that similar HCSs can also be prepared from other materials, such as glucose, starch and rice carbohydrates [27, 28]. These findings substantially widen the system category of carbon-based anode materials toward mass storage.

6.2.3 Alloys

It is well known that many metals and alloys can store a large quantity of lithium through the electrochemical alloying reaction; for example, $Li_{4.4}Sn$ gives a lithium storage capacity of 992 mAh g^{-1}, nearly triple that of conventional graphite (in the form of LiC_6, with a theoretical capacity of 372 mAh g^{-1}), and $Li_{4.4}Si$ can even provide a capacity of 4200 mAh g^{-1}. These materials exhibit a great potential in substituting graphite to become the next-generation anode materials used in high-energy lithium-ion batteries. However, the significant volume expansion during lithiation/delithiation cycle, which leads to the pulverization of electrode materials and very rapid capacity decay, brings a huge problem to their practical applications [14, 29, 30]. To conquer this problem, much work has been done, and one of the most promising methods is the introduction of carbon matrix into a nanocomposite electrode material. For example, Derrien et al. [31] have demonstrated the use of Sn–C composite (i.e., Sn nanoparticles dispersed in the carbon matrix, which acts as a buffer) can effectively relieve the strain associated with volume variations of tin while preventing the aggregation of Sn nanoparticles upon cycling and hence leads to a much improved anode material with large capacity of ca. 500 mAh g^{-1} over several hundred cycles.

When referring to carbon matrix, hollow carbon spheres can always be considered a favorable candidate, and recent work [30] has proved that by encapsulating tin nanoparticles into hollow carbon spheres with a uniform size, anode materials with high specific capacity and good cycling performance will be obtained, as explained in detail below.

6.2.4 Transition Metal Oxides

After 2000, the use of transition metal oxides as anode materials for lithium-ion batteries became more and more popular [32]. However, most of the materials consisting of transition metal oxides suffer from a high overpotential (voltage difference between the working voltage and thermodynamic equilibrium voltage, and the value is about 1 V) when undergoing conversion reactions for both lithiation and delithiation process. Therefore, the theoretical lithiation capacity can be achieved only when the thermodynamic equilibrium voltage of the material for conversion reaction is higher than 1 V [33]. Cr_2O_3 and MnO, with their high lithium storage capacity and relatively low thermodynamic equilibrium voltage, are thus more suitable as anode materials for lithium battery with enhanced energy densities [33, 34].

6.3 Cathode Materials

In a lithium-ion battery, cathode materials are usually oxides of transition metals that can undergo oxidation to higher valences when lithium ions are removed [6, 35], while maintaining their structural stability over a wide range of composition (for example, from fully charged states to completely discharged states). Among all the cathode materials, $LiCoO_2$ is the first cathode material that has been used in commercialized lithium-ion batteries and is still widely used. With decades of study, the research on the cathode materials for lithium-ion batteries has focused on the following three types of materials: a layered, structured hexagonal oxide (e.g., $LiCoO_2$), a spinel structured oxide (e.g., $LiMn_2O_4$), and an olivine structured oxide LiFePO$_4$), as shown in Table 6.1. In addition, transition metal sulfides such as titanium disulfide (TiS$_2$) have also been adopted in the cathode of lithium batteries.

6.3.1 Layered Structured Hexagonal Oxide

As one of the most successful cathode materials, LiCoO$_2$ forms a distorted rock-salt structure, the same as α-NaFeO$_2$, in which the cations order in alternating (111) planes [36]. This crystalline structure provides planes of lithium ions through which lithiation and delithiation can occur. Currently, LiCoO$_2$ suffers from two major problems. On the one hand, the availability of cobalt is relatively lower than that of other transition metals such as manganese, nickel, and iron, and this leads to a much higher cost. On the other hand, the structure of LiCoO$_2$ is not as stable as that of other cathode materials (e.g., LiFePO$_4$), and it undergoes performance degradation or even failure when overcharged [37, 38]. Therefore, it still is highly desired to search for alternative materials to replace LiCoO$_2$. One of the candidates is LiNiO$_2$, which shares the same layered structure as LiCoO$_2$. Compared with LiCoO$_2$, LiNiO$_2$ is lower in cost and higher in energy density; however, it also suffers from serious structural instability [39, 40]. Element doping of cobalt can effectively increase the degree of ordering, thus enhancing its structural stability; the same holds for LiMnO$_2$, which can achieve a higher capacity as well as a better rate capability through the addition of cobalt and nickel to form a composition of Li(Ni$_{1/3}$Mn$_{1/3}$Co$_{1/3}$)O$_2$ [41–43].

6.3.2 Spinel Structured Oxide

Another promising cathode material is LiMn$_2$O$_4$, which forms a spinel structure, and this structure enables manganese to occupy the octahedral sites while lithium predominantly occupies the tetrahedral sites [44]. LiMn$_2$O$_4$ has a lower cost and enhanced safety compared with LiCoO$_2$, but it also has some limitations [45]. One of its limitations comes from the phase change during the charge/discharge cycling, which will then be responsible for its capacity fade [46, 47]. Doping has been widely used to improve the electrochemical performance of LiMn$_2$O$_4$. For example, the addition of iron may lead to an enhancement in the charge plateau of $LiMn_2O_4$ at high voltages, the addition of cobalt may help to stabilize the spinel structure of $LiMn_2O_4$, thus bettering the capacity retention upon cycling, and the addition of nickel can improve the capacity of $LiMn_2O_4$ by decreasing its lattice parameters [48–50]. Other doping atoms, such as aluminum, can also be used to improve the electrochemical performance of $LiMn_2O_4$.

6.3.3 Olivine Structured Oxide

Lithium phosphates ($LiMPO_4$) with an olivine structure have attracted extensive interests as potential cathode materials for lithium-ion batteries. In the crystalline structure, phosphorous occupies tetrahedral sites, while the transition metal (M) occupies octahedral sites and lithium forms one-dimensional chains along the [010] direction [51]. The most widely studied phosphate is $LiFePO_4$. Owing to its high power density and long cycle life, $LiFePO_4$ has a great potential in manufacturing lithium-ion batteries to power the next-generation EVs and HEVs. However, the electronic conductivity of $LiFePO_4$ is relatively low (ca. 10^{-9} S cm^{-1} for the pure $LiFePO_4$). In order to overcome this shortcoming, many methods have been used, among which the introduction of hierarchical 3D mixed conducting networks is one of the best choices in enhancing the power and rate performance of $LiFePO_4$ [52], and this work will be further discussed in Section 6.5.3.

6.4 Nanostructured Electrode Materials

Although lithium-ion batteries have achieved great commercial success, further application of these batteries is now limited by their performance (e.g., charge/discharge rate) due to the fact that most lithium-ion batteries are based on micrometer-sized electrode materials, thus poor in their kinetics, lithium-ion intercalation capacities, and structural stability; limits will always exist if no improvement is made on the intrinsic diffusivity of Li ions in the solid state (ca. 10^{-8} cm^2 s^{-1}) or other material properties [8]. Therefore, in order to meet the demands of next-generation HEVs and clean energy storage, people may still face challenges in developing new electrode materials with high power density (viz. high rates), high energy density, longer cycle life, and improved safety [14]. As a result, people naturally turn to nanomaterials, for the sake of size effects, ultrahigh specific surface areas and other attractive features they possess. For example, reduced dimensions of nanomaterials may accelerate the intercalation/deintercalation rates of Li ions, hence enabling the high-power characteristic of the battery. However, using nanomaterials does not necessarily mean taking a panacea. Before using nanomaterials, it is extremely important to understand effects, both positive and negative, of nanomaterials on the performance of lithium-ion batteries.

6.4.1 Advantages of Nanomaterials

Generally, the greatest obstacle to the use of lithium-ion batteries in HEVs or EVs lies in its relatively low power density, which results from kinetic problems in solid-state electrode materials, i.e., the slow Li^+ and e^- diffusion rate. And the mean diffusion (or storage) time, τ_{eq}, can be expressed by the diffusion coefficient, D, and the diffusion length, L, as shown in Eq. 6.3:

$$\tau_{eq} = \frac{L^2}{2D} \tag{6.3}$$

Clearly, the value of τ_{eq} can be reduced through two approaches: One is to increase the value of D, and this can be achieved through doping foreign atoms; the other approach is to reduce the value of L, and this is what nanomaterials are for [53]. Although the first method can improve the mixed conduction, only limited rate-performance enhancement can be observed, and sometimes this method may even bring about unstable crystal structures [14]. However, through nanostructuring, the crystal structure of electrode material remains the same, while the diffusion distance of Li ions can be greatly shortened. Thus, τ_{eq} can be reduced dramatically. In this way, fast insertion/extraction of lithium ions in the electrode material can be realized on the condition that no deterioration will come to the structure or performance of the material. Moreover, it has been reported that electrode materials inactive toward Li insertion may become active when "going nano." For example, low Li diffusion rate of rutile TiO_2 along the ab-plane ($D_{ab} \sim 10^{-15}$ cm^2 s^{-1}) often makes Li insertion into rutile extremely hard. However, nano-sized rutile TiO_2

(10–40 nm) is able to reversibly accommodate Li up to $Li_{0.5}TiO_2$ (168 mAh g^{-1}) at 1–3 V vs. Li$^+$/Li with excellent capacity retention on cycling compared with its micrometer-sized counterparts [54]. The difference in rutile TiO_2 is mainly related to the drastic decrease in the diffusion time τ_{eq}, which is evidently caused by the shortening of Li diffusion path. This is the same for β-MnO_2, which, when made mesoporous, will allow reversible lithium intercalation without destruction of the rutile structure [55]. In addition, with a large surface area, there will be a higher contact area between the material and the electrolyte. Hence, more lithium ions can be quickly absorbed onto and stored in the fine particles; meanwhile, the specific current density of active material can be significantly reduced, which will then enable a high lithium-ion flux across the interface. As a result, material performance such as capacity and high rate performance (or high power) can be improved. It can also be concluded from the above example that nano-sized rutile TiO_2 with a specific surface area of ca. 110 m^2 g^{-1} also exhibits an excellent high rate performance (100 mAh g^{-1} at 10 C and 70 mAh g^{-1} at 30 C, where 1 C = 336 mA g^{-1}) [54].

Another advantage brought about by nanomaterials is the enhanced structural stability. When material has a particle radius r_p larger than the critical nucleation radius r_c for that phase, structural transition to thermodynamically undesirable structures may take place. Therefore, by using nanoparticles with $r_p < r_c$, it is possible to eliminate such transitions. For example, layered $LiMnO_2$ suffers from serious structural instability during the Li insertion/extraction process, which is responsible for its cycling capacity fade. In order to overcome such difficulties, materials with nanocrystalline structures are introduced, for a much higher Li-intercalation capacity than convention materials resulting from an easily accommodated lattice stress due to Jahn–Teller distortion [56]. In nanoparticles, it is widely accepted that the smaller the particles are, the more atoms these particles will have at the surface. Since the charge accommodation occurs mainly at or very near the surface, the need for diffusion of Li$^+$ in the solid phase of nanomaterials may be reduced, which will then account for an enhancement in the charge and discharge rate of the electrode as well as a reduction in the volumetric change and lattice stress caused by repeated Li insertion and expulsion.

Besides, using nanomaterials will enable some lithium-storage mechanisms available for mass storage. One such new mechanism is the so-called "conversion" mechanism [32], which is first found in transition metal oxides, then in fluorides, sulfides, and nitrides [33, 57, 58], and the mechanisms can described by Eq. 6.4:

$$MX + yLi^+ + ye^- \rightleftarrows Li_yX + M, \tag{6.4}$$

where X = O, S, F, or N and M = Fe, Co, Ni, Cr, Mn, Cu, and so on.

As can be seen, the mechanism is mainly related to the reversible in situ formation and decomposition of Li_yX upon Li uptake and release, and the reversible capacities in these systems are usually in the range of 400–1100 mAh g^{-1}. It is reported that electrodes made of CoO nanoparticles can achieve a specific capacity of 700 mAh g^{-1} with almost 100% capacity retention for up to 100 charge/discharge cycles [32]. In addition to conversion mechanism, other mechanisms such as interfacial Li storage [59] and nanopore Li storage [24] have also been proposed, and more research work will follow this direction in the future.

Nanostructured electrode materials also possess other merits, such as the change of electrode potential (or the thermodynamics of the reaction) [60], and the more extensive range of solid-solution-existing composition [61]; these merits, along with other above-mentioned advantages, provide infinite possibilities to nanomaterials.

6.4.2 Disadvantages of Nanomaterials

For nanomaterials, the excess surface free energy should be taken into consideration for the chemical potential, as shown in Eq. 6.5:

$$\mu_0(r) = \mu_0(r = \infty) + 2(\gamma/r)V, \tag{6.5}$$

where $2(\gamma/r)V$ gives the excess surface free energy, γ is the effective surface tension, V is the partial molar volume, and r is the effective grain radius.

Although the excess surface free energy contributes to the high electroactivity toward Li storage, it also results in several disadvantages. First, nanoparticles have low thermodynamic stability, therefore tend to agglomerate due to their high surface energy, and this makes them difficult to be dispersed or mixed with carbon black and binder to produce electrodes. It is also noted that the high electrolyte/electrode surface area may raise the risk of secondary reactions involving electrolyte decomposition between electrode and electrolyte, and this may be responsible for the low coulombic efficiency and poor cycle life of the battery and even lead to serious safety problems. In addition, nanoparticles are often difficult to synthesize, and the uniformity of their particle sizes may also be difficult to control. Moreover, since the density of nanopowder is always less than that of the same material formed from micrometer-sized particles, a decrease in energy density may happen during the nanostructuring process of electrode materials. In view of all problems in the use of nanomaterials, it is necessary to find a solution to overcome the shortcomings of nanosized materials while retaining their merits. In this context, the construction of hierarchically nanostructured electrode materials has been provided to be a solution to such problem.

6.5 Hierarchically Nanostructured Electrode Materials

It is known that nanostructured electrode materials, compared with its micro-sized counterparts, have a more enhanced kinetics for both Li ions and electrons due to a shorter diffusion length, a better accommodation of the strain caused by Li insertion/extraction and some new lithium storage mechanisms [14]; therefore, it can surpass the micro-sized materials in many electrochemical properties such as capacity, charge/discharge rate, and cycling performance. However, micro-sized materials are easy to operate, and the cost of their production is usually affordable, whereas for nanomaterials, a lot of problems limit their practical applications, such as low thermodynamic stability, high activity toward surface reactions, and difficulties in handling. Besides, nanostructuring process often requires more time and more complicated procedures, which will then raise the cost of manufacture [8, 14]. Since the properties of nanomaterials and properties of micro-sized materials are complementary to each other, if a material can integrate all merits of the both it will be highly possible to obtain an electrode material with superior performance. Concerning factors in all the aspects, nano/micro hierarchical structured materials become the best choice for novel structures of electrodes.

Nano/micro hierarchical structures often consist of nanometer-sized building blocks and micro- or submicrometer-sized assemblies, and as stated above, they have been chosen as the best system for lithium-ion batteries because they can take advantages of both the components of nano and micro. The former can greatly shorten diffusion times and provide possible new Li storage mechanisms, while the latter guarantees good stability and makes the materials easy to handle [14, 62]. On the basis of their morphology, commonly reported nano/micro hierarchical structured materials include sphere-like nano/micro hierarchical structures, flower-like nano/micro hierarchical structures, and hierarchical 3D mixed conducting networks. The following part will discuss these structures.

6.5.1 Sphere-Like Nano/Micro Hierarchical Structures for Electrode Materials

Sphere-like nano/micro hierarchical structures are often at micro-scale or submicron-scale, and they can be synthesized from various nanometer-sized building blocks such as nanoparticles and nanorods. As one of the most promising hierarchical structures for lithium-ion batteries, its large quantity of nanochannels and nanopores will enable a fast migration for both Li ions and electrons, whereas the sphere-like microstructure can enhance its stability and bring about a better dispersivity. For example, through a mediated polyol process, highly ordered superstructures of V_2O_5 can be self-assembled, in which nanoparticles interconnect to form nanorods, which will then circle around to form hollow microspheres, as shown in Fig. 6.3a,b [63]. The self-assembled V_2O_5 hollow microspheres with uniform hedgehog-like morphology have desirable electrochemical properties such as high capacity and remarkable cycle reversibility. It should be noted that V_2O_5 is not a cathode material for traditional lithium-ion batteries but a promising cathode material for lithium-based batteries with lithium-containing anodes.

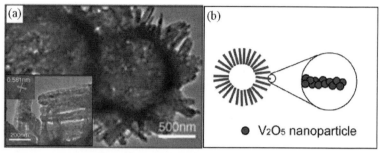

Figure 6.3 (a) TEM image of the as-prepared V_2O_5 hollow microspheres. Inset gives a high magnification transmission electron microscopy (TEM) image of a V_2O_5 microsphere and a HRTEM picture taken from the edge of a nanorod surface. (b) Sketch of V_2O_5 hollow microspheres (all reproduced with permission from [63], copyright 2005, Wiley-VCH Verlag GmbH & Co. KGaA).

Another example in this context is the tin nanoparticles encapsulated hollow carbon spheres (TNHCs), with a uniform size (ca. 500 nm), in which several tin nanoparticles (diameter ca. 100 nm) are encapsulated in one hollow carbon sphere (thickness ca. 20 nm), as shown in Fig. 6.5a,b [30]. With SiO_2 spheres (ca. 400 nm) and Na_2SnO_3 as the starting materials, SnO_2 coated SiO_2 spheres can be synthesized; then after a etching step by NaOH, followed by the treatment with glucose at 190°C, SnO_2/C double shell sphere is obtained, and the final products can be prepared after the heat treatment in the Nitrogen atmosphere at 700°C. In the composite microspheres, the content of Sn is up to 74 wt%, which provides a high theoretical specific capacity (up to 831 mAh g^{-1}), and the volume ratio of the Sn nanoparticles and the void space encapsulated is about 1:3, which can accommodate the volume expansion during the lithiation process (it can be seen in Fig 6.4c that even in case of the formation of $Li_{4.4}Sn$, there still exists enough free space (about 17% of the total space) in the hollow carbon spheres. Consequently, this type of Sn-based nanocomposites have an extremely high specific capacity of larger than 800 mAh g^{-1} in the first 10 cycles, and ca. 550 mAh g^{-1} even after 100 cycles.

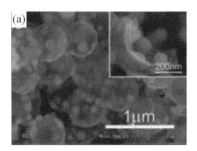
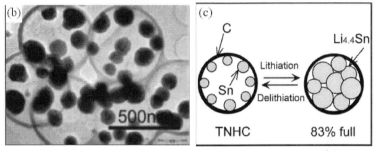

Figure 6.4 (a) TEM image of TNHCs. Inset shows a close view of a single broken carbon spherical shell studded with tin particles. (b) SEM image of TNHCs. (c) Sketch of the lithiation and delithiation process of TNHCs (all reproduced with permission from [30], copyright 2008, Wiley-VCH Verlag GmbH & Co. KGaA).

Other sphere-like hierarchical structured materials include mesoporous microspheres [55]. It has been reported that mesoporous anatase TiO_2 sub-micrometer spheres, including mesoporous TiO_2 spheres that have an average particle size of 300 nm and a pore size in the range of 3–30 nm (as shown in Fig 6.5a), seem to be more promising for anode materials of lithium-ion batteries [64, 65]. They can be synthesized by a two-step method: first, calcine the Ti–Cd precursors in the air to offer TiO_2–$CdSO_4$ composite, followed by a washing step to give the final product. With a BET specific surface area ca. 131 m^2 g^{-1} and a porosity ca. 48%, this material exhibits a favorable performance at low current rates, whereas at high current rates above 10 C, the performance becomes worse due to the poor electronic characteristics of the material.

Figure 6.5 (a) SEM images of mesoporous TiO$_2$ spheres. Inset shows the high-magnification SEM image of surface structures of a single TiO$_2$ sphere. (b) HRTEM image of a part of a mesoporous TiO$_2$ sphere. Inset exhibits the corresponding SAED pattern of the sphere (all reproduced with permission from [65], copyright 2006, The Royal Society of Chemistry).

6.5.2 Flower-Like Nano/Micro Hierarchical Structures for Electrode Materials

Nature can always inspire people in the design of new materials, and that is why people turn to structures in nature when developing hierarchical nano/microstructures for lithium-ion batteries. For example, hierarchical flower-like SnO$_2$ nano/microstructure material has been synthesized via facile solvent-induced and surfactant-assisted self-assembly method followed by a suitable thermal treatment [66]. As can be seen in Fig. 6.6, this flower-like structure consists of 3D interconnected SnO$_2$ nanoparticles and nanopores, and in its use of gas sensors for detecting carbon monoxide and hydrogen, as-prepared SnO$_2$ nano/microstructure shows an admirable sensitivity and extremely low detecting limit (5 ppm), as well as good reproducibility and short response/recovery times. After coating with carbon, both SnO$_2$–C composite and Sn–C composite are obtained. Benefiting from the outside carbon coating layer as well as the in situ formed matrix of Li$_2$O during Li uptake, the obtained SnO$_2$–C composite shows a very high reversible capacity (~700 mAh g^{-1} after 20 cycles, as shown in Fig. 6.7) and extremely high coulombic efficiency in the initial few cycles; when compared with said bare SnO$_2$ nanostructure as well as the Sn–C nanocomposite, this SnO$_2$–C composite exhibits a much enhanced cycling behavior, thus gaining its application potential as a novel anode material in lithium-ion batteries.

Hierarchically Nanostructured Electrode Materials | 241

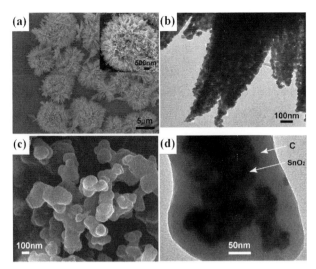

Figure 6.6 Flower-like nano/micro hierarchical structures of SnO_2 ((a) and (b)) and SnO_2–C composites ((c) and (d)). (a) SEM image of as-obtained SnO_2. Inset shows a close view of a single flower-like structure. (b) TEM image of "petals" of a SnO_2 flower (all reproduced with permission from [66], copyright 2009, American Chemical Society).

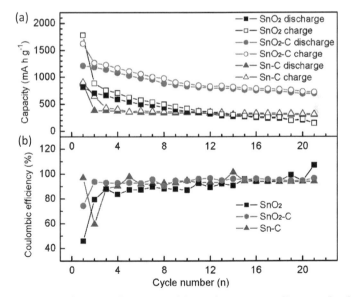

Figure 6.7 Cycling performance (a) and corresponding coulombic efficiency profiles (b) of the SnO_2, SnO_2–C, and Sn–C samples at a rate of C/5 between voltage limits of 0.02 and 3.0 V [66].

6.5.3 Hierarchical 3D Mixed Conducting Networks

Hierarchical 3D mixed conducting network consists of transport channels for both electrons and Li ions, which has shown an exceeding electrochemical performance compared with other electrode materials through enhancing the electronic conductivity of the material.

An example of this structure involves with the above anatase TiO_2 sub-micrometer spheres (see Fig. 6.5), the performance of which becomes poor at high current rates. In order to obtain anode materials for high-energy and high-power lithium-ion batteries, an optimization has to be made on the structure design of electrode materials, and the introduction of hierarchical 3D mixed conducting networks on nano/micron scale can be an effective way to carry out such an optimization. For example, with said mesoporous TiO_2 spheres as the starting material, TiO_2:RuO_2 nanocomposite can be synthesized through RuO_2 coating, and the prepared nanocomposite shows superior high rate capability when used as anode material for lithium-ion batteries [34]. As can be seen in Fig. 6.8, the nanoscopic network structure is composed of a dense net of metalized mesopores that allow both Li^+ and e^- to migrate; thereby, the effective diffusion length is reduced. Moreover, this network with mesh size of about 10 nm can be overlapped by another net with similar structure on the micron-scale formed by the composite of the mesoporous particles and the conductive admixture. Besides, RuO_2 or Li_xRuO_2 formed during Li insertion also allows for quick Li permeation, and this can be proved by the fact that at a very high charge/discharge rate of 30 C, the specific charge capacity of the composite is 91 mAh g^{-1}, nine times larger than that of the original mesoporous TiO_2 spheres (10 mAh g^{-1}).

The concept of hierarchical 3D mixed conducting networks has also been successfully used in constructing high rate cathode materials such as $LiFePO_4$. It has been found that benefiting from the combining use of carbon and nanometer-sized RuO_2 as electronic interconnects, the kinetics and rate capability of the composite of C–$LiFePO_4$–RuO_2 are significantly improved [67].

Bearing in mind that nanoporous carbon itself is already a mixed conducting 3D network, we propose and realize an alternative optimized nanostructure design of electrode materials for high-power and high-energy lithium batteries by combining the advantages of nanoporous carbon and nanometer-sized active particles. Superior cathode materials are obtained by dispersing nanometer-sized particles of active materials into a nanoporous carbon matrix. In view of its facility in terms of synthesis and porous structure, the design is superior to either directly coating nanoparticles with thin carbon layers or inserting nanoparticles into a nonporous carbon matrix. For example, through dispersing LiFePO$_4$ nanoparticles into a nanoporous carbon matrix (abbreviated LFP-NP@NPCM, see Fig. 6.9), an optimized nanostructure with much enhanced power and rate performance can be realized, and the obtained materials are promising for both lithium-ion batteries and supercapacitors [52].

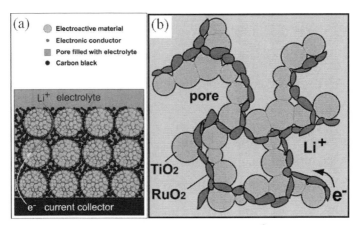

Figure 6.8 (a) Sketch of TiO$_2$:RuO$_2$ nanocomposites. (b) Schematic illustration of the self-wired path of deposited RuO$_2$ nanoparticles (all reproduced with permission from [34], copyright 2007, Wiley-VCH Verlag GmbH & Co. KGaA).

In the context of constructing hierarchical 3D mixed conducting networks, another approach is the introduction of 3D carbon nanotube (CNT) networks into active materials, such as CuO–CNTs composite. Through a simple solution method, self-assembled CuO and CuO–CNTs nanomicrospheres with an average size of 2.5 μm can be easily synthesized. It has been demonstrated that the as-formed CuO–CNTs nanomicrospheres contain uniformly distributed CNTs [68]. Furthermore, electrochemical experiments have confirmed that these composite spheres exhibit a superior cycling performance (good retention of capacity on cycling) and rate performance compared with using CuO spheres alone as anode materials in lithium-ion batteries. The reason for the improvement in the performance lies in the 3D network of CNTs, which has dual functions in enabling the transport of both electrons and Li ions.

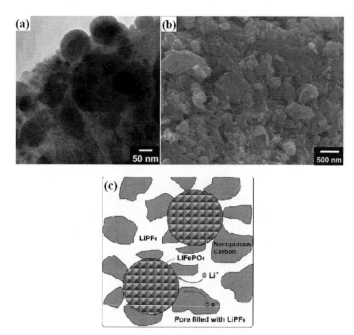

Figure 6.9 TEM (a) and SEM (b) images of LFP-NP@NPCM nanocomposite. (c) Schematic illustration of the LFP-NP@NPCM nanocomposite for Li storage (all reproduced with permission from [52], copyright 2009, Wiley-VCH Verlag GmbH & Co. KGaA).

Considering all the merits that the hierarchical 3D mixed conducting networks have, such as negligible diffusion times, much enhanced local conductivities, and possibly faster transfer reactions, such networks can provide excellent power performance, thus making them especially suitable for EVs and HEVs [14].

6.6 Conclusions

The introduction of new technologies always brings about new opportunities as well as new challenges to the industry, and the use of new materials may always be the breakthrough point of the technological innovation. Since the invention of the first voltaic pile, the development of the battery industry has been lasted for hundreds of years, with more and more novel batteries coming into being to serve the world. And the lithium-ion battery, though with a history of decades, yet has exhibited its great potential for a wide range of applications. However, to meet the ever-increasing demand and extend its applications to other industrial fields, improvements based on old materials and developments of new materials are necessary to be made, and nanomaterials, with their enhanced kinetics and activity, can always be taken into consideration for high-performance energy devices. With the help of optimized hierarchically nano/micro composite, it is believed that the problems caused by the disadvantages of conventional nanomaterials, such as low thermodynamic stability, high secondary reactions, and difficulties in synthesis, will be solved neatly while retaining their intrinsic favorable properties. However, exploring new materials and solving existing problems require a lot of interdisciplinary collaborations from researchers all over the world, and their success will surely make contributions to the development of sustainable energy accompanied by its economics.

References

1. Koksbang, R., et al. (1996). Cathode materials for lithium rocking chair batteries, *Solid State Ionics*, **84**(1–2): pp. 1–21.
2. Scrosati, B. (1992). Lithium rocking chair batteries: An old concept? *J. Electrochem. Soc.*, **139**(10): pp. 2776–2781.
3. Koksbang, M. Y. S. R., et al. (1997). Rocking-chair batteries based on $LiMn_2O_4$ and V_6O_{13}, *J. Power Sources*, **68**(2): pp. 726–729.
4. Armand, M. (1980). *Materials for Advanced Batteries*, eds. D. W. Murphy, J. Broodhead, and B. C. H. Steele., New York: Plenum Press.
5. Armand, M., and Tarascon, J.-M. (2008). Building better batteries, *Nature*, **451**(7179), pp. 652–657.
6. Whittingham, M. S. (2004). Lithium batteries and cathode materials. *Chem. Rev.*, **104**(10): pp. 4271–4301.
7. Maier, J. (2005). Nanoionics: Ion transport and electrochemical storage in confined systems. *Nat. Mater.*, **4**(11): pp. 805–815.

8. Bruce, P. G., Scrosati, B., and Tarascon, J.-M. (2008). Nanomaterials for rechargeable lithium batteries. *Angew. Chem. Int. Ed.*, **47**(16): pp. 2930–2946.

9. Arico, A. S., et al. (2005). Nanostructured materials for advanced energy conversion and storage devices. *Nat. Mater.*, **4**(5): pp. 366–377.

10. *World Secondary Lithium-ion Battery Markets*, Frost & Sullivan Industrial Report, Published by Frost & Sullivan on October 15, 2007, pp. 1–131.

11. Arora, P., and Zhang, Z. (2004). Battery separators. *Chem. Rev.*, **104**(10): pp. 4419–4462.

12. Winter, M., and Brodd, R. J. (2004). What are batteries, fuel cells, and supercapacitors? *Chem. Rev.*, **104**(10): pp. 4245–4270.

13. Wang, Y., and Cao, G. (2008). Developments in nanostructured cathode materials for high-performance lithium-ion batteries. *Adv. Mater.*, **20**(12): pp. 2251–2269.

14. Guo, Y.-G., Hu, J.-S., and Wan, L.-J. (2008). Nanostructured materials for electrochemical energy conversion and storage devices. *Adv. Mater.*, **20**(15): pp. 2878–2887.

15. Whittingham, M. S. (1976). Electrical energy storage and intercalation chemistry. *Science*, **192**(4244): pp. 1126–1127.

16. Nishi, Y. (2001). The development of lithium ion secondary batteries. *Chem. Rec.*, **1**(5): pp. 406–413.

17. Basu and Samar (1981). *Rechargeable battery*, US 4304825 (A), Bell Telephone Laboratories, Incorporated (Murray Hill, NJ) United States, 1981, pp. 1–6.

18. Yamaura, J., et al. (1993). High voltage, rechargeable lithium batteries using newly-developed carbon for negative electrode material. *J. Power Sources*, **43**(1–3): pp. 233–239.

19. Li, H., et al. (2009). Research on advanced materials for Li-ion batteries. *Adv. Mater.*, **21**(45): pp. 4593–4607.

20. Aurbach, C. A. D., et al. (2002). Advances in lithium-ion batteries, eds. W.A.V. Schalkwijk and B. Scrosati. New York: Kluwer Academic/Plenum Publishers.

21. Watanabe, I., and Yamaki, J.-I. (2006). Thermalgravimetry-mass spectrometry studies on the thermal stability of graphite anodes with electrolyte in lithium-ion battery. *J. Power Sources*, **153**(2): pp. 402–404.

22. Dahn, J. R., et al. (1995). Mechanisms for lithium insertion in carbonaceous materials. *Science*, **270**(5236): pp. 590–593.

23. Azuma, H., et al. (1999). Advanced carbon anode materials for lithium ion cells. *J. Power Sources*, **81–82**: pp. 1–7.
24. Wang, Q., et al. (2001). Monodispersed hard carbon spherules with uniform nanopores. *Carbon*, **39**(14): pp. 2211–2214.
25. Wang, Q., et al. (2002). Novel spherical microporous carbon as anode material for Li-ion batteries. *Solid State Ionics*, **152–153**: pp. 43–50.
26. Hu, J., Li, H., and Huang, X. (2007). Electrochemical behavior and microstructure variation of hard carbon nano-spherules as anode material for Li-ion batteries. *Solid State Ionics*, **178**(3–4): pp. 265–271.
27. Sun, X., and Li, Y. (2004). Colloidal carbon spheres and their core/shell structures with noble-metal nanoparticles. *Angew. Chem. Int. Ed.*, **43**(5): pp. 597–601.
28. Cui, X., Antonietti, M., and Yu, S.-H. (2006). Structural effects of iron oxide nanoparticles and iron ions on the hydrothermal carbonization of starch and rice carbohydrates. *Small*, **2**(6): pp. 756–759.
29. Huggins, R. A. (1999). Lithium alloy negative electrodes. *J. Power Sources*, **81–82**: pp. 13–19.
30. Zhang, W.-M., et al. (2008). Tin-nanoparticles encapsulated in elastic hollow carbon spheres for high-performance anode material in lithium-Ion batteries. *Adv. Mater.*, **20**(6): pp. 1160–1165.
31. Derrien, G., et al. (2007). Nanostructured Sn-C composite as an advanced anode material in high-performance lithium-ion batteries. *Adv. Mater.*, **19**(17): pp. 2336–2340.
32. Poizot, P., et al. Nano-sized transition-metal oxides as negative-electrode materials for lithium-ion batteries. *Nature*, 2000, **407**: pp. 496–499.
33. Li, H., Balaya, P., and Maier, J. (2004). Li-storage via heterogeneous reaction in selected binary metal fluorides and oxides. *J. Electrochem. Soc.*, **151**(11): pp. A1878–A1885.
34. Guo, Y. G., et al. (2007). Superior electrode performance of nanostructured mesoporous TiO_2 (anatase) through efficient hierarchical mixed. *Adv. Mater.*, **19**(16): pp. 2087–2091.
35. Guyomard, D. (2000). Advanced cathode materials for lithium batteries. In Osaka, T., and Datta, M. (Eds.) *New Trends in Electrochemical Technology. Volume 1: Energy Storage Systems for Electronics.* Amsterdam: Gordon and Breach. pp. 253–350.
36. Antolini, E. (2004). $LiCoO_2$: Formation, structure, lithium and oxygen nonstoichiometry, electrochemical behaviour and transport properties. *Solid State Ionics*, **170**(3–4): pp. 159–171.

37. Belova, D., and Yang, M.-H. (2008). Investigation of the kinetic mechanism in overcharge process for Li-ion battery. *Solid State Ionics*, **179**(27–32): pp. 1816–1821.

38. Belov, D., and Yang, M.-H. (2007). Failure mechanism of Li-ion battery at overcharge conditions. *J. Solid State Electrochem.*, **12**(7–8): pp. 885–894.

39. Yamada, A., Chung, S. C., and Hinokuma, K. (2001). Optimized LiFePO$_4$ for lithium battery cathodes. *J. Electrochem. Soc.*, **148**(3): pp. A224–A229.

40. Liu, H., Yang, Y., and Zhang, J. (2007). Reaction mechanism and kinetics of lithium ion battery cathode material LiNiO$_2$ with CO$_2$. *J. Power Sources*, **173**(1): pp. 556–561.

41. Stewart, S. G., Srinivasan, V., and Newman, J. (2008). Modeling the performance of lithium-ion batteries and capacitors during hybrid-electric-vehicle operation. *J. Electrochem. Soc.*, **155**(9): pp. A664–A671.

42. Martha, S.K., et al. (2009). A short review on surface chemical aspects of Li batteries: A key for a good performance. *J. Power Sources*, **189**(1): pp. 288–296.

43. Xu, H. Y., Wang, Q. Y., and Chen, C. H. (2008). Synthesis of Li[Li$_{0.2}$Ni$_{0.2}$Mn$_{0.6}$]O$_2$ by radiated polymer gel method and impact of deficient Li on its structure and electrochemical properties. *J. Solid State Electrochem.*, **12**(9): pp. 1173–1178.

44. Thackaray, M. M., Kock, A. D., and David, W. I. F. (1993). Synthesis and structural characterization of defect spinels in the lithium-manganese- oxide system. *Mater. Res. Bull.*, **28**(10): pp. 1041–1049.

45. Pasquier, A. D., Huang, C. C., and Spitler, T. (2009). Nano Li$_4$Ti$_5$O$_{12}$–LiMn$_2$O$_4$ batteries with high power capability, improved cycle-life. *J. Power Sources*, **186**(2): pp. 508–514.

46. Liu, Q., et al. (2007). Phase conversion and morphology evolution during hydrothermal preparation of orthorhombic LiMnO$_2$ nanorods for lithium ion battery application. *J. Power Sources*, **173**(1): pp. 538–544.

47. Molenda, J., et al. (2007). Electrochemical and high temperature physicochemical properties of orthorhombic LiMnO$_2$. *J. Power Sources*, **173**(2): pp. 707–711.

48. Shigemura, H., et al. (2001). Structure and electrochemical properties of LiFe$_x$Mn$_{2-x}$O$_4$ ($0 \leq x \leq 0.5$) spinel as 5 V electrode material for lithium batteries. *J. Electrochem. Soc.*, **148**(7): pp. A730–A736.

49. Arora, P., Popov, B. N. and White, R. E. (1998). Electrochemical investigations of cobalt-doped $LiMn_2O_4$ as cathode material for lithium-ion batteries. *J. Electrochem. Soc.*, **145**(3): pp. 807–815.

50. Fang, T.-T. and Chung, H.-Y. (2008). Reassessment of the electronic-conduction behavior above Verwey-like transition of Ni^{2+}- and Al^{3+}-doped $LiMn_2O_4$. *J. Am. Ceram. Soc.*, **91**(1): pp. 342–345.

51. Padhi, A. K., Nanjundaswamy, K. S., and Goodenough, J. (1997). Phospho-olivines and positive-electrode materials for rechargeable lithium batteries. *J. Electrochem. Soc.*, **144**(4): pp. 1188–1194.

52. Wu, X.-L., et al. (2009). $LiFePO_4$ Nanoparticles embedded in a nanoporous carbon matrix: superior cathode material for electrochemical energy-storage devices. *Adv. Mater.*, **21**(25–26): pp. 2710–2714.

53. Okubo, M., et al. (2007). Nanosize effect on high-rate Li-ion intercalation in $LiCoO_2$ electrode. *J. Am. Chem. Soc.*, **129**(23): pp. 7444–7452.

54. Hu, Y.-S., et al. (2006). High lithium electroactivity of nanometer-sized rutile TiO_2. *Adv. Mater.*, **18**(11): pp. 1421–1426.

55. Jiao, F., and Bruce, P. G. (2007). Mesoporous crystalline β-MnO_2—a reversible positive electrode for rechargeable lithium batteries. *Adv. Mater.*, **19**(5): pp. 657–660.

56. Robertson, A. D., Armstrong, A. R., and Bruce, P. G. (2001). Layered $Li_xMn_{1-y}Co_yO_2$ intercalation electrodesinfluence of ion exchange on capacity and structure upon cycling. *Chem. Mater.*, **13**(7): pp. 2380–2386.

57. Pereira, N., et al. (2003). Electrochemistry of Cu_3N with lithium. *J. Electrochem. Soc.*, **150**(9): pp. A1273–A1280.

58. Chen, J., et al. (2005). α-Fe_2O_3 nanotubes in gas sensor and lithium-ion battery applications. *Adv. Mater.*, **17**(5): pp. 582–586.

59. Jamnik, J., and Maier, J. (2003). Nanocrystallinity effects in lithium battery materials Aspects of nano-ionics. Part IV. *Phys. Chem. Chem. Phys.*, **5**(23): pp. 5215–5220.

60. Balaya, P., et al. (2006). Nano-ionics in the context of lithium batteries. *J. Power Sources*, **159**(1): pp. 171–178.

61. Meethong, N., et al. (2007). Size-dependent lithium miscibility gap in nanoscale $Li_{1-x}FePO_4$. *Electrochem. Solid-State Lett.*, **10**(5): pp. A134–A138.

62. Jiang, C., Hosono, E., and Zhou, H. (2006). Nanomaterials for lithium ion batteries. *Nano Today*, **1**(4): pp. 28–33.

63. Cao, A.-M., et al. (2005). Self-assembled vanadium pentoxide (V_2O_5) hollow microspheres from nanorods and their application in lithium-ion batteries. *Angew. Chem. Int. Ed.*, **44**(28): pp. 4391–4395.
64. Hu, J.-S., et al. (2004). Interface assembly synthesis of inorganic composite hollow spheres. *J. Phys. Chem. B*, **108**(28): pp. 9734–9738.
65. Guo, Y.-G., Hu, Y.-S., and Maier, J. (2006). Synthesis of hierarchically mesoporous anatase spheres and their application in lithium batteries. *Chem. Commun.*, (26): pp. 2783–2785.
66. Jiang, L.-Y., et al. (2009). SnO_2-based hierarchical nanomicrostructures facile synthesis and their applications in gas sensors and lithium-ion batteries. *J. Phys. Chem. C*, **113**(32): pp. 14213–14219.
67. Hu, Y.S., et al. (2007). Improved electrode performance of porous $LiFePO_4$ using RuO_2 as an oxidic nanoscale interconnect. *Adv. Mater.*, **19**(15): pp. 1963–1966.
68. Zheng, S.-F., et al. (2008). Introducing dual functional CNT networks into CuO nanomicrospheres toward superior electrode materials for lithium-ion batteries. *Chem. Mater.*, **20**(11): pp. 3617–3622.

Chapter 7

Ionic Liquid-Assisted Fabrication of Graphene-Based Electroactive Composite Materials

Pia Damlin, Bhushan Gadgil, and Carita Kvarnström

*Turku University Centre for Materials and Surfaces (MATSURF),
Laboratory of Materials Chemistry and Chemical Analysis,
University of Turku, FI-20014, Finland*

pia.damlin@utu.fi, carita.kvarnstrom@utu.fi

7.1 Introduction

Graphene has attracted great attention due to its unique structure and properties such as high electrical conductivity, high thermal conductivity, extraordinary elasticity, and stiffness. To take full advantage of its properties for applications, integration of graphene in polymer matrices to form advanced functional composites is one promising and currently attracting route. Various graphene composites with insulating or conducting polymers (CPs) have been prepared through non-covalent or covalent approaches, leading to improved properties (tensile strength, elastic modulus, electrical and thermal conductivity, etc.) at very low loading. Conducting polymers are organic polymers that possess

---

Electrochemical Nanofabrication: Principles and Applications (Second Edition)
Edited by Di Wei
Copyright © 2016 Pan Stanford Publishing Pte. Ltd.
ISBN 978-981-4613-86-6 (Hardcover), 978-981-4613-87-3 (eBook)
www.panstanford.com

conjugated chain structures and electrical conductivities. In their neutral state, they are semiconductors or insulators, but compared to conventional polymers they can be doped in order to achieve their highly conductive states, making them promising materials in a wide range of electrochemical applications.

This chapter focuses on the IL-assisted fabrication of graphene/CP composite structures. The application of these composite materials is briefly mapped in the field of supercapacitors and as electrochromic materials. The motivation for incorporation of graphene (carbon materials) into organic electrochromic devices lies in the improvement of the cycling stability and optical contrast. Furthermore, in supercapacitor applications it is an effective approach in order to boost the mechanical and electrochemical cycling stability. Primary focus is on graphene fillers produced from graphite by scalable top-down techniques such as chemical synthesis and electrochemical exfoliation utilizing ionic liquids (ILs) as electrolytes. The use of ILs will be addressed in the synthesis of organic electroactive composite materials as they open up new opportunities for the influence of film properties.

7.2 Conducting Polymers

Conductivity in conjugated polymers was discovered several decades ago [1]. Since then, there has been major advances in the synthesis, characterization and applicability of conjugated polymers for different device applications. They all have in common a conjugated bond structure that allows for electron delocalization and charge transport along the polymer backbone resulting in unique electrical, optical and magnetic properties. Molecular structures for some commonly used CPs are shown in Fig. 7.1.

CPs can be synthesized by chemically initiated polymerization, vapor-phase methods, photochemical polymerization or by electrochemical polymerization where the choice of polymerization method depends to a large extent on the final application in mind [2, 3]. Chemical polymerization usually produces powdery materials and is the preferred technique for large-scale production of CPs. The major disadvantages of chemical polymerization are the limited choice of suitable oxidants, introduction of impurities with the oxidant and difficulties to control the oxidation strength of the reaction solution, leading to poor control of the degree of chemical degradation during synthesis. The obtained materials also show poor processing properties why functionalization often has to be made in order to improve their solubility [4]. Thereafter films can be made using spin coating or other solution casting methods on a variety of conductive or non-conductive substrates.

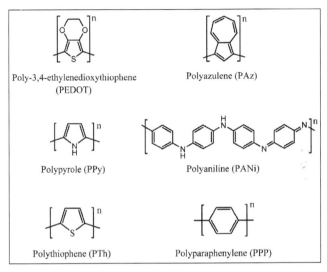

Figure 7.1 Molecular structures of some commonly utilized conducting polymers in their pristine state.

Electrochemical polymerization is a fast and simple, widely used method to synthesize different conducting polymers. Electrodeposition enables film formation on surfaces with complicated patterns as well as control of film thickness [5]. Furthermore, the subsequent growth of the polymer film and the charging reactions can be followed in situ [6]. Parameters such as solvent media, electrolyte, electrochemical method used for polymerization and monomer material, all have a profound effect on film morphology, charge transfer and transport properties. Different from the powdery products prepared through chemical approaches, this method enables an easy, one-step deposition of the film directly at the surface of the electrode substrate that can be further applied for electrochemical purposes.

7.3 Ionic Liquids

Room-temperature molten salts, or better known as room-temperature ionic liquids (RTILs), are liquids at ambient or even lower temperatures. ILs are not new materials although their importance has increased dramatically over the past decade. In fact, they have been known for almost 100 years all since the study of Walden on electrical conductivity of ethylammonium nitrate was reported [7]. The interest in ILs increased significantly in the 1970s by the discovery of pyridinium and imidazolium haloaluminate salts [8]. Haloaluminate ILs are, however, highly water sensitive as a result of the strong Lewis acid groups (e.g., $AlCl_3$) making them highly reactive under ambient conditions. The new generation air and moisture stable imidazolium-based ILs was reported in 1992 [9] and most of the currently employed ILs consists of imidazolium or pyridinium cations and halide-containing anions. The structures for some cations and anions found in common non-haloaluminate ILs are shown in Fig. 7.2.

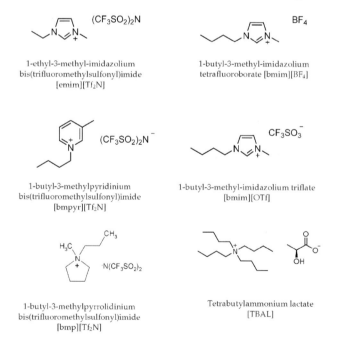

Figure 7.2 Structures and abbreviations for some cations and anions found in non-haloaluminate ILs.

Concerns have risen over the potential toxicity as well as low biodegradability of some of these traditionally employed ILs. To overcome these drawbacks, ILs have been synthesized from renewable, non-toxic bio-resources, thus meeting one of the main principles of green chemistry. Among these sources, natural products (amino acids, vitamins, etc.) with an interesting molecular diversity have the potential to be converted into ILs by means of green procedures, such as simple ion exchange and/or acid–base reactions. A group of amino acid–based ILs, in which amino acids act as the cations or anions, have been reported since 2005 [10–12]. These ILs showed excellent biodegradability and low toxicity to filamentous fungi [13] and the freshwater crustacean *Daphnia magna* [14]. In this respect, also biomass derived ionic liquids will be good candidates. The synthesis, starting from cheap commercial tetrabutylammonium hydroxide (TBAOH) and different acids derived from biomass, are shown in Scheme 7.1 [15]. Antimicrobial toxicity screening was included showing that the ILs were non-toxic to all tested fungi strains up to 1 mM concentration.

$$TBA^+OH^- + R\text{-}COOH \xrightarrow{H_2O,\ 100°C,\ 24h} TBA^+\ R\text{-}COO^-$$
(yield 93-97%)

Scheme 7.1 Formation of TBA ILs.

Another reported class of bio-ILs is the group of cholinium-based ILs (choline, an essential nutrient) with amino acids as the anion [16, 17]. Reported applications of such ILs have focused on their activities for the pretreatment of lignocellulosic biomass and as catalysts [18, 19]. Recently gel-based electrolytes, choline chloride–based eutectic solvents, were also used as electrolyte for the fabrication of environmentally friendly supercapacitors on paper [20].

7.3.1 Ionic Liquids in Electrochemistry

Ionic liquids can be used in a wide range of applications owing to their unusual physical and chemical properties [21]. They show high thermal stability, negligible vapor pressure, non-flammability, excellent electrochemical stability and high ionic conductivity. This, combined with their advantages of being non-volatile and biodegradable makes them an ideal non-aqueous solvent for the electrochemical production of graphene flakes (which will be further discussed in Section 7.4.3). ILs have been proposed as green alternatives to conventional organic solvents in a range of applications, including organic synthesis, catalysis and liquid–liquid extractions, just to mention a few [22].

From an electrochemical point of view, the relatively wide potential window, high conductivity, and thermal stability are properties making ILs an interesting alternative to conventional electrolytes. Table 7.1 shows the electrochemical window for four ILs with different anion–cation combinations at platinum electrode. The width of the window varies for different ILs ranging from 2 to 6 V depending on the ion combinations. Also the electrode substrate influence why larger potential windows are often reported for GC electrodes than for Pt as a result of the catalytic activity of the latter [23]. Even if many ILs are considered hydrophobic, they do absorb a certain amount of water from the surrounding atmosphere. The effect of residual water on the potential window has been studied in ILs containing the imidazolium cation and fluoroanions ($BMIMBF_4$ and $BMIMPF_6$). Addition of more than 3 wt% of water caused a decrease in the potential window of over 2 V [24]. Also the effect of very small amount of water (15 to 58 ppm) has been studied for trimethylpropylammonium-bis(trifluoromethylsulfonyl)imide (TMPA-TFSI) [25]. For GC, no significant effects were observed before addition of 0.59 wt% shrinking the cathodic limiting potential from −3.2 to −2.7 V. However, for a Pt electrode, a significant effect could be observed decreasing the usable cathodic potential from −3.2 to −1.8 V. These studies indicate the importance of removing water from the ILs, which due to their nonvolatile nature can be easily accomplished by vacuum drying.

The potentiality of using ionic liquids in electrosynthesis of organic materials has been investigated for electropolymerization of conducting polymers. In this case, the IL functions as both electrolyte and solvent media and will have a big influence on film morphology and on its electrochemical properties. As the thermodynamics and kinetics of reactions carried out in ILs are different to those in conventional organic electrolytes, this opens up new interesting research areas in the field of CPs. During the last few years conjugated polymers like PPy [27, 28], PTh [29, 30],

PANi [31, 32], PPP [33, 34], PEDOT [35, 36], and PAz [37, 38] have been electropolymerized in different non-haloaluminate ILs. For some CPs, electropolymerization in ILs results in films showing enhanced stability, improved charging capacity and morphology. However, it is important to point out that in ILs, the electron transfer rates and solvation mechanism is different compared to conventional electrolytes why not all polymers benefit from electropolymerization in ILs. Differences in ion transport also have an effect on the charge balancing occurring during doping/de-doping of CPs.

Table 7.1 Ionic liquids with different anion–cation combinations together with their potential windows on platinum [26].

Ionic liquid	Abbreviation	Electrochemical window [V] (vs. NHE)
1-Ethyl-3-methylimidazolium tetrafluoroborate	[EMIM][BF$_4$]	–2.6 to 2.1
Butyltrimethylammonium bis(trifluoromethylsulfonyl) imide	[BMA][Tf$_2$N]	–3.2 to 2.9
1-Ethyl-1-methylpyrrolidinium bis(trifluoromethylsulfonyl) imide	[EMIM][Tf$_2$N]	–2.5 to 2.8
1-Ethyl-3-methylimidazolium bis(trifluoromethylsulfonyl) imide	[EMIM][Tf$_2$N]	–2.1 to 2.6

7.4 Graphene

A natural source for graphene is graphite. Usually graphene sheets are stacked neatly in graphite where every other layer is positioned identically to each other [39]. In graphite, there are no chemical bonds between the individual graphene sheets. Due to the aromatic structure of graphene, there is a π-electron network throughout the graphene sheet that interacts with the π-electrons of graphene sheets above and below. These π–π-interactions are quite weak but due to large surface area of the sheets, the combined forces keep the individual sheets tightly stacked to each other, making their exfoliation difficult. On the other hand, because of the lack of strong chemical bonds between the sheets, they can quite easily slide with respect to each other. In fact, this property has been used in some graphite-based lubricants.

The strength of graphene is slightly higher than of carbon nanotubes due to extra stress in nanotubes caused by tight rolling. While graphene is mechanically very strong, it is also lightweight and flexible. Graphene can be rolled without any damage to the structure although the applied stress causes observable changes in its electrical properties [40]. The strong and flexible material is ideal for many applications on its own, if large enough sheets are available, but usually graphene sheets are just a few micrometers wide. Nevertheless, these graphene micro sheets have been used to enhance the mechanical strength of many polymeric films which will be discussed in Section 7.5. In addition to its mechanical properties, the electrical properties of graphene are excellent. In case of graphene, the two-dimensional material is constructed from alternating single and double bonds between carbon atoms. This creates a "frictionless" surface where electrons move freely, not just along one chain, but through the whole graphene sheet. In fact, the mobility of charge carriers in graphene is extremely good, 15000 $cm^2V^{-1}s^{-1}$, making electrons behave like massless Dirac fermions [41]. Despite this, graphene is a zero-band gap semiconductor rather than a conductor. This means that at charge neutrality point, e.g., at undoped stage, graphene has zero conductivity. However, any kind of a doping, positive or negative, chemical or electrical, makes graphene conducting.

7.4.1 Preparation of Graphene

There are plenty of methods to produce graphene and graphene-like materials and new ways are continuously invented. The preparation methods can be divided in two categories: top-down and bottom-up methods [42–45]. In the top-down methods, natural or synthetic graphite is used as precursor and various physical or chemical routes are used in order to peel off single graphene sheets from graphite. Bottom-up methods, on the other hand, usually make use of small carbon containing molecules or some other carbon source and build up graphene structures by joining carbon atoms together to grow up graphene.

In the bottom-up category, chemical vapor deposition (CVD) is the most used technique, since it can produce large area graphene sheets, with lateral dimensions up to several centimeters and with controlled number of graphene layers. While graphene can be prepared by mechanical cleavage of graphite or CVD, the exfoliation of graphite oxide has provided an affordable route to bulk quantities of graphene-based materials.

7.4.2 Production of Graphene Oxide

Although the yield and quality of graphene produced by the CVD process has improved considerably, most of the graphene produced lately has been via reduction of a graphene oxide (GO) aqueous dispersion. The solution processable GO is obtained by harsh chemical oxidation of graphite, nowadays mostly by applying the modified Hummers method, in which potassium permanganate is used as the main oxidant [46]. From this monolayer material GO is produced by exfoliation of graphite oxide. Regardless of the oxidation method and synthesis conditions, the GO produced contains more or less the same kind of functional groups but their abundance may vary significantly from one method to another. Also the starting material, graphite flake size, affects the efficiency of the oxidation process [47]. On the basis of theoretical calculations, chemical properties of GO and spectroscopic data, different structural models have been proposed for GO [48, 49]. Still today the exact structure of GO remains unclear and new models or interpretations to the old ones are made.

7.4.2.1 Post treatment of GO

As physicochemical properties of GO differ significantly from pristine graphene, most often GO is reduced to form materials known as reduced graphene oxide (rGO) [50]. The reduction processes follows via chemical or electrochemical reduction, thermal annealing of surface functional groups (carbonyl, hydroxyl, epoxy, carboxyl, etc.) of GO and in this way one recovers, at least partially, the excellent properties of graphene. Throughout the reduction process, the brown colored dispersion of GO in water turns black and the reduced sheets aggregate and form precipitates. This is due to the removal of oxygen atoms from the GO sheets making the final product less hydrophilic after reduction. The proposed scheme of GO reduction to rGO is shown in Fig. 7.3.

Each reduction method has their advantages and disadvantages. The most common method for rGO production is by chemical reduction of GO using different reducing agents of which the most efficient ones can reduce GO at room temperature or slightly elevated temperatures. There is a wide variety of different reducing agents to choose from, of which some are toxic (hydrazine) while others are environmentally friendly (ascorbic acid). In a recent review, 50 different reducing agents were covered emphasizing on their reduction mechanism and efficiency [51]. Much less attention has been put into the electrochemical reduction method unless it brings several advantages. It is an extremely simple, fast and environmentally friendly method, which furthermore offers control on the reduction process of GO, in both films and solution phase, by tuning of the applied voltages. Electrochemical reduction has been carried out in a range of different electrolytes like aqueous solutions [52], eutectic melts [53], ionic liquids [54] and organic solvents [55, 56]. While electrochemistry alone cannot give details about structural changes in thin films, it is often applied in combination with a spectroscopic technique [57]. Raman spectroscopy is an efficient method to identify and characterize various carbon nanostructures, while IR-spectroscopy gives information about the presence of oxygen-containing functional groups. The combination of these techniques gives the possibility to compare the potential induced changes in the oxygen-containing functional groups, as well as reorganization in the GO nanostructure as the electrochemical reduction of GO progresses [58, 59]. Furthermore, the change of electrolyte from water to organic media allows extension of the negative potential range used for the reduction of GO. According to such in situ spectroelectrochemical results, it could be stated that the reduction process can be effectively controlled by the choice of reduction potential and electrolyte medium [60, 61].

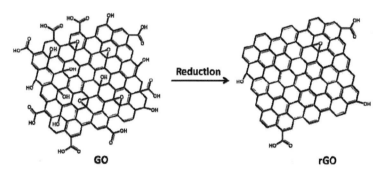

Figure 7.3 Proposed scheme for the reduction of GO to rGO.

7.4.3 Electrochemical Exfoliation of Graphite

The aforementioned top-down methods are unable to produce graphene in bulk at a reasonable cost and in an environmentally friendly manner. For widespread engineering applications of graphene-based technology, simple processes for its production in big scale are essential. The electrochemical approach has the advantage of being a single-step and at ambient conditions easy to operate method, where the yield and properties of graphene flakes can be tuned by controlling the electrolysis parameters and electrolytes. Compared to the chemical methods, the electrochemical approaches have not yet been extensively explored, however, gaining increased interest in the past few years as reckoned method for the production of graphene flakes [62].

The electrochemical method utilizes an aqueous or non-aqueous electrolyte and an electrical current (cathodic or anodic) to drive expansion, intercalation and ultimately exfoliation of a graphite electrode (e.g., rod, plate or wire). Figure 7.4 gives an illustration of electrochemical synthesis of graphene flakes involving oxidation (a) or reduction (b) of a host graphite working electrode leading to intercalation of cations or anions from the electrolyte.

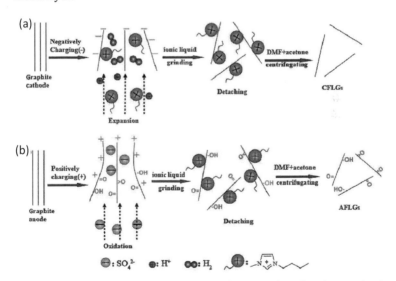

Figure 7.4 Illustration of electrochemical approaches for the synthesis of graphene flakes using (a) oxidation or (b) reduction of a graphite electrode. Reprinted with permission from Ref. [63]. Copyright © 2013, Royal Society of Chemistry.

When using the anodic oxidation as approach (Fig. 7.4a), a positive potential is applied to the graphite rod in order to oxidize it, thereby allowing intercalation of anions leading to structural expansion and ultimate exfoliation of graphene flakes [64]. However, if the applied potential is not properly controlled, side reactions will take place resulting in release of O_2 and CO_2 leading ultimately to the exfoliation of GO flakes [65, 66]. In this case the method suffers from the same drawback as in the chemical approach, it is hard to transform back the defect structure to sp^2 in the carbon atom.

The cathodic reduction of the graphite electrode is another electrochemical mean of preparing graphene (Fig. 7.4b). The advantage of the cathodic reduction technique includes the absence of strong oxidizing conditions preventing the generation of irreversible sp^3 defects. However, compared to the anodic approach, just a few papers can be found on the cathodic reduction of graphene mainly due to the limitation of water electrolysis in aqueous electrolytes. In one paper, inspired by the lithium rechargeable battery concept, a non-aqueous electrolyte (propylene carbonate) containing Li^+ ions was used [67]. With this concept, highly conductive few-layer graphene flakes could be produced with high yield. The results from this work using cathodic reduction showed promise towards scaled-up production of graphene. However, there will be questions concerning the cost and environmental friendliness of using non-aqueous electrolytes. Ionic liquids are another alternative non-aqueous solvent that has shown remarkable tendency to intercalate graphitic electrodes and yield gram-scale quantities of carbon nanostructures and graphene sheets. The first published work on graphene electrosynthesis from graphite electrode using ILs goes back to 2008 using 1:1 IL ([BMIM][PF_6])/water volume ratios [68]. This line of research was continued by Lu et al. focusing on the influence of water on the exfoliation mechanism [69]. They showed that by changing the water/IL ([BMIM][BF_4]) ratio, several nanomaterials can be made in different shapes and sizes. Furthermore, by controlling this ratio it was possible to generate fluorescent nanomaterials to be used in the ultraviolet and visible range. In a recent work by Najafabadi et al. [26], an electrolyte composed of IL and acetonitrile as co-solvent (1:50 volume ratio) was used for the electrochemical exfoliation of an iso-molded graphite anode.

Four different ILs were tested (Table 7.1) focusing on electrolyte coloration during electro-exfoliation and the role of water in the exfoliation process. It was found that switching from BF_4^- to more oxygenated anions such as bis(trifluoromethylsulfonyl)imide (BTA) enhanced the exfoliation yield and the electrolyte coloration to be due to diverse reactions involving the IL (Fig. 7.5). Additionally, it was shown that presence of water is unnecessary for the electrochemical graphite exfoliation, the co-solvent acetonitrile was proven as a more advantageous alternative.

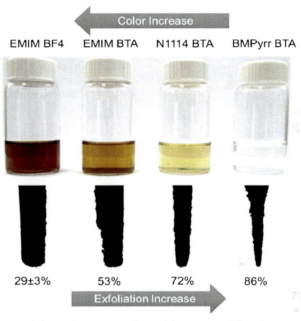

Figure 7.5 Color changes and exfoliation yield after 4 h of electrochemical graphite exfoliation at 7 V using 0.1 M IL/acetonitrile. Reprinted with permission from Ref. [26]. Copyright © 2014, Elsevier Ltd.

7.5 Graphene/CP Composites

Given the unique structure and properties of CPs (light weight, good tensile strength and flexibility) they have been shown as promising electrode materials for a range of different applications such as supercapacitors, sensors, solar cells, batteries, electrochromic cells and actuators [70–75]. However, CPs exhibit large volume changes during charging/discharging processes leading to decrease in electrochemical performance and stability. One approach to effectively improve the mechanical and electrochemical cycling stability of CPs is the incorporation of various carbon materials, such as activated carbons, carbon monolith, template porous carbon, and carbon nanotubes (CNTs) [76–80]. However, in such composites, the total capacitance value is usually limited by the microstructure of the carbon material. In case of CNTs also the cost effectiveness is a limitation.

Recently, the preparation of graphene–CP composites has attracted considerable interest. These composites exhibit a remarkable improved stability due to the synergetic combination of the excellent conducting and mechanical properties of graphene sheets and the high pseudo-capacitance of CPs [81]. In comparison to CNTs, the interfacial surface area in graphene–CP is very high why there is no need to add surplus graphene to the composite. Moreover, graphene has higher mobility and shows good mechanical properties and conductivity compared to traditional porous carbon materials and CNTs.

Graphene-CP composites have mainly been fabricated using non-covalent strategies starting with unreduced (GO/CPs) or reduced graphene oxide (rGO/CPs) via simple mixing or by in situ polymerization using either the chemical or electrochemical polymerization route [82, 83]. In situ polymerization involves the polymerization of monomer in a system containing the graphene material where the morphology can be controlled by tuning the polymerization conditions. Among CP, PANi and PPy have been the mostly investigated for the preparation of composite materials. For example, PANi on GO sheets has been prepared by chemical oxidative polymerization where the morphology of the composites was heavily affected by the ratio of aniline monomer to GO. Two morphologies of PANI nanowires were produced, randomly connected PANi nanowires from homogeneous nucleation in bulk solution and aligned PANi nanowires from heterogeneous nucleation on the GO nanosheet (Fig. 7.6) [84]. In order to obtain high-quality composites by the chemical polymerization method, it is important to control the pH value of the polymerization solution since over-acidification of the reaction will cause congestion of the GO sheets.

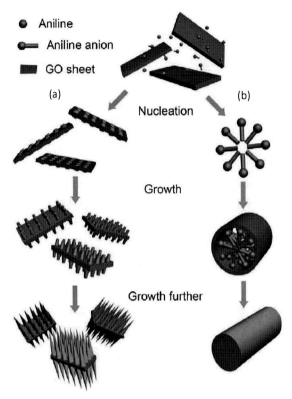

Figure 7.6 Schematic illustration of nucleation and growth mechanism of PANi nanowires: (a) heterogeneous nucleation on GO nanosheets; (b) homogeneous nucleation in bulk solution. Reprinted with permission from Ref. [84]. Copyright © 2010, American Chemical Society.

Graphene/CPs can also be produced by in situ electrochemical polymerization. Compared to the powdery products often obtained through chemical approaches the electrochemical polymerization yields composite films directly on the electrode substrate. A graphene/PANi composite film was prepared using in situ anodic electrochemical polymerization utilizing a graphene paper as electrode (Fig. 7.7). The mechanical and electrochemical properties of the composite material could be easily tuned by adjusting the polymerization time, as it changes the PANi content [85]. This graphene-based composite paper electrode showed excellent flexibility and favorable tensile strength.

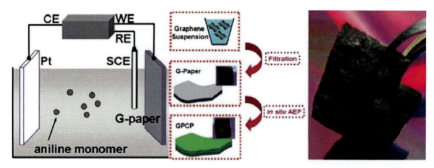

Figure 7.7 Schematic representation for the production and electrodeposition of aniline monomer on graphene paper. Reprinted with permission from Ref. [85]. Copyright © 2009, American Chemical Society.

For GO/CP composites, a further reduction process of GO by chemical, thermal or electrochemical means is essential to exhibit the conductivity of graphene [86]. Chemical reduction, using hydrazine, is by far the most used technique for the reduction of GO. However, this reducing agent not only is hazardous to the environment but also compromises on the conductivity of graphene. This is due to a nitrogen-containing ring structure, pyrazole, formed at the edges of rGO increasing the resistance of graphene [87, 88]. Furthermore, in the case of composite structures with CP, polymer degradation can arise from the reducing agents and high reduction temperatures, if using thermal reduction of GO. A more preferable method for composite fabrication, in which one avoids the post-processing, would be to use rGO. However, direct coating of CPs on rGO sheets is difficult because the dispersion of rGO sheets in water is poor. To prevent the irreversible aggregation of rGO during reduction, non-covalent and covalent functionalization of rGO platelets with various organic molecules has become an attractive alternative. However, the presence of possible stabilizers often compromises on the electrical properties of its composites with conducting polymers. As all of these above-mentioned methods involve time-consuming fractional fabrication of composites and/or toxic chemicals for GO reduction, the search for a simple, fast, green, and scalable method to prepare graphene–CP composites is highly desirable. In the one-pot electrodeposition method, shown in Fig. 7.8, graphene/CP composite films can be made through direct cyclic voltammetry (CV) electrolysis in a bath containing GO and the monomer material both with [89] and without presence of additional dopants [90, 91]. The resulting composite, where GO will be incorporated as dopant, can then be subjected to post-electrochemical reduction to form an rGO/CP film. GO in electrochemically prepared PPy and PEDOT composite films has been electrochemically reduced in aqueous electrolytes using CV or a constant potential method [92–96].

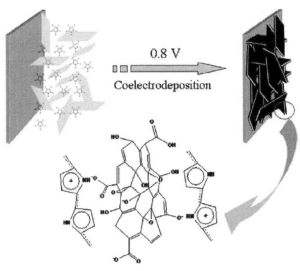

Figure 7.8 A typical one-step co-electrodeposition method for the synthesis of GO/PPy nanocomposites. Reprinted with permission from Ref. [91]. Copyright © 2012, Royal Society of Chemistry.

Furthermore, for some monomer materials the CV scan can be extended to anodic and cathodic potentials during the electrodeposition process ensuring both monomer oxidation and GO reduction, thus eliminating the need for the post-electrochemical reduction step [96–98].

As discussed in Section 7.4.3, graphene can also be produced by the electrochemical exfoliation of graphite using ILs or other solvent combinations as electrolyte. Saxena et al. reported on electropolymerization of EDOT using both electrochemically exfoliated graphene (fluoro alkyl phosphate–based ionic liquid functionalized graphene (ILFG)) and rGO as electrolytes [99].

The formed PEDOT-ILFG composite showed higher electrochemical activity and improved ionic/electronic conductivity in comparison to PEDOT-rGO. A more intact carbon structure with a significantly lower level of oxidation was obtained for ILFG compared to rGO. The IL not only had the ability to fortify the structure of graphene but also facilitated charge transport through the bulk of the film by providing less impeded pathways. In Fig. 7.9, the interactions taking place between PEDOT upon formation of a composite together with ILFG and rGO are compared. In case of PEDOT-rGO the carboxylate ions on the reduced graphene oxide function as the charge compensating anions and link to the oxidized thiophene rings of PEDOT. For PEDOT-ILFG there are more interactions taking place as also the ones between the imidazolium cation and graphene sheets has to be accounted for.

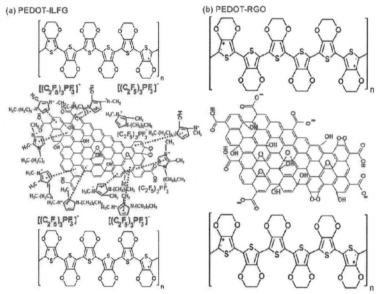

Figure 7.9 Schematic comparison of the interactions prevalent in (a) PEDOT-ILFG and (b) PEDOT-RGO nanocomposites. Reprinted with permission from Ref. [99]. Copyright © 2011, American Chemical Society.

In CPs, a number of variables during electropolymerization, such as the polymerization method and nature of dopant and electrolyte, to name a few, strongly affect the morphology of the film which plays an important role in their electrochemical performance. Many ILs show good compatibility with CPs and electrochemical synthesis using ILs as dopant and electrolyte has in several cases resulted in polymer films with improved surface morphology, higher electroactivity and improved cycling stability [100, 101]. PEDOT/GO composite films electrochemically polymerized and reduced utilizing ILs as electrolyte media displayed an enhanced electrochemical performance after electrochemical reduction of the composite. Such composite films will be free of water, valuable for their long term utilization, and additionally complete reduction of incorporated GO could be guaranteed due to the broad electrochemical window of the IL [102].

7.6 Applications of Composite Materials

Given the unique structure and properties of graphene and CPs, these materials have shown important impact in a broad range of applications. CPs are well suited for application where flexibility, processability, light-weight, optical properties, and conductivity are important. Carbon nanomaterials, particularly graphene, have been recognized as promising active materials due to their outstanding conductivity and large surface area. For graphene/CP composite films, the goal of combining the materials has been both to obtain a mechanically more robust material and to combine the attractive properties of the individual components to obtain a superior material. As discussed above, graphene/CP composite materials can be synthesized by a range of different methods. In this section, electropolymerization of graphene/CP composite electrode materials and the direct use of such electrodes in the field of supercapacitors and electrochromic devices will be briefly summarized.

7.6.1 Supercapacitors

The rise of ubiquitous, wearable, and flexible electronics brings challenges in the demand for new energy storage solutions [103]. Supercapacitors, also known as electrochemical capacitors (ECs) or ultracapacitors, are energy storage devices that can be used in many applications, especially when long cycle life, high power capability, and environmental friendliness is required. Compared to batteries, ECs typically have lower energy densities but significantly longer cycle lives and can be rapidly charged and discharged [104]. An important advantage of supercapacitors in universal and disposable electronics is that they can be prepared from non-toxic materials [105]. Thus, supercapacitors act as a perfect complement for lithium ion batteries and fuel cells and when used in cooperation, they are considered to be promising power supplies for versatile applications. Supercapacitors are used, for example, as temporary energy storage devices in renewable energy systems such as wind power plants, as high peak power sources in hybrid electric vehicles or as an interim power source in autonomous energy harvesting circuits when the primary source is not available [106, 107].

Depending on the charge storage mechanism, supercapacitors can be classified into two types: electrical double layer capacitors (EDLC) and pseudocapacitors [108]. EDLCs store and release energy based on the accumulation of charges at the interface between a porous electrode, typically a carbonaceous material with high surface area, and the electrolyte. In pseudocapacitors, the mechanisms rely on fast and reversible Faradaic redox reactions at the surface and/or in the bulk.

CPs are interesting pseudocapacitive electrode materials that exhibit high redox capacitance and fast switching between redox states. They are also lightweight, show good tensile strength and are flexible, making them among the most potential pseudocapacitor materials when thinking about high-performance portable and flexible supercapacitor applications [109, 110]. Among CPs, PANi and PPy have been paid the most intensive attention because of their fast electrochemical reactions, low cost, and high specific capacitances. The common disadvantage of CPs, which has mitigated against their use as sole supercapacitor materials, is their poor cycling stability. CP electrodes exhibit large volume changes during charging/discharging processes, leading to decrease in electrochemical performance and in stability. One commonly applied strategy used to further improve the mechanical and electrical properties of CPs has been the formation of composites together with a wide variety of materials like metal nanoparticles and different carbon materials [111–113]. Carbon-based materials are of great interest owing to their abundance, stability, and environmental friendliness [114, 115]. Graphene, is an obvious choice for this application offering high specific surface area, high intrinsic electrical conductivity, and good chemical and thermal stability [116]. The porous layer structures of graphene not only facilitate the transport of electrolyte ions within the electrodes and hence improve the specific capacitance, but also control the morphology of CPs in the graphene/polymer composite.

Graphene oxide (GO) has become the most common starting material for graphene-based applications as it can be produced in large quantities as well-separated GO sheets. For many applications, the reduction, using chemical or electrochemical reduction or thermal annealing, of GO is desired in order to restore the graphitic structure. Of these reduction methods, the electrochemical reduction is advantageous as it requires no oxidant or toxic reducing agents (e.g., hydrazine) and can be operated at room temperature [50]. Unfortunately, with the reduction of GO (rGO), oxygen-containing groups largely decrease and sheets tend to restack as a graphite-like structure that shows low dispersibility in solvents. Aggregation reduces the available surface area and limit electron and ion transport, giving unsatisfactory capacitive performance [117]. Thus, one applied strategy has been the post-reduction of CP/GO composites in order to prepare uniform CP/rGO materials as it avoids the restacking of rGO sheets.

In the literature, composite materials of CPs together with graphene and GO have received significant attention in supercapacitor applications [118–120]. In the following, an overview of current research on the development of CP/graphene-based composite electrodes prepared by the electrodeposition method for ECs will be given. Different from the powdery products prepared through chemical approaches, this method enables an easy, one-step deposition of the composite film directly on the surface of the current collector electrode. Furthermore, there is no need of binders/additives as in the chemical approach to enhance the mechanical strength. Such electrodes suffer from sluggish rate of ion transport during the redox reaction due to the high interfacial resistance and the inherent resistance of the binder materials, and hence to poor device performance.

Fabrication of freestanding thin-film composite electrodes by electrodeposition was first reported by Wang et al. [85]. In this work, freestanding and flexible graphene composite paper was used as the working electrode on which a PANi film was formed by an in situ anodic electropolymerization technique

(Fig. 7.7). The infusion of PANi into the voids of the graphene paper not only improved its mechanical properties but also the ion accessibility. This graphene-based composite paper electrode showed an electrochemical capacitance of 233 F/g, twice the value obtained for a pure graphene electrode.

A micro-supercapacitor was fabricated by electrodepositing PANi nanorods on the surfaces of rGO patterns [121]. This micro-supercapacitor possesses a high specific capacitance of 970 F/g at a relatively high discharge current density of 2.5 A/g, showing retention of 90% after 1700 cycles (Fig. 7.10).

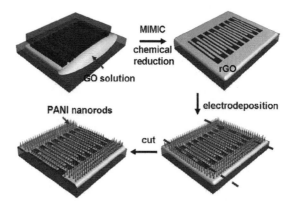

Figure 7.10 Fabrication of rGO/PANI microelectrodes. Reprinted with permission from Ref. [121]. Copyright © 2012, John Wiley and Sons.

Further, Davies et al. [122] have developed flexible, uniform graphene/PPy composite films using a pulsed electropolymerization for supercapacitor electrodes. This flexible supercapacitor film could achieve high energy and power densities, due to the favorable nucleation of the PPy chains at defects sites on the graphene surface. One work also compared graphene composites of PANi, PPy, or PEDOT [118]. rGO/PANi exhibited a specific capacitance of 361 F/g at a current density of 0.3 A/g. The composites consisting of rGO and PPy or PEDOT displayed specific capacitances of 248 or 108 F/g at the same current density. Moreover, the composites showed improved electrochemical stability. Taking the rGO/PANi composite as an example, it kept about 82% of its initial capacitance after a 1000 cycle test, which is much better than that of pristine PANi fibers (68% after 600 charge/discharge cycles).

Recently, Zhang et al. [123] have demonstrated a flexible composite membrane of reduced graphene oxide and polypyrrole nanowire (rGO–PPy-NWs) via in situ reduction. A symmetric supercapacitor has been fabricated by direct coupling of two membrane electrodes, without the use of any binder or conductive additive. The supercapacitor achieved a large areal capacitance (175 mF/cm^2) and excellent cycling stability. The in situ reduction of GO in the composite dispersion with PPy-NWs renders the formation of the rGO–PPy composite foam via self-assembly, as shown in Fig. 7.11.

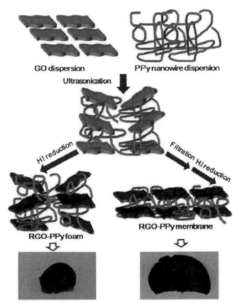

Figure 7.11 Illustration of the formation processes of rGO–PPy foam and rGO–PPy membrane. Reprinted with permission from Ref. [123]. Copyright © 2013, Royal Society of Chemistry.

The redox process occurring during charging/discharging of a CP is a combination of charge injection into the polymer backbone from the current collector and ion exchange from the electrolyte to maintain charge neutrality in the film. The rate of this charging/discharging is influenced by the mass transport of ions through the film and therefore it is highly dependent on film morphology and type of electrolyte. Therefore, besides the proper choice of active electrode materials, appropriate choice of electrolytes is also an important factor to consider in improving the performance of ECs. The electrolyte will not only determine the operating voltage but it is also important for the charging/discharging rates (power density). As energy density is proportional to the square of the voltage, a wide operating electrochemical window can substantially improve the energy density of supercapacitors. In any electrochemical process, the safe or withstanding working potential window is limited due to the electrochemical stability of the electrolytes and the generation of gaseous products at cathode and anodes (positive or negative electrodes) during the voltammetric measurements. In aqueous systems, the operating voltage is usually limited to about 1 V, organic and ionic liquid–based electrolytes can improve the working potential window at least 1.5–2 times.

A substantial part of the work on composite materials in ECs report specific capacitances using acid and basic aqueous electrolytes. Lately, ILs have been evaluated for ECs as they offer the promise of improved safety over organic electrolytes because of their low vapor pressure and low flammability. Unlike organic solvents, the high boiling point of ILs makes the encapsulation of the device easier and avoids electrolyte evaporation during testing, thereby further enhancing the cyclability of the device. They also possess moderate intrinsic conductivity and many are stable over a broad temperature and potential range [8]. In CP-based electrochemical devices, an improved cycling stability has been obtained with ILs compared to organic electrolytes [124, 125]. ILs has also been used for the electrochemical exfoliation of graphene sheets from graphite and for preparing IL-attached graphene nanosheets simply by adding IL into GO prior to reduction [62, 126–128]. Such composites have been used as supercapacitor electrodes showing good rate capability, high specific capacitance (147.5 F/g) and cycle performance [127]. However, until now no reports could be found on utilizing ILs both for the electrodeposition of CP/graphene onto flexible electrode substrates and as electrolyte for the EC devices.

7.6.2 Electrochromic Composite Materials

Many conducting polymers such as polyaniline, polyazulene, polypyrrole, and polythiophene derivatives including PEDOT have been electrosynthesized in various ILs [129, 130].

It has been observed that the use of ionic liquids as electrolytes leads to improved electrochemical activities, altered surface morphologies, and stable optoelectronic properties, in comparison to the conventional solvent medias [131]. Lately, Gadgil et al. reported the use of a common IL to successfully electrodeposit a viologen functionalized polythiophene (PTh-V) derivative [130]. The as-formed film was cycled for its electrochromic response by repetitive charging/discharging scans and found to sustain its electroactivity after several cycles. Electrochromic devices (ECDs) based on PEDOT as electrochromic material synthesized in ILs has been found to show better device performance [132]. This enhanced performance is owing to the IL, which provides highly porous morphology and both the ionic charge and the transport channel necessary for the devices to efficiently operate, facilitating charge injection/extraction at the electrode/active-layer interface.

In recent years, graphene has emerged as a promising component in electrochromic materials [133]. ILs have been used to improve the dispersion of graphene, producing high stable IL-modified graphene composite [134]. Conducting polymers are often added as an optical component in order to improve optoelectronic and electrochemical properties of the resulting composite with graphene. For an electrochemical fabrication of these composite films, the electropolymerizable monomer is mixed with exfoliated graphene or a GO dispersion in ILs and electrocodeposition is carried out. Recently, Sindhu and co-workers fabricated PEDOT-graphene films by electropolymerizing EDOT monomers in IL functionalized graphene (IFGR) sheets, obtained by electrochemical exfoliation of graphite rods [135]. The as-prepared PEDOT-IFGR films revealed good electrochemical properties like better ionic and electrical conductivity, significant band gap, and excellent electrochromic properties. The composite absorbs in the whole visible and near-IR spectral region. The PEDOT-IFGR film showed electrochromic response between transmissive blue and darkish grey, high contrast, and excellent switching stability. In another study, Saxena et al. performed electropolymerization of the EDOT monomer in presence of chemically reduced GO (rGO) or exfoliated IL functionalized graphene (ILFG) [99]. The corresponding PEDOT-rGO or PEDOT-ILFG nanocomposite films were obtained possessing exclusively different morphologies. While defect free graphene nanosheets were achieved with a size of a few nanometers in PEDOT-ILFG, PEDOT-rGO films appeared as GO sheets with wrinkled/disordered polymer structures (Fig. 7.12). The comparative studies showed that electroactivity, ionic/electronic conductivity, switching kinetics and electrochromic efficiency are all superior in PEDOT-ILFG in comparison to PEDOT-rGO films (Fig. 7.13). This improved performance can be attributed by the ability of the ionic liquid to not only fortify the structure of graphene but also facilitate electron/charge transport through the bulk of the film, further reiterates the role of ILs and graphene in ECDs.

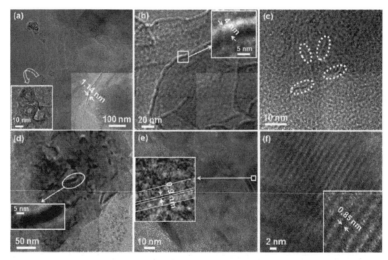

Figure 7.12 HRTEM micrographs of (a) neat ILFG showing the crumpled sheet-like texture (inset on right-hand side shows the thickness of the nanosheet), (b) neat RGO with the inset showing a blown up view of a fold on the nanosheet, (c) the PEDOT-RGO nanocomposite; the ellipses encircle the streak-like structures typical of RGO and are surrounded by the amorphous polymer, (d) PEDOT-ILFG nanocomposite, the elongated shapes (as in the inset) are characteristic of ILFG, (e) coexisting crystalline and amorphous phases in PEDOT-ILFG (inset is a magnified view of a quasi-ordered arrangement of lattice fringes), and (f) a relatively defect free crystallite of ILFG in PEDOT-ILFG, inset shows the fringe separation of 0.85 nm. Reprinted with permission from Ref. [99]. Copyright © 2011, American Chemical Society.

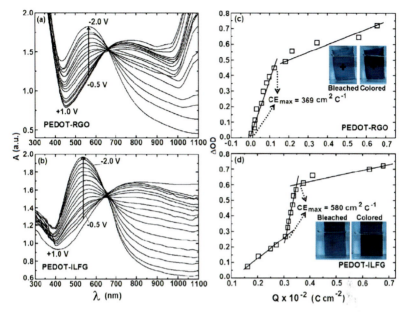

Figure 7.13 Absorption spectra of (a) PEDOT-RGO and (b) PEDOT-ILFG nanocomposite films recorded under different dc potentials in the IL. Change in optical density versus inserted charge density plots for (c) PEDOT-RGO and (d) PEDOT-ILFG nanocomposite films at a monochromatic wavelength of 550 nm. Insets of (c) and (d) are the same films in dark (−2.0 V) and pale (+1.0 V) blue states. Reprinted with permission from Ref. [99]. Copyright © 2011, American Chemical Society.

In some cases, graphene or GO is modified with ILs and subsequently reduced, if necessary, to get rGO and further dispersed with monomer and chemically polymerized to yield polymer-rGO or polymer-graphene hybrid composites [136]. Hybrid materials of rGO and PEDOT were prepared by poly(ionic liquid)-mediated hybridization by Tung et al. [137]. Poly(ionic liquid)s (PILs) are found to be preferentially physically adsorbed onto the rGO platelets, and also promoted PEDOT growth on rGO platelets through effective molecular interaction of PIL with PEDOT chains (Fig. 7.14). The composite showed delay in the thermal degradation due to a strong interaction between PEDOT matrix and PIL:rGO. Further, the PEDOT-PIL:rGO composites dispersed in AN and coated on PET film forming 500 nm-thick conductive layers, showed light transmittance of 86.0% at 550 nm, which was similar to the optical transmittance of the PET film itself. Authors claimed that this hybrid composite can work efficiently in optoelectronic devices, especially where transparency, electrical conductivity, and stability properties are required.

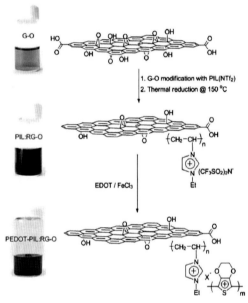

Figure 7.14 The process for preparation of PIL:RG-O hybrids with conducting polymer PEDOT using poly(ionic liquids). Reprinted with permission from Ref. [137]. Copyright © 2011, Elsevier B.V.

References

1. Chiang, C. K., Fincher Jr, C., Park, Y., Heeger, A., Shirakawa, H., Louis, E., Gau, S., and MacDiarmid, A. G. (1977). Electrical conductivity in doped polyacetylene, *Phys. Rev. Lett.*, **39**, pp. 1098.

2. Inzelt, G. (1987). Chemical and electrochemical syntheses of conducting polymers. In: *Conducting Polymers: a New Era in Electrochemistry,* 2nd ed., Chapter 4 (Springer Science & Business Media) pp. 149–171.

3. Bernier P., Bidan G., and Lefrant, S. (1999). *Advances in Synthetic Metals: Twenty Years of Progress in Science and Technology* (Elsevier Science & Technology, Amsterdam, The Netherlands).

4. Roncali, J. (1992). Conjugated poly(thiophenes): Synthesis, functionalization, and applications, *Chem. Rev.,* **92**, pp. 711–738.
5. Gadgil, B., Damlin, P., Ääritalo, T., Kankare, J., and Kvarnström, C. (2013). Electrosynthesis and characterization of viologen cross linked thiophene copolymer, *Electrochim. Acta,* **97**, pp. 378–385.
6. Nalwa, H. S. (2001). *Advanced Functional Molecules and Polymers: Electronic and Photonic Properties,* **3** (Gordon and Breach Science Publishers, Amsterdam, The Netherlands).
7. Walden, P. (1914). Über die Molekulargröße und elektrische Leitfähigkeit einiger geschmolzener Salze, *Bull. Acad. Imper. Sci. St. Petersburg,* **8**, pp. 405–422.
8. Hapiot, P., and Lagrost, C. (2008). Electrochemical reactivity in room-temperature ionic liquids, *Chem. Rev.,* **108**, pp. 2238–2264.
9. Wilkes, J. S., and Zaworotko, M. J. (1992). Air and water stable 1-ethyl-3-methylimidazolium based ionic liquids, *J. Chem. Soc. Chem. Commun.,* pp. 965–967.
10. Fukumoto, K., Yoshizawa, M., and Ohno, H. (2005). Room temperature ionic liquids from 20 natural amino acids, *J. Am. Chem. Soc.,* **127**, pp. 2398–2399.
11. Tao, G., He, L., Liu, W., Xu, L., Xiong, W., Wang, T., and Kou, Y. (2006). Preparation, characterization and application of amino acid-based green ionic liquids, *Green Chem.,* **8**, pp. 639–646.
12. He, L., Tao, G., Parrish, D. A., and Shreeve, J. M. (2009). Slightly viscous amino acid ionic liquids: Synthesis, properties, and calculations, *J. Phys. Chem. B,* **113**, pp. 15162–15169.
13. Petkovic, M., Ferguson, J. L., Gunaratne, H. N., Ferreira, R., Leitao, M. C., Seddon, K. R., Rebelo, L. P. N., and Pereira, C. S. (2010). Novel biocompatible cholinium-based ionic liquids—toxicity and biodegradability, *Green Chem.,* **12**, pp. 643–649.
14. Nockemann, P., Thijs, B., Driesen, K., Janssen, C. R., Van Hecke, K., Van Meervelt, L., Kossmann, S., Kirchner, B., and Binnemans, K. (2007). Choline saccharinate and choline acesulfamate: Ionic liquids with low toxicities, *J. Phys. Chem. B,* **111**, pp. 5254–5263.
15. Ferlin, N., Courty, M., Gatard, S., Spulak, M., Quilty, B., Beadham, I., Ghavre, M., Haiß, A., Kümmerer, K., Gathergood, N., and Bouquillon, S. (2013). Biomass derived ionic liquids: Synthesis from natural organic acids, characterization, toxicity, biodegradation and use as solvents for catalytic hydrogenation processes, *Tetrahedron,* **69**, pp. 6150–6161.
16. García-Suárez, E. J., Moriel, P., Menéndez-Vázquez, C., Montes-Morán, M. A., and García, A. B. (2011). Carbons supported bio-ionic liquids:

Stability and catalytic activity, *Micropor. Mesopor. Mater.*, **144**, pp. 205–208.

17. Liu, Q., Hou, X., Li, N., and Zong, M. (2012). Ionic liquids from renewable biomaterials: Synthesis, characterization and application in the pretreatment of biomass, *Green Chem.*, **14**, pp. 304–307.

18. Moriel, P., García-Suárez, E. J., Martínez, M., García, A. B., Montes-Morán, M. A., Calvino-Casilda, V., and Bañares, M. A. (2010). Synthesis, characterization, and catalytic activity of ionic liquids based on biosources, *Tetrahedron Lett.*, **51**, pp. 4877–4881.

19. Hou, X., Li, N., and Zong, M. (2013). Renewable bio ionic liquids-water mixtures-mediated selective removal of lignin from rice straw: Visualization of changes in composition and cell wall structure, *Biotechnol. Bioeng.*, **110**, pp. 1895–1902.

20. Pettersson, F., Keskinen, J., Remonen, T., von Hertzen, L., Jansson, E., Tappura, K., Zhang, Y., Wilén, C.-E., and Österbacka, R. (2014). Printed environmentally friendly supercapacitors with ionic liquid electrolytes on paper, *J. Power Sources*, **271**, pp. 298–304.

21. Rogers, R. D., and Seddon, K. R. (2003). Ionic liquids--solvents of the future? *Science,* **302**, pp. 792–793.

22. Gutowski, K. E., Broker, G. A., Willauer, H. D., Huddleston, J. G., Swatloski, R. P., Holbrey, J. D., and Rogers, R. D. (2003). Controlling the aqueous miscibility of ionic liquids: Aqueous biphasic systems of water-miscible ionic liquids and water-structuring salts for recycle, metathesis, and separations, *J. Am. Chem. Soc.*, **125**, pp. 6632–6633.

23. Sahami, S., and Osteryoung, R. A. (1983). Voltammetric determination of water in an aluminum chloride-Nn-butylpyridinium chloride ionic liquid, *Anal. Chem.*, **55**, pp. 1970–1973.

24. Schröder, U., Wadhawan, J. D., Compton, R. G., Marken, F., Suarez, P. A., Consorti, C. S., de Souza, R. F., and Dupont, J. (2000). Water-induced accelerated ion diffusion: Voltammetric studies in 1-methyl-3-[2,6-(S)-dimethylocten-2-yl]imidazolium tetrafluoroborate, 1-butyl-3-methylimidazolium tetrafluoroborate and hexafluorophosphate ionic liquids, *N. J. Chem.*, **24**, pp. 1009–1015.

25. Ohno, H. (2005). Electrochemical windows of room-temperature ionic liquids. In: *Electrochemical Aspects of Ionic Liquids,* Matsumoto, H., eds., Chapter 4 (John Wiley & Sons, Inc., Hoboken, NJ, USA) pp. 35–54.

26. Najafabadi, A. T., and Gyenge, E. (2014). High-yield graphene production by electrochemical exfoliation of graphite: Novel ionic

liquid (IL)–acetonitrile electrolyte with low IL content, *Carbon,* **71**, pp. 58–69.
27. Pickup, P., and Osteryoung, R. (1984). Electrochemical polymerization of pyrrole and electrochemistry of polypyrrole films in ambient temperature molten salts, *J. Am. Chem. Soc.,* **106**, pp. 2294–2299.
28. Sekiguchi, K., Atobe, M., and Fuchigami, T. (2003). Electrooxidative polymerization of aromatic compounds in 1-ethyl-3-methylimidazolium trifluoromethanesulfonate room-temperature ionic liquid, *J. Electroanal. Chem.,* **557**, pp. 1–7.
29. Janiszewska, L., and Osteryoung, R. A. (1987). Electrochemistry of polythiophene and polybithiophene films in ambient temperature molten salts, *J. Electrochem. Soc.,* **134**, pp. 2787–2794.
30. Pringle, J. M., Forsyth, M., MacFarlane, D. R., Wagner, K., Hall, S. B., and Officer, D. L. (2005). The influence of the monomer and the ionic liquid on the electrochemical preparation of polythiophene, *Polymer,* **46**, pp. 2047–2058.
31. Koura, N., Ejiri, H., and Takeishi, K. (1993). Polyaniline secondary cells with ambient temperature molten salt electrolytes, *J. Electrochem. Soc.,* **140**, pp. 602–605.
32. Innis, P. C., Mazurkiewicz, J., Nguyen, T., Wallace, G. G., and MacFarlane, D. (2004). Enhanced electrochemical stability of polyaniline in ionic liquids, *Curr. Appl. Phys.,* **4**, pp. 389–393.
33. Kobryanskii, V. M., and Arnautov, S. A. (1992). Electrochemical synthesis of poly(p-phenylene) in an ionic liquid, *Macromol. Chem. Phys.,* **193**, pp. 455–463.
34. Wagner, M., Kvarnström, C., and Ivaska, A. (2010). Room temperature ionic liquids in electrosynthesis and spectroelectrochemical characterization of poly(para-phenylene), *Electrochim. Acta,* **55**, pp. 2527–2535.
35. Randriamahazaka, H., Plesse, C., Teyssié, D., and Chevrot, C. (2003). Electrochemical behaviour of poly(3,4-ethylenedioxythiophene) in a room-temperature ionic liquid, *Electrochem. Commun.,* **5**, pp. 613–617.
36. Damlin, P., Kvarnström, C., and Ivaska, A. (2004). Electrochemical synthesis and in situ spectroelectrochemical characterization of poly(3,4-ethylenedioxythiophene) (PEDOT) in room temperature ionic liquids, *J. Electroanal. Chem.,* **570**, pp. 113–122.
37. Österholm, A., Kvarnström, C., and Ivaska, A. (2011). Ionic liquids in electrosynthesis and characterization of a polyazulene-fullerene composite, *Electrochim. Acta,* **56**, pp. 1490–1497.
38. Österholm, A., Damlin, P., Kvarnström, C., and Ivaska, A. (2011). Studying electronic transport in polyazulene–ionic liquid systems

using infrared vibrational spectroscopy, *Phys. Chem. Chem. Phys.,* **13**, pp. 11254–11263.

39. Chung, D. (2002). Review graphite, *J. Mater. Sci.,* **37**, pp. 1475–1489.
40. Kim, K. S., Zhao, Y., Jang, H., Lee, S. Y., Kim, J. M., Kim, K. S., Ahn, J., Kim, P., Choi, J., and Hong, B. H. (2009). Large-scale pattern growth of graphene films for stretchable transparent electrodes, *Nature,* **457**, pp. 706–710.
41. Novoselov, K., Geim, A. K., Morozov, S., Jiang, D., Katsnelson, M., Grigorieva, I., Dubonos, S., and Firsov, A. (2005). Two-dimensional gas of massless Dirac fermions in graphene, *Nature,* **438**, pp. 197–200.
42. Allen, M. J., Tung, V. C., and Kaner, R. B. (2009). Honeycomb carbon: A review of graphene, *Chem. Rev.,* **110**, pp. 132–145.
43. Park, S., and Ruoff, R. S. (2009). Chemical methods for the production of graphenes, *Nat. Nanotechnol.,* **4**, pp. 217–224.
44. Weiss, N. O., Zhou, H., Liao, L., Liu, Y., Jiang, S., Huang, Y., and Duan, X. (2012). Graphene: An emerging electronic material, *Adv. Mater.,* **24**, pp. 5782–5825.
45. Ferrari, A. C., Bonaccorso, F., Fal'ko, V., Novoselov, K. S., Roche, S., Boggild, P., Borini, S., Koppens, F. H. L., Palermo, V., Pugno, N., Garrido, J. A., Sordan, R., Bianco, A., Ballerini, L., Prato, M., Lidorikis, E., Kivioja, J., Marinelli, C., Ryhanen, T., Morpurgo, A., Coleman, J. N., Nicolosi, V., Colombo, L., Fert, A., Garcia-Hernandez, M., Bachtold, A., Schneider, G. F., Guinea, F., Dekker, C., Barbone, M., Sun, Z., Galiotis, C., Grigorenko, A. N., Konstantatos, G., Kis, A., Katsnelson, M., Vandersypen, L., Loiseau, A., Morandi, V., Neumaier, D., Treossi, E., Pellegrini, V., Polini, M., Tredicucci, A., Williams, G. M., Hee Hong, B., Ahn, J., Min Kim, J., Zirath, H., van Wees, B. J., van der Zant, H., Occhipinti, L., Di Matteo, A., Kinloch, I. A., Seyller, T., Quesnel, E., Feng, X., Teo, K., Rupesinghe, N., Hakonen, P., Neil, S. R. T., Tannock, Q., Lofwander, T., and Kinaret, J. (2015). Science and technology roadmap for graphene, related two-dimensional crystals, and hybrid systems, *Nanoscale,* **7**, pp. 4598–4810.
46. Hummers Jr, W. S., and Offeman, R. E. (1958). Preparation of graphitic oxide, *J. Am. Chem. Soc.,* **80**, pp. 1339–1339.
47. Botas, C., Álvarez, P., Blanco, C., Santamaría, R., Granda, M., Ares, P., Rodríguez-Reinoso, F., and Menéndez, R. (2012). The effect of the parent graphite on the structure of graphene oxide, *Carbon,* **50**, pp. 275–282.
48. Dreyer, D. R., Park, S., Bielawski, C. W., and Ruoff, R. S. (2010). The chemistry of graphene oxide, *Chem. Soc. Rev.,* **39**, pp. 228–240.

49. Johari, P., and Shenoy, V. B. (2011). Modulating optical properties of graphene oxide: Role of prominent functional groups, *ACS Nano,* **5**, pp. 7640–7647.
50. Pei, S., and Cheng, H. (2012). The reduction of graphene oxide, *Carbon,* **50**, pp. 3210–3228.
51. Chua, C. K., and Pumera, M. (2014). Chemical reduction of graphene oxide: A synthetic chemistry viewpoint, *Chem. Soc. Rev.,* **43**, pp. 291–312.
52. Guo, H., Wang, X., Qian, Q., Wang, F., and Xia, X. (2009). A green approach to the synthesis of graphene nanosheets, *ACS Nano,* **3**, pp. 2653–2659.
53. Dilimon, V. S., and Sampath, S. (2011). Electrochemical preparation of few layer-graphene nanosheets via reduction of oriented exfoliated graphene oxide thin films in acetamide–urea–ammonium nitrate melt under ambient conditions, *Thin Solid Films,* **519**, pp. 2323–2327.
54. Fu, C., Kuang, Y., Huang, Z., Wang, X., Du, N., Chen, J., and Zhou, H. (2010). Electrochemical co-reduction synthesis of graphene/Au nanocomposites in ionic liquid and their electrochemical activity, *Chem. Phys. Lett.,* **499**, pp. 250–253.
55. Harima, Y., Setodoi, S., Imae, I., Komaguchi, K., Ooyama, Y., Ohshita, J., Mizota, H., and Yano, J. (2011). Electrochemical reduction of graphene oxide in organic solvents, *Electrochim. Acta,* **56**, pp. 5363–5368.
56. Kauppila, J., Kunnas, P., Damlin, P., Viinikanoja, A., and Kvarnström, C. (2013). Electrochemical reduction of graphene oxide films in aqueous and organic solutions, *Electrochim. Acta,* **89**, pp. 84–89.
57. Nalwa, H. S. (2001). Processing and spectroscopy. In: *Advanced Functional Molecules and Polymers,* Kvarnstrom, C., Neugebauer, H., and Ivaska, A., eds., Chapter 2 (Gordon and Breach Publishers) pp. 139.
58. Acik, M., Mattevi, C., Gong, C., Lee, G., Cho, K., Chhowalla, M., and Chabal, Y. J. (2010). The role of intercalated water in multilayered graphene oxide, *ACS Nano,* **4**, pp. 5861–5868.
59. Acik, M., Lee, G., Mattevi, C., Pirkle, A., Wallace, R. M., Chhowalla, M., Cho, K., and Chabal, Y. (2011). The role of oxygen during thermal reduction of graphene oxide studied by infrared absorption spectroscopy, *J. Phys. Chem. C,* **115**, pp. 19761–19781.
60. Viinikanoja, A., Wang, Z., Kauppila, J., and Kvarnström, C. (2012). Electrochemical reduction of graphene oxide and its in situ spectroelectrochemical characterization, *Phys. Chem. Chem. Phys.,* **14**, pp. 14003–14009.

61. Viinikanoja, A., Kauppila, J., Damlin, P., Suominen, M., and Kvarnström, C. (2015). In situ FTIR and Raman spectroelectrochemical characterization of graphene oxide upon electrochemical reduction in organic solvents, *Phys. Chem. Chem. Phys.*, **17**, pp. 12115–12123.

62. Low, C. T. J., Walsh, F. C., Chakrabarti, M. H., Hashim, M. A., and Hussain, M. A. (2013). Electrochemical approaches to the production of graphene flakes and their potential applications, *Carbon*, **54**, pp. 1–21.

63. Mao, M., Wang, M., Hu, J., Lei, G., Chen, S., and Liu, H. (2013). Simultaneous electrochemical synthesis of few-layer graphene flakes on both electrodes in protic ionic liquids, *Chem. Commun.*, **49**, pp. 5301–5303.

64. Wang, G., Wang, B., Park, J., Wang, Y., Sun, B., and Yao, J. (2009). Highly efficient and large-scale synthesis of graphene by electrolytic exfoliation, *Carbon*, **47**, pp. 3242–3246.

65. Morales, G. M., Schifani, P., Ellis, G., Ballesteros, C., Martínez, G., Barbero, C., and Salavagione, H. J. (2011). High-quality few layer graphene produced by electrochemical intercalation and microwave-assisted expansion of graphite, *Carbon*, **49**, pp. 2809–2816.

66. You, X., Chang, J., Ju, B. K., and Pak, J. J. (2011). An electrochemical route to graphene oxide, *J. Nanosci. Nanotechnol.*, **11**, pp. 5965–5968.

67. Wang, J., Manga, K. K., Bao, Q., and Loh, K. P. (2011). High-yield synthesis of few-layer graphene flakes through electrochemical expansion of graphite in propylene carbonate electrolyte, *J. Am. Chem. Soc.*, **133**, pp. 8888–8891.

68. Liu, N., Luo, F., Wu, H., Liu, Y., Zhang, C., and Chen, J. (2008). One-step ionic-liquid-assisted electrochemical synthesis of ionic-liquid-functionalized graphene sheets directly from graphite, *Adv. Funct. Mater.*, **18**, pp. 1518–1525.

69. Lu, J., Yang, J., Wang, J., Lim, A., Wang, S., and Loh, K. P. (2009). One-pot synthesis of fluorescent carbon nanoribbons, nanoparticles, and graphene by the exfoliation of graphite in ionic liquids, *ACS Nano*, **3**, pp. 2367–2375.

70. Stenger-Smith, J. D., Webber, C. K., Anderson, N., Chafin, A. P., Zong, K., and Reynolds, J. R. (2002). Poly(3,4-alkylenedioxythiophene)-based supercapacitors using ionic liquids as supporting electrolytes, *J. Electrochem. Soc.*, **149**, pp. A973–A977.

71. Thomas, S. W., Joly, G. D., and Swager, T. M. (2007). Chemical sensors based on amplifying fluorescent conjugated polymers, *Chem. Rev.*, **107**, pp. 1339–1386.

72. Okuzaki, H., Suzuki, H., and Ito, T. (2009). Electrically driven PEDOT/PSS actuators, *Synth. Met.,* **159**, pp. 2233–2236.
73. Lock, J. P., Lutkenhaus, J. L., Zacharia, N. S., Im, S. G., Hammond, P. T., and Gleason, K. K. (2007). Electrochemical investigation of PEDOT films deposited via CVD for electrochromic applications, *Synth. Met.,* **157**, pp. 894–898.
74. Nogueira, A. F., Longo, C., and De Paoli, M.-. (2004). Polymers in dye sensitized solar cells: Overview and perspectives, *Coord. Chem. Rev.,* **248**, pp. 1455–1468.
75. Novák, P., Müller, K., Santhanam, K., and Haas, O. (1997). Electrochemically active polymers for rechargeable batteries, *Chem. Rev.,* **97**, pp. 207–282.
76. Frackowiak, E., and Béguin, F. (2001). Carbon materials for the electrochemical storage of energy in capacitors, *Carbon,* **39**, pp. 937–950.
77. Qu, D., and Shi, H. (1998). Studies of activated carbons used in double-layer capacitors, *J. Power Sources,* **74**, pp. 99–107.
78. He, X., Wang, T., Qiu, J., Zhang, X., Wang, X., and Zheng, M. (2011). Effect of microwave-treatment time on the properties of activated carbons for electrochemical capacitors, *N. Carbon Mater.,* **26**, pp. 313–319.
79. Zhang, W., Huang, Z., Cao, G., Kang, F., and Yang, Y. (2012). A novel mesoporous carbon with straight tunnel-like pore structure for high rate electrochemical capacitors, *J. Power Sources,* **204**, pp. 230–235.
80. Liu, C. G., Fang, H. T., Li, F., Liu, M., and Cheng, H. M. (2006). Single-walled carbon nanotubes modified by electrochemical treatment for application in electrochemical capacitors, *J. Power Sources,* **160**, pp. 758–761.
81. Sun, Y., and Shi, G. (2013). Graphene/polymer composites for energy applications, *J. Polym. Sci. Part B Polym. Phys.,* **51**, pp. 231–253.
82. Kim, H., Abdala, A. A., and Macosko, C. W. (2010). Graphene/polymer nanocomposites, *Macromolecules,* **43**, pp. 6515–6530.
83. Zhang, Q., Li, Y., Feng, Y., and Feng, W. (2013). Electropolymerization of graphene oxide/polyaniline composite for high-performance supercapacitor, *Electrochim. Acta,* **90**, pp. 95–100.
84. Xu, J., Wang, K., Zu, S., Han, B., and Wei, Z. (2010). Hierarchical nanocomposites of polyaniline nanowire arrays on graphene oxide sheets with synergistic effect for energy storage, *ACS Nano,* **4**, pp. 5019–5026.
85. Wang, D., Li, F., Zhao, J., Ren, W., Chen, Z., Tan, J., Wu, Z., Gentle, I., Lu, G. Q., and Cheng, H. (2009). Fabrication of graphene/polyaniline

composite paper via in situ anodic electropolymerization for high-performance flexible electrode, *ACS Nano,* **3**, pp. 1745–1752.

86. Kuila, T., Mishra, A. K., Khanra, P., Kim, N. H., and Lee, J. H. (2013). Recent advances in the efficient reduction of graphene oxide and its application as energy storage electrode materials, *Nanoscale,* **5**, pp. 52–71.

87. Gao, W., Alemany, L. B., Ci, L., and Ajayan, P. M. (2009). New insights into the structure and reduction of graphite oxide, *Nat. Chem.,* **1**, pp. 403–408.

88. Park, S., Hu, Y., Hwang, J. O., Lee, E., Casabianca, L. B., Cai, W., Potts, J. R., Ha, H., Chen, S., Oh, J., Kim, S. O., Kim, Y., Ishii, Y., and Ruoff, R. S. (2012). Chemical structures of hydrazine-treated graphene oxide and generation of aromatic nitrogen doping, *Nat. Commun.,* **3**, pp. 638.

89. Deng, M., Yang, X., Silke, M., Qiu, W., Xu, M., Borghs, G., and Chen, H. (2011). Electrochemical deposition of polypyrrole/graphene oxide composite on microelectrodes towards tuning the electrochemical properties of neural probes, *Sens. Actuators B: Chem.,* **158**, pp. 176–184.

90. Österholm, A., Lindfors, T., Kauppila, J., Damlin, P., and Kvarnström, C. (2012). Electrochemical incorporation of graphene oxide into conducting polymer films, *Electrochim. Acta,* **83**, pp. 463–470.

91. Zhu, C., Zhai, J., Wen, D., and Dong, S. (2012). Graphene oxide/polypyrrole nanocomposites: One-step electrochemical doping, coating and synergistic effect for energy storage, *J. Mater. Chem.,* **22**, pp. 6300–6306.

92. Yang, Y., Wang, C., Yue, B., Gambhir, S., Too, C. O., and Wallace, G. G. (2012). Electrochemically synthesized polypyrrole/graphene composite film for lithium batteries, *Adv. Energy Mater.,* **2**, pp. 266–272.

93. Chang, H., Chang, C., Tsai, Y., and Liao, C. (2012). Electrochemically synthesized graphene/polypyrrole composites and their use in supercapacitor, *Carbon,* **50**, pp. 2331–2336.

94. Si, P., Chen, H., Kannan, P., and Kim, D. (2011). Selective and sensitive determination of dopamine by composites of polypyrrole and graphene modified electrodes, *Analyst,* **136**, pp. 5134–5138.

95. Lindfors, T., Österholm, A., Kauppila, J., and Pesonen, M. (2013). Electrochemical reduction of graphene oxide in electrically conducting poly(3,4-ethylenedioxythiophene) composite films, *Electrochim. Acta,* **110**, pp. 428–436.

96. Feng, X., Li, R., Ma, Y., Chen, R., Shi, N., Fan, Q., and Huang, W. (2011). One-step electrochemical synthesis of graphene/polyaniline

composite film and its applications, *Adv. Funct. Mater.*, **21**, pp. 2989–2996.

97. Tang, Y., Wu, N., Luo, S., Liu, C., Wang, K., and Chen, L. (2012). One-step electrodeposition to layer-by-layer graphene–conducting-polymer hybrid films, *Macromol. Rapid Commun.*, **33**, pp. 1780–1786.

98. Lim, Y., Tan, Y. P., Lim, H. N., Tan, W. T., Mahnaz, M., Talib, Z. A., Huang, N. M., Kassim, A., and Yarmo, M. A. (2013). Polypyrrole/graphene composite films synthesized via potentiostatic deposition, *J. Appl. Polym. Sci.*, **128**, pp. 224–229.

99. Saxena, A. P., Deepa, M., Joshi, A. G., Bhandari, S., and Srivastava, A. K. (2011). Poly (3, 4-ethylenedioxythiophene)-ionic liquid functionalized graphene/reduced graphene oxide nanostructures: Improved conduction and electrochromism, *ACS Appl. Mater. Interfaces*, **3**, pp. 1115–1126.

100. Pringle, J. M., Forsyth, M., MacFarlane, D. R., Wagner, K., Hall, S. B., and Officer, D. L. (2005). The influence of the monomer and the ionic liquid on the electrochemical preparation of polythiophene, *Polymer*, **46**, pp. 2047–2058.

101. Damlin, P., Kvarnström, C., and Ivaska, A. (2004). Electrochemical synthesis and in situ spectroelectrochemical characterization of poly(3,4-ethylenedioxythiophene) (PEDOT) in room temperature ionic liquids, *J. Electroanal. Chem.*, **570**, pp. 113–122.

102. Damlin, P., Suominen, M., Heinonen, M., and Kvarnström, C. (2015). Non-covalent modification of graphene sheets in PEDOT composite materials by ionic liquids, *Carbon*, **93**, pp. 533–543.

103. Zhan, Y., Mei, Y., and Zheng, L. (2014). Materials capability and device performance in flexible electronics for the Internet of Things, *J. Mater. Chem. C*, **2**, pp. 1220–1232.

104. Kötz, R., and Carlen, M. (2000). Principles and applications of electrochemical capacitors, *Electrochim. Acta*, **45**, pp. 2483–2498.

105. Jost, K., Stenger, D., Perez, C. R., McDonough, J. K., Lian, K., Gogotsi, Y., and Dion, G. (2013). Knitted and screen printed carbon-fiber supercapacitors for applications in wearable electronics, *Energy Environ. Sci*, **6**, pp. 2698–2705.

106. Lehtimäki, S., Li, M., Salomaa, J., Pörhönen, J., Kalanti, A., Tuukkanen, S., Heljo, P., Halonen, K., and Lupo, D. (2014). Performance of printable supercapacitors in an RF energy harvesting circuit, *Int. J. Elec. Power*, **58**, pp. 42–46.

107. Schlichting, A., Tiwari, R., and Garcia, E. (2012). Passive multi-source energy harvesting schemes, *J. Int. Mater. Syst. Struct.*, **23**, pp. 1921–1935.

108. Pieta, P., Obraztsov, I., D'Souza, F., and Kutner, W. (2013). Composites of conducting polymers and various carbon nanostructures for electrochemical supercapacitors, *ECS J. Solid State Sci. Technol.*, **2**, pp. M3120–M3134.

109. Nyholm, L., Nyström, G., Mihranyan, A., and Strømme, M. (2011). Toward flexible polymer and paper-based energy storage devices, *Adv. Mater.*, **23**, pp. 3751–3769.

110. Shown, I., Ganguly, A., Chen, L., and Chen, K. (2015). Conducting polymer-based flexible supercapacitor, *Energy Sci. Eng.*, **3**, pp. 2–26.

111. Wang, Y., Li, H., and Xia, Y. (2006). Ordered whiskerlike polyaniline grown on the surface of mesoporous carbon and its electrochemical capacitance performance, *Adv. Mater.*, **18**, pp. 2619–2623.

112. Kovalenko, I., Bucknall, D. G., and Yushin, G. (2010). Detonation nanodiamond and onion-like-carbon-embedded polyaniline for supercapacitors, *Adv. Funct. Mater.*, **20**, pp. 3979–3986.

113. Snook, G. A., Kao, P., and Best, A. S. (2011). Conducting-polymer-based supercapacitor devices and electrodes, *J. Power Sources*, **196**, pp. 1–12.

114. Simon, P., and Gogotsi, Y. (2008). Materials for electrochemical capacitors, *Nat. Mater.*, **7**, pp. 845–854.

115. Zhang, L. L., and Zhao, X. (2009). Carbon-based materials as supercapacitor electrodes, *Chem. Soc. Rev.*, **38**, pp. 2520–2531.

116. Choi, H., Jung, S., Seo, J., Chang, D. W., Dai, L., and Baek, J. (2012). Graphene for energy conversion and storage in fuel cells and supercapacitors, *Nano Energy*, **1**, pp. 534–551.

117. Meng, Y., Wang, K., Zhang, Y., and Wei, Z. (2013). Hierarchical porous graphene/polyaniline composite film with superior rate performance for flexible supercapacitors, *Adv. Mater.*, **25**, pp. 6985–6990.

118. Zhang, J., and Zhao, X. (2012). Conducting polymers directly coated on reduced graphene oxide sheets as high-performance supercapacitor electrodes, *J. Phys. Chem. C*, **116**, pp. 5420–5426.

119. Yu, C., Ma, P., Zhou, X., Wang, A., Qian, T., Wu, S., and Chen, Q. (2014). All-solid-state flexible supercapacitors based on highly dispersed polypyrrole nanowire and reduced graphene oxide composites, *ACS Appl. Mater. Interfaces*, **6**, pp. 17937–17943.

120. Kumar, N. A., and Baek, J. (2014). Electrochemical supercapacitors from conducting polyaniline–graphene platforms, *Chem. Commun.*, **50**, pp. 6298–6308.

121. Xue, M., Li, F., Zhu, J., Song, H., Zhang, M., and Cao, T. (2012). Structure-based enhanced capacitance: In situ growth of highly ordered polyaniline nanorods on reduced graphene oxide patterns, *Adv. Funct. Mater.,* **22**, pp. 1284–1290.
122. Davies, A., Audette, P., Farrow, B., Hassan, F., Chen, Z., Choi, J., and Yu, A. (2011). Graphene-based flexible supercapacitors: Pulse-electropolymerization of polypyrrole on free-standing graphene films, *J. Phys. Chem. C,* **115**, pp. 17612–17620.
123. Zhang, J., Chen, P., Oh, B. H., and Chan-Park, M. B. (2013). High capacitive performance of flexible and binder-free graphene–polypyrrole composite membrane based on in situ reduction of graphene oxide and self-assembly, *Nanoscale,* **5**, pp. 9860–9866.
124. Stenger-Smith, J. D., Lai, W. W., Irvin, D. J., Yandek, G. R., and Irvin, J. A. (2012). Electroactive polymer-based electrochemical capacitors using poly(benzimidazo-benzophenanthroline) and its pyridine derivative poly(4-aza-benzimidazo-benzophenanthroline) as cathode materials with ionic liquid electrolyte, *J. Power Sources,* **220**, pp. 236–242.
125. Österholm, A. M., Shen, D. E., Dyer, A. L., and Reynolds, J. R. (2013). Optimization of PEDOT films in ionic liquid supercapacitors: Demonstration as a power source for polymer electrochromic devices, *ACS Appl. Mater. Interfaces,* **5**, pp. 13432–13440.
126. Kim, J., and Kim, S. (2014). Surface-modified reduced graphene oxide electrodes for capacitors by ionic liquids and their electrochemical properties, *Appl. Surf. Sci.,* **295**, pp. 31–37.
127. Kim, J., and Kim, S. (2014). Preparation and electrochemical property of ionic liquid-attached graphene nanosheets for an application of supercapacitor electrode, *Electrochim. Acta,* **119**, pp. 11–15.
128. Lehtimäki, S., Suominen, M., Damlin, P., Tuukkanen, S., Kvarnström, C., and Lupo, D. (2015). Preparation of supercapacitors on flexible substrates with electrodeposited PEDOT/graphene composites, *ACS Appl. Mater. Interfaces,* **7**, pp. 22137–22147.
129. Lehtimäki, S., Suominen, M., Damlin, P., Tuukkanen, S., Kvarnström, C., and Lupo, D. (2015). Preparation of supercapacitors on flexible substrates with electrodeposited PEDOT/graphene composites, *ACS Appl. Mater. Interfaces,* **7**, pp. 22137–22147.
129. Wei, D. (2011). Synthesis of organic electroactive materials in ionic liquids. In: *Electrochemical Nanofabrication: Principles and Applications,* Wagner, M, Carita, K., Ivaska, A., eds., Chapter 3 (Pan Stanford Publishing) pp. 117–143.

130. Gadgil, B., Damlin, P., Ääritalo, T., and Kvarnström, C. (2014). Electrosynthesis of viologen cross-linked polythiophene in ionic liquid and its electrochromic properties, *Electrochim. Acta,* **133**, pp. 268–274.

131. Lu, W., Fadeev, A. G., Qi, B., Smela, E., Mattes, B. R., Ding, J., Spinks, G. M., Mazurkiewicz, J., Zhou, D., Wallace, G. G., MacFarlane, D. R., Forsyth, S. A., and Forsyth, M. (2002). Use of ionic liquids for pi-conjugated polymer electrochemical devices, *Science,* **297**, pp. 983–987.

132. Lu, W., Fadeev, A. G., Qi, B., and Mattes, B. R. (2004). Fabricating conducting polymer electrochromic devices using ionic liquids, *J. Electrochem. Soc.,* **151**, pp. H33–H39.

133. Gadgil, B., Damlin, P., Heinonen, M., and Kvarnström, C. (2015). A facile one step electrostatically driven electrocodeposition of polyviologen–reduced graphene oxide nanocomposite films for enhanced electrochromic performance, *Carbon,* **89**, pp. 53–62.

134. Tunckol, M., Durand, J., and Serp, P. (2012). Carbon nanomaterial–ionic liquid hybrids, *Carbon,* **50**, pp. 4303–4334.

135. Siju, C., Raja, L., Shivaprakash, N., and Sindhu, S. (2015). Gray to transmissive electrochromic switching based on electropolymerized PEDOT-ionic liquid functionalized graphene films, *J. Solid State Electrochem.,* pp. 1–10.

136. Mittal, V. (2014). Functional polymer nanocomposites with graphene: A review, *Macromol. Mater. Eng.,* **299**, pp. 906–931.

137. Tung, T. T., Kim, T. Y., Shim, J. P., Yang, W. S., Kim, H., and Suh, K. S. (2011). Poly(ionic liquid)-stabilized graphene sheets and their hybrid with poly(3,4-ethylenedioxythiophene), *Org. Electron.,* **12**, pp. 2215–2224.

Chapter 8

Chemically Converted Graphene: Functionalization, Nanocomposites, and Applications

Li Niu, Yuanyuan Jiang, Yizhong Lu, Shiyu Gan, Fenghua Li, and Dongxue Han

Engineering Laboratory for Modern Analytical Techniques, c/o State Key Laboratory of Electroanalytical Chemistry, Changchun Institute of Applied Chemistry, Changchun 130022, Jilin, China

lniu@ciac.ac.cn

Graphene, a two-dimensional, monoatomic thick building block of all graphitic forms, has emerged as one of the most exciting materials of research of the 21st century and received worldwide attention due to its fascinating properties and great potential for various applications [1]. This single-atom-thick sheet of carbon atoms arrayed in a honeycomb pattern is the world's thinnest and strongest material, as well as an excellent conductor of both electricity and heat. It is commonly accepted by the researchers that this two-dimensional material is considered to be even more promising than other nanostructured carbon allotropes, such as, one-dimensional nanotubes and zero-dimensional fullerenes.

Electrochemical Nanofabrication: Principles and Applications (Second Edition)
Edited by Di Wei
Copyright © 2016 Pan Stanford Publishing Pte. Ltd.
ISBN 978-981-4613-86-6 (Hardcover), 978-981-4613-87-3 (eBook)
www.panstanford.com

Nowadays, a "gold rush" has been triggered all over the world for the functionalization of graphene and exploiting the possible application of graphene-based nanocomposites. This chapter gives a brief overview of the recent research concerning physical and chemical functionalization of graphene, and their applications in various fields.

8.1 Introduction

The direct observation of mechanically exfoliated graphene monolayer by Novoselov et al. in 2004 has sparked the exponential growth of graphene research in both the scientific and engineering communities [2]. Graphene, a 2D single layer of sp^2 hybridized carbon atoms with the hexagonal packed structure, has shown many unique properties, such as the quantum hall effect [3], high carrier mobility at room temperature [4], large theoretical specific surface area [5, 6], good optical transparency [6], and excellent thermal conductivity [7]. To further exploit these properties in various applications, efforts have been made to prepare high quality graphene and its derivative, ranging from top-down exfoliation of graphite (by means of intercalation, oxidation, and/or sonication) to bottom-up epitaxial growth (such as chemical vapor deposition or organic synthesis) [8]. In particular, the production of graphene by chemical reduction of graphene oxide (GO) has the characteristic of low-cost, facile preparation and can realize the mass production. The chemically converted graphene (CCG) possesses many reactive oxygen-containing groups for further functionalization [9]. Besides, the abundant structural defects in CCG are advantageous for electrochemical and other applications [10]. Therefore, in this chapter, we will focus on the functionalization of the chemically converted graphene.

Pristine graphene has bad solubility in most solvents and is apt to stack back to graphite through the strong cohesive van der Waals energy of the π-stacked layer, which greatly limits its applications [11]. To improve the dispersibility of graphene and endow it with more attractive properties, various functional materials have been attached to this carbon backbone by covalent or noncovalent functionalization [12]. To date, graphene-based nanocomposites have been successfully made with polymers, metal or metal oxide nanoparticles, organic molecules including biomaterials, etc. These graphene-based nanocomposites have been intensively explored in applications such as electrochemical sensors, fuel cells, supercapacitors, batteries, photocatalysis, transparent electrodes, biosensors and so on [13]. In this critical review, we will discuss the recent progress for synthesis of graphene and its derivatives, as well as graphene-based nanocomposites, with particular emphasis placed on their fabrication/synthesis methods, properties and applications.

8.2 Graphene-Based Nanocomposite

Considering the unexceptionable performance of graphene and its bad dispersibility in aqueous and organic solvents, enormous efforts have been made to synthesize graphene-based nanocomposites. It is widely accepted that appropriate combination of graphene with some polymers, organic molecules or other nanoscale materials can improve the dispersibility of graphene, leading to the development of multifunctional nanocomposites with enhanced performance. However, several key issues should be taken into account for preparing graphene-based nanocomposites. First, graphene should exist in the form of individual nanosheets in the corresponding hybrids in order to effectively exhibit the excellent property of graphene. Second, the combination of graphene with other materials should be forceful enough during the experiments. Third, the size, morphology and component of the nanocomposites should be well controlled for the reproduction of the performance. In this chapter, we will discuss three main kinds of functionalized graphene composites: graphene–polymer composites, graphene-nanoparticle composites, and graphene–organic molecule composites.

8.2.1 Graphene–Polymer Composites

Based on the kind of interaction between graphene sheets and polymers, the graphene–polymer composite can be mainly classified into two types: graphene-filled polymer composites and polymer-functionalized graphene.

8.2.2 Graphene-Filled Polymer Composites

Carbon-based materials, such as carbon nanotubes (CNTs) and amorphous carbon, are traditional fillers for polymer matrices to enhance their electronic, mechanical and thermal properties. CNTs were regarded as one of the most efficient filler materials; however, the cost was relatively high [14, 15]. Because of their good conductivity, thermal stability, and excellent mechanical strength, graphene and its derivatives have shown a great potential as filler materials for polymer composites. Similar to the conventional polymer reinforcing methods, the most common synthesis strategies applied for the fabrication of graphene–polymer composites are solution mixing [16, 17], melt blending [18, 19], and in situ polymerization [20–22].

Solution mixing. Solution mixing is the most commonly used method to fabricate polymer-based composites provided the polymers and graphene or its derivatives are readily soluble in common aqueous and organic solvents. Graphene oxide, for its abundant hydrophilic oxygen-containing groups on the surface, can be directly mixed with water-soluble polymers, such as poly(vinyl alcohol) (PVA), at various concentrations [17, 23, 24]. However, graphene sheets have bad dispersibility in both aqueous and organic solvents, even hydrophilic GO had bad solubility in organic solvents. To solve this problem, ultrasonication is usually used to produce short-time metastable dispersions of GO or graphene materials, which are then quickly mixed with polymers solutions, such as poly(methyl methacrylate) (PMMA) [25], polyurethane (PU) [26], and polyaniline (PANI) [27]. However, re-aggregation of the graphene sheets might occur during the fabrication process. Another way to improve the solubility of graphene materials is by modifying the graphene sheets with functional molecules [28]. Our group reported the preparation of tunably loaded chemically converted GO/epoxy resin nanocomposites through two-phase extraction of the mixture of GO aqueous and the epoxy resin (Fig. 8.1). Great improvements in mechanical properties such as compressive failure strength and toughness were achieved for this chemically converted GO/epoxy resin for a 0.0375 wt% loading of chemically converted GO sheets in epoxy resin by 48.3% and 1185.2%, respectively [29].

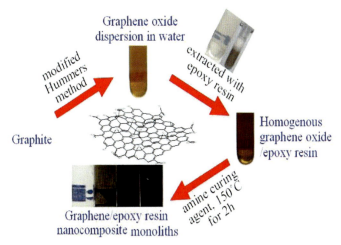

Figure 8.1 Procedures to create the chemically converted GO/epoxy nanocomposite monoliths. Reprinted from reference [29] by permission of The Royal Society of Chemistry.

Melt blending. Melt blending is another effective method for the preparation of polymer composites with graphene. This technique uses a high temperature and shear force to blend the filler and the melt of a thermoplastic polymer. Melt blending is more economical and suitable for mass production in factory and is environment-friendly without the requirement of common solvents. Kim and Macosko studied this technique systemically; they prepared thermally expanded CCG/polycarbonate (PC) composite by melt blending using an extruder at 250°C [30]. Graphene–polylactic acid (PLA) [31] and graphene–polypropylene (PP) [32] were also prepared by this method.

In situ polymerization. This technique starts with the dispersion of GO or graphene in monomer, followed by the polymerization of the monomers. In situ polymerization is another often used method to prepare graphene-filled polymer composites, such as those with epoxy and polyaniline. In a typical process for preparation of graphene–epoxy composite, a curing agent is added to initiate the polymerization after the graphene filler is mixed with epoxy resins under high-shear forces [33]. In the polymerization of PANI, which is an oxidative process, an oxidative agent such as ammonium persulfate is used to trigger the polymerization [34]. Alternatively, graphene–PANI composite can also be prepared by the in situ anodic electropolymerization [35]. Other composites such as PMMA-GO [36], PP-GO [37], and polyethylene (PE)-graphite nanosheets [38] have also been prepared by this technique.

8.2.3 Polymer-Functionalized Graphene

Instead of being used as fillers to enhance the performance of polymers, graphene and its derivatives can be applied as 2D templates for polymer decoration via covalent or noncovalent functionalization. The polymers here are used to improve the solubility of the graphene derivatives, and offer additional functionality to the resulting graphene–polymer composites.

Based on the van der Waals force, π–π stacking, or electrostatic interaction, the noncovalent functionalization is easier to carry out than covalent functionalization and provides effective means to tailor the electronic property and solubility of the graphene sheets without destroying the chemical structure of the graphene sheets. The first example for the noncovalent functionalization of graphene sheets was the reduction of exfoliated graphite by hydrazine in the presence of poly(sodium 4-styrenesulfonate) (PSS) [39], in which the backbone of PSS interacted with graphene through π–π interaction, and the hydrophilic sulfonate side groups improved the dispersibility of the composite in water. Later on, graphene-based materials with poly(2,5-bis(3-sulfonatopropoxy)-1,4-ethynylphenylene-alt-1,4-ethynylphenylene) sodium salt (PPE-SO_3^-) polyelectrolyte were successfully obtained and exhibited high conductivity and good dispersibility in water [40]. The PPE-SO_3^- molecule exhibited interesting optoelectronic property, endowing the resulting graphene-based composite (PPE-SO_3^--G) with potential applications in a variety of optoelectronic devices (Fig. 8.2).

The covalent functionalization of graphene derivatives is mainly based on the reactions between the oxygenated groups on the GO or CCG surface and the functional groups of the polymers. Graphene sheets covalently functionalized with biocompatible poly-L-lysine (PLL) were synthesized through the reaction of epoxy groups on GO and amino groups on PLL in the presence of KOH. The advantages of PLL, such as good biocompatibility, flexible molecular backbone, plentiful active amino groups, and relative good solubility in water, were all attached to the graphene–PLL composite [41]. The carboxylic groups on GO were also reacted with the amino groups in the six-armed polyethylene glycol (PEG) to form NGO–PEG composite. The as-prepared NGO–PEG had good biocompatibility and could form stable dispersion in various biological solutions, and was further used in delivery of water-insoluble cancer drugs [42]. Sometimes, additional process are needed to modify GO or graphene with desired functional groups before the grafting of polymers.

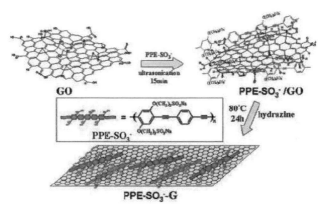

Figure 8.2 The schematic preparation process of the PPE-SO$_3^-$-G composite. Reprinted with permission from reference [40]. Copyright 2010 American Chemical Society.

8.2.4 Graphene–Nanoparticle Composites

For its high conductivity, large specific area and extended 2D planar structure, CCG sheets can serve as a useful substrate for immobilizing nanoparticles (NPs). Resent research mainly focuses on some kinds of graphene-based NPs, such as graphene–metal NPs, graphene–metal oxide NPs, graphene–quantum dots hybrids, and other graphene-based NPs. These graphene-NPs composites have been widely employed in a broad range of areas, such as catalysis, sensing, surface-enhanced Raman scattering (SERS), electronics, optics, energy conversion and storage, and so on. The fabrication methods are generally classified as the ex situ hybridization, and in situ growth. The ex situ hybridization involves the solution mixing of graphene-based nanosheets and commercially available or pre-synthesized nanomaterials. Before mixing, surface modification of the NPs and/or graphene sheets is usually carried out in favor of forceful combination through either chemical bonding or noncovalent interaction. In situ growth is the most widely used method for preparing graphene-NPs composites. Usually the salts containing the metal ions were added to the GO or graphene solution before the addition of reducing agent to obtain the graphene-NPs composites. Other methods, such as layer-by-layer (LBL) deposition method is also developed.

8.2.5 Composites with Metal Nanoparticles

Graphene can usually improve the sensing or catalytic activity of metal nanoparticles (MNPs) for its high conductivity and large specific area. Hydrophobic and static interactions are the main driving forces of adsorbing MNPs on graphene surface. Most of this kind of composites reported in literature are graphene-supported noble MNPs, including Au, Ag, Pt, and Pd. Besides, MNPs of transition metals, such as Fe, Co, and Cu, were also supported on graphene for various applications. Some graphene-bimetallic hybrids were also reported.

In situ growth: In situ growth is the most widely used method for preparing graphene-NPs composites. Usually the salts containing the metal ions were added to the GO or graphene solution, then reducing agent were added along with heat treatment to obtain the graphene-NPs composites. The oxygen-containing groups on the surface of both GO and graphene can facile the nucleation of the MNPs. In this sense, graphene sheets act as a supporter and stabilizer for MNPs.

The first graphene-Au NPs composite was prepared by the reduction of $AuCl_4^-$ with $NaBH_4$ in an octadecylamine (ODA) solution containing graphene. The Au NPs anchored on ODA-functionalized graphene were readily dispersible in low-polarity solvents, e.g., tetrahydrofuran (THF) medium [43]. Later on, a general approach for the preparation of graphene–metal NPs nanocomposites (Au, Pt, Pd) in a water-ethylene glycol system was reported. The water-ethylene glycol system was used considering the facile exfoliation nature of graphite oxide sheets in water and the mild reducibility of the ethylene glycol. The metal NPs formed first and were adsorbed on the surface of GO sheets, which can prevent the restacking of the later reduced graphene [44].

Since graphene sheets contain much less residual oxygen containing functional groups, their solubility is poor in some polar solvents, such as water. Capping agents or some hydrophilic functional groups are widely used to improve the dispersibility and control the morphology of the graphene-MNPs. 3,4,9,10-Perylene tetracarboxylic acid (PTCA)-modified graphene sheets were used as template for immobilizing Au NPs by the reduction of $HAuCl_4$ with the protection of amino-terminated ionic liquid (IL-NH_2). Because of additional carboxylic groups from PTCA, the resulting CCG/PTCA/Au-IL composites were water soluble and had a dense coverage of Au NPs [45].

Besides the spherical nanoparticles attached on graphene, anisotropic metal nanostructures, such as Au nanorods and Au square sheets, have also been successfully prepared on graphene film. For example, Au nanorods was anchored on graphene films or seed-modified graphene films by using a shape-directing surfactant, cetyl-trimethylammonium bromide (CTAB), to facilitate the one-dimensional growth of the nanorods [46]. Recently, Zhang's group synthesized square-like Au nanosheets with an edge length of 200–500 nm and a thickness of 2.4 nm (~16 Au atomic layers) on graphene sheets, as shown in Fig. 8.3. Hexagonal close packed (hcp) Au nanostructure was directly synthesized by wet chemical method for the first time [47].

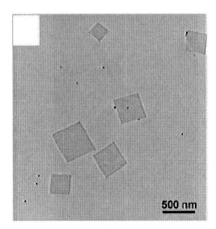

Figure 8.3 TEM image of 2.4 nm-thick Au nanosheets on GO surface (scale bar, 500 nm). The black dots are nanoparticle by products, that is, spherical Au nanoparticles. Reprinted from reference [47] by permission of Nature Publishing Group.

In addition to the single-phased metal structures, some two-phased metals are also reported. $Au_{core}Pd_{shell}$ (Au@Pd) bimetallic nanoparticles of sub-10 nm dispersed on graphene were successfully synthesized by a simple one-step reducing method. The as-synthesized Au@Pd-graphene hybrid exhibited extraordinary peroxidase activity as catalysts [48]. Graphene-supported Pt–Au alloy NPs were also successfully synthesized by simultaneous ethylene glycol reduction [49]. Pt-on-Pd bimetallic nanodendrites on polyvinylpyrrolidone (PVP)-modified graphene sheets were fabricated by a seed-mediated sequential process, in which the Pd seeds were anchored on graphene by reducing H_2PdCl_4 with HCOOH, and then Pt nanodendrites on the Pd seeds were subsequently formed by further reducing K_2PtCl_4 with ascorbic acid [50]. The PVP-protected graphene improved the solubility of the graphene sheets and provided high nucleation density of the Pd seeds. Graphene-supported PtPd alloy nanocubes (PtPd/RGO) were also synthesized via a facile and versatile one-pot hydrothermal synthetic strategy in dimethylformamide (DMF) solution with PdI_4^- and PtI_4^- as precursors [51]. DMF here acted as both solvent and reductant of GO in the reaction process, and I^- acted as a structure-directing agent (Figs. 8.4 and 8.5). The reported process proved to be a universal technique for preparing other graphene-supported alloy nanocrystals.

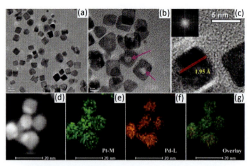

Figure 8.4 (a–c) High-resolution TEM micrographs of the PtPd alloy nanocubes supported on the RGO surface at different magnifications. The inset in (c) shows the FFT pattern of an individual PtPd nanocrystal. (d) The high-angle annular dark-field (HAADF)-STEM image of PtPd/RGO, and the corresponding elemental mapping of (e) Pt, (f) Pd, and (g) the overlay. Reprinted with permission from reference [51]. Copyright 2013 American Chemical Society.

Figure 8.5 Schematic illustration of the procedure for one-pot hydrothermal synthesis of RGO-supported PtPd alloy nanocubes. Stage 1: The reduction of GO and the nucleation of nanocrystals. Stage 2: The growth process of alloy nanocubes. Reprinted with permission from reference [51]. Copyright 2013 American Chemical Society.

Electrodeposition method is another efficient way applied in the preparation of graphene-MNPs. It is considered to be a controllable and effective technique for synthesizing metal or alloy NPs. Recently, the electrochemical reduction of GO has also drawn increasing attention due to its fast and green nature. In general, the electrochemical synthesis of graphene is carried out via two steps: First, GO was immobilized on the electrode by solution deposition methods, and then subjected to electrochemical reduction. Then metal NPs can be further loaded on this graphene-modified electrode by electrodeposition. The resulting metal NPs are located on the surface of the graphene film, that is, graphene sheets act like "agglomerate." Besides, the film thickness is difficult to control in the above-mentioned method. Our group reported a one-step facile method to synthesize electrochemically reduced GO and Pd NPs nanocomposite (ERGO-Pd) in the mixed solution of GO colloid and Na_2PdCl_4 (Fig. 8.6). The as-prepared nanocomposite was very "clean" as a result of reductant-free and surfactant-free synthesis process. Pd NPs and graphene sheets could be alternately attached onto the electrode, forming an interpenetrating network. The as-synthesized Pd NPs had a homogeneous and dense distribution on the graphene. The ERGO-Pd remarkably improved the electrocatalytic performance for formic acid oxidation. This facile and controllable method can readily be applied in preparation of other graphene-MNPs [52].

The photochemical reduction has also been applied for synthesis of graphene–metal NPs. In this case, graphene–metal NPs nanocomposites were prepared by the irradiant white light instead of the chemical reducing agent. In one report, noble metal salts adsorbed on ultrasonication exfoliated graphene flakes, which were functionalized with ionic surfactants, were reduced by irradiation with white light [53]. Microwave reduction was also demonstrated to be a useful way. Graphene-supported Pt catalysts were formed by co-reduction of GO and Pt salt in ethylene glycol solution under microwave irradiation conditions. The reduction mechanism was found to be a synergistic co-reduction whereby the presence of Pt ions led to a faster reduction of GO, and the presence of defect sites on the reduced GO served as anchor points for the heterogeneous nucleation of Pt [54].

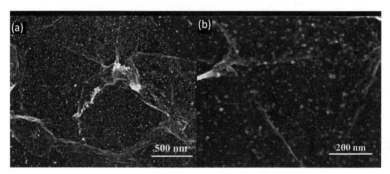

Figure 8.6 SEM images of the electrodeposited ERGO-Pd composite. Reprinted from reference [52]. Copyright 2012, with permission from Elsevier.

Besides forming composite with nanoparticles, GO can also act as structure-directing agent in the formation of graphene-MNPs composite. Yizhong Lu et al. reported that GO could act as a structure-directing agent for the formation of PtPd alloy concave nanocubes enclosed by high index facets (C–PtPd–RGO, Fig. 8.7). In contrast, only cubic PtPd alloy nanocrystals were obtained in the absence of GO. This work highlighted the unique role of GO in the formation of metal nanocrystals as not only a catalyst support but also a structure- and/or morphology-directing agent, due to the presence of various functional groups on GO sheets [55].

Ex situ hybridization. The ex situ hybridization is also widely used for preparation of graphene-NPs. Electrostatic interaction is usually the driving force for self-assembling MNPs on CCG sheets. Some binding molecules are often used in order to combine both graphene sheets and MNPs, thus improving the MNPs density anchored on graphene. Zhang and Lin reported that pre-synthesized PtAu alloy NPs captured on poly(diallyldimethylammonium chloride) (PDDA) were mixed with as-prepared graphene sheets to form the graphene-PtAu alloy nanocomposites. They found that the key is the introduction of PDDA, which not only acted as "nanoreactors" for the preparation of PtAu alloy NPs, but also facilitated the dense and uniform loading of PtAu on graphene [56]. Liu and Deng reported a bovine serum albumin (BSA) modified graphene (BSA-graphene) in which BSA acted as both reductant and stabilizer. Graphene-BSA-MNPs nanocomposites were prepared by mixing BSA-graphene with pre-synthesized Au, Pt, Pd or Ag nanostructures (Figs. 8.8 and 8.9). The protein here acted as a "universal glue" and turned the BSA-graphene into a general platforms towards the efficient assembly of NPs with varying sizes, shapes, compositions, and surface properties [57].

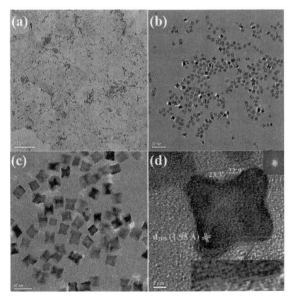

Figure 8.7 (a, b) TEM images of C–PtPd–RGO at different magnifications. (c, d) HRTEM images of the as-synthesized C–PtPd–RGO at different magnifications. The inset in the top right in (d) shows the FFT pattern of the concave nanocube and the inset in the bottom right is a zoom-in image of the selected area marked by a red rectangle in (d). Reprinted from reference [55] by permission of The Royal Society of Chemistry.

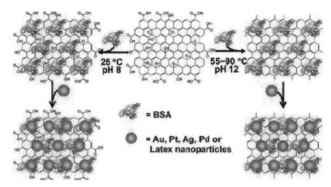

Figure 8.8 Scheme of protein-based decoration and reduction of GO, leading to a general nanoplatform for nanoparticle assembly. Reprinted with permission from reference [57]. Copyright 2010 American Chemical Society.

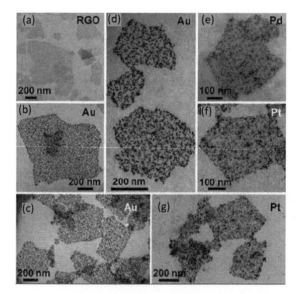

Figure 8.9 TEM images of (a) BSA-RGO, (b, c, d) AuNP-BSA-RGO, (e) PdNP-BSA-RGO, and (f, g) PtNP-BSA-RGO. Reprinted with permission from reference [57]. Copyright 2010 American Chemical Society.

Other methods: Layer-by-layer (LBL) deposition and solid reaction are used to prepare graphene-MNPs sometimes. Zhu and Dong et al. reported a LBL self-assembly method for constructing a graphene/platinum NPs three-dimensional hybrids base on the electrostatic interaction between positively charged immidazolium salt-based ionic liquid-functionalized graphene and negative charged citrate-stabilized Pt NPs [58]. The graphene multilayer supported Au NPs were also assembled by the electrostatic interaction of positively charged 4-dimethylaminopyridine (DMAP) coated Au NPs and negatively charged GO [59]. Lu and Chen et al. reported a GO-Ag composite synthesized via a dry approach. They decorated GO sheets with aerosol Ag nanocrystals using an electrostatic force directed assembly technique at room temperature [60].

8.2.6 Composites with Metal Oxide Nanoparticles

Owing to their desirable optical, electronic, magnetic, and catalytic properties, metal oxides NPs have many applications in the fields of sensors, energy, photochemistry, electronics and biomedicine, etc. The development of graphene–metal oxide nanoparticle (MONPs) provides an important milestone to further improve application performance of metal oxides. In order to construct high-quality graphene–MONPs hybrids, some aspects should be taken into account. First, GO or graphene sheets should have good solubility in the solvents throughout the experiments, so that individual sheets can be maintained to maximize their performance in the composites. Second, the presence of some oxygen-containing groups and defects are favorable for the anchor of MONPs on graphene sheets. Besides, the morphology of the metal oxide should be well controlled to optimize its property. A variety of metal compounds have been investigated for synthesizing their composites with CCG, such as SnO_2 [61–63], TiO_2 [64–67], ZnO [68], Fe_3O_4 [69–70], Co_3O_4 [71, 72], MnO_2 [73–75], and $Ni(OH)_2$ [76, 77]. The most widely used methods for synthesizing graphene–MONPs are in situ growth and ex situ hybridization.

The in situ growth can give rise to uniform surface coverage of nanocrystals by controlling the nucleation sites on GO/graphene via surface functionalization. For example, $TiCl_3$ aqueous solution was added into anionic sulfate surfactant (i.e., sodium dodecyl sulfate)-functionalized graphene sheets dispersion. After the self-assembly of surfactant with the metal oxide precursor, and the in situ crystallization of metal oxide, the self-assembled TiO_2–graphene hybrid (rutile and anatase) was obtained. The as-synthesized TiO$_2$–graphene hybrid materials showed enhanced Li-ion insertion/extraction kinetics, especially at high charge/discharge rates [64]. Liang and Guo developed a facile one-step solution-based process to in situ synthesize SnO$_2$/graphene nanocomposites by using the mixture of dimethyl sulfoxide (DMSO) and H$_2$O as both solvent and reactant. The reduction of GO and the in situ formation of SnO$_2$ NPs were realized in one step [62]. Liang and Dai reported a two-step method for synthesizing graphene–Co$_3$O$_4$ nanocrystals [72]. In the first step, Co(OAc)$_2$ aqueous solution was added to GO/ethanol dispersion and then the solution was stirred for 10 h at 80°C. In the second step, the mixture was subjected to hydrothermal reaction at 150°C for 3 h. NH$_4$OH was added after Co(OAc)$_2$ addition in the first step for the synthesis of Co$_3$O$_4$/nitrogen doped graphene. The Co$_3$O$_4$/nitrogen doped graphene hybrid exhibits similar catalytic activity but superior stability to commercial Pt/C in alkaline solution. A three-dimensional (3D) N-doped graphene aerogel supported Fe$_3$O$_4$ (Fe$_3$O$_4$/N-graphene) was also synthesized by a hydrothermal in situ growth method with Fe(II) ions and GO as precursors and polypyrrole (PPy) as nitrogen source [78].

Ex situ hybridization is also widely used for preparation of graphene–MONPs. Peak and Honma et al. prepared SnO$_2$ sol by hydrolysis of SnCl$_4$ with NaOH, and then a graphene dispersion was mixed with the SnO$_2$ sol in ethylene glycol to form graphene–SnO$_2$ composites [63]. Pre-synthesized GO powder and TiO$_2$ colloidal suspension in ethanol were mixed and sonicated to produce GO–TiO$_2$ nanocomposites. UV-irradiation of the GO–TiO$_2$ led to the photocatalytic reduction of GO; thus graphene–TiO$_2$ nanocomposites were formed [65]. Lee and Park reported highly photoactive, low-bandgap graphene-wrapped anatase TiO$_2$ NPs, which could highly enhance the photocatalytic activity of TiO$_2$ under visible light irradiation [79]. The scheme of the preparation process and the SEM images of the graphene wrapped TiO$_2$ were shown in Fig. 8.10. In this work, synthesis steps of graphene-wrapped TiO$_2$ NPs were conducted as follows: Bare amorphous TiO$_2$ NPs were prepared by a sol-gel method, afterwards the surface of amorphous TiO$_2$ NPs was modified by 3-aminopropyl-trimethoxysilane (APTMS) and then wrapped by GO nanosheets via electrostatic interaction. Graphene-wrapped anatase TiO$_2$ NP was obtained through a one-step GO reduction and TiO$_2$ crystallization via hydrothermal treatment.

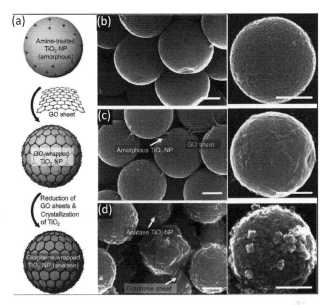

Figure 8.10 (a) Schematic illustration of synthesis steps for graphene-wrapped anatase TiO_2 NPs. (b) SEM images of bare, amorphous TiO_2 NPs. (c) SEM images of GO-wrapped amorphous TiO_2 NPs. The weight ratio of GO to TiO_2 was 0.02:1. (d) SEM images of graphene-wrapped anatase TiO_2 NPs. Scale bar: 200 nm. Reprinted from reference [79] by permission of John Wiley & Sons Ltd.

8.2.7 Graphene Quantum Dots Hybrids

Quantum dots (QDs), as a new class of fluorescent probes with higher sensitivity, better photostability, and chemical stability than conventional organic fluorophore markers, have attracted wide interest due to their size-dependent optical absorption and potential applications in various fields. Cao and Liu reported for the first time a one-step method to prepare graphene-GdS hybrids directly from GO in DMSO with Cd(II) salt. In this approach, DMSO served as a solvent, a source of sulfur and a reductant, thus the reduction of GO and the deposition of CdS on graphene occurred simultaneously [80]. The single-layer graphene sheets homogeneously decorated with CdS QDs were shown in the TEM image in Fig. 8.11. Zhang's group developed a facile method to directly synthesize red-emission Au nanodots (NDs) by the photochemical reduction of $HAuCl_4$ in the presence of

1-octadecanethiol (ODT) molecules [81]. Using graphene sheets as templates, the in situ synthesized Au NDs were self-assembled into particle chains which aligned along the (100) directions of the graphene lattice. This alignment was directed by thiol groups of the self-assembled ODT molecules on graphene. The pattern of in situ synthesized AuNDs on graphene is shown in Fig. 8.11.

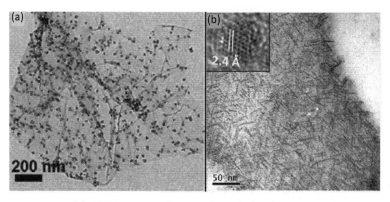

Figure 8.11 (a) TEM image of a graphene/CdS sheet sparsely coated with CdS QDs, showing natural wrinkles of a single graphene sheet. (b) TEM image of ODT-capped Au NDs synthesized in situ and assembled on graphene surface. Inset is a HRTEM image of an Au ND. Reprinted from reference [80] and [81] by permission of John Wiley & Sons Ltd.

ZnS nanodots highly dispersed on the surface of graphene sheets were also successfully fabricated by a facile one-step hydrothermal method by using Na_2S as reducing agent as well as sulfide source [82]. A composite of graphene and CdSe QDs (graphene-CdSe QDs) was fabricated via π–π stacking of aromatic structures between pre-synthesized graphene and pyridine capped CdSe QDs [83]. The as-prepared graphene-CdSe QDs can be readily used as flexible and transparent optoelectronic films. Later on, a linker-free graphene/CdSe NPs was produced by directly anchoring CdSe NPs onto graphene [84].

8.2.8 Composite with Other Nanoparticles

Other graphene-NPs nanocomposites were also reported. For example, Lee and Kung et al. reported a graphene-Si NPs nanocomposites by a solution mixing method. The as-prepared graphene-NPs supported by a 3-D network of graphite exhibited high Li ion storage capacities and cycling stability. There were also some reports about synthesis and applications of graphene–SiO$_2$ composites [85–88]. Transparent and electrically conductive composite silica films were fabricated on glass and hydrophilic SiO$_x$/silicon substrates by incorporation of individual GO sheets into silica sols followed by spin-coating, chemical reduction, and thermal curing. The electrical conductivity of the films compared favorably to those of composite thin films of carbon nanotubes in silica [85]. Graphene composite with transition metal complex were also prepared. Cao and Lu et al. reported an in situ controllable growth of Prussian blue nanocubes on GO by a wet chemical procedure [89]. Jiang et al. in our group developed a facile electrochemical deposition method to grow Prussian blue (PB) on graphene matrix [90]. We carried out a cyclic voltammetry scanning on graphene-modified electrode in a freshly deaerated mixed solution containing FeCl$_3$, K$_3$[Fe(CN)$_6$], KCl, and HCl. The process is controllable by varying the electrochemical deposition conditions. The PB-Graphene/GCE had admirable electrocatalytic performance towards both the reduction of H$_2$O$_2$ and the oxidation of hydrazine.

8.2.9 Graphene Composite with Organic Molecules

Some organic molecules, except for organic polymers, were also used for graphene functionalization. The procedures used for preparing these composites are similar to those applied for fabricating graphene/polymer composites. The preparation method can be mainly divided as noncovalent compositing, and covalent grafting.

Noncovalent compositing. Small organic molecules can usually act as stabilizers for dispersing GO or graphene. Surfactants were used to exfoliate graphite in water with the ultrasonication via noncovalent compositing with graphene. A sodium dodecyl benzene sulfonate (SDBS) surfactant was successfully applied for this purpose [91]. Xu and Shi et al. reported a supermolecular assembly between graphene and cationic 5,10,15,20-tetrakis (1-methyl-4-pyridinio) porphyrin (TMPyP) through the $\pi-\pi$ interaction. They found an unexpected large flattening of the TMPyP molecules adsorbed on graphene, which greatly accelerated the incorporation of Cd^{2+} ions into TMPyP rings. Thus the graphene-TMPyP composite could act as an optical probe for rapid and selective detection of Cd^{2+} ions in aqueous media [92]. Other graphene-porphyrin/phthalocyanine composites were also successfully prepared by noncovalent method [93–98]. Single stranded DNA–graphene nanocomposite was also fabricated by noncovalent method [99]. Our group also synthesized 3,4,9,10-perylene tetracarboxylic acid (PTCA)-functionalized graphene [45] and Congo red-functionalized graphene via the π–π interaction [100]. We also found that except forming composite with some molecules like FePc through π–π interaction (Fig. 8.12), partly reduced graphene can also improve the dispersibility of indissolvable FePc, which is helpful for further applications of the composite [101].

Besides forming composites with two-dimensional graphene, organic molecules can also form composite with other forms of graphene materials, such as three-dimensional graphene. In one research work of our group, after lyophilization and pyrolysis treatment of the VB12 and GO mixture, we successfully obtained a novel Co-N complex on three-dimensional graphene framework. The as-prepared composite exhibited admirable electrocatalytic performance for oxygen reduction reaction [102].

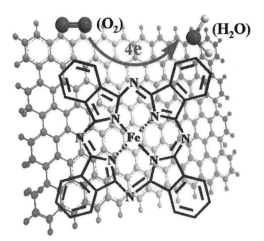

Figure 8.12 Schematic illustration of the interaction between graphene and FePc and the ORR process on g-FePc. Reprinted from reference [101] by permission of The Royal Society of Chemistry.

Covalent grafting The functionalization of graphene with organic groups has been developed for several purposes. One important aim is improvement in dispersibility of graphene, especially in common organic solvents. In addition, some organic functional groups can offer new properties that could be combined with the property of graphene. The covalent functionalization of graphene with organic molecules includes two general routes: The most common route is the formation of covalent bonds between organic functional oxygen containing groups of GO or graphene; another route is the formation of covalent bonds between free radicals or dienophiles and C=C bonds of graphene.

Our group reported a facile method to obtain ionic liquid-functionalized graphene (IL-graphene) through an epoxide nucleophilic ring-opening reaction between GO and amine-terminated ionic liquid when catalyzed by KOH (Fig. 8.13a). The resulting IL-graphene can be stably dispersed in water, DMF, and DMSO without any stabilizers [103]. Negatively charged glucose oxidase (GOD) was further immobilized onto the composite matrix simply by ionic exchange (Fig. 8.13b). The novel IL-graphene-GOD bionanocomposite can act as a desirable glucose sensor [104].

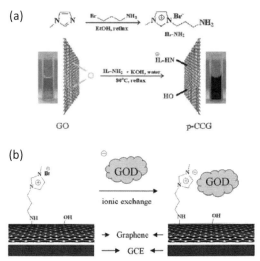

Figure 8.13 (a) Scheme of the preparation of IL-graphene. Reprinted from reference [103] by permission of The Royal Society of Chemistry. (b) Illustration showing the construction of the IL-graphene-GOD bionanocomposite. Reprinted from reference [104]. Copyright 2012, with permission from Elsevier.

Some simple organic chromophores such as porphyrins, phthalocyanines, and azobenzene with very interesting optoelectronic properties have been covalently attached on graphene sheets [105–109]. Xu and Chen et al. reported the first covalently bonded and organic soluble graphene hybrid with porphyrin via the formation of amide bonds between the carboxylic groups of GO and amino-porphyrin (Fig. 8.14) [110]. James M. Tour's group reported a surfactant-wrapped chemically converted graphene sheets functionalized by treatment with diazonium salts (Fig. 8.15). The diazonium-functionalized graphene sheets with high amounts of varying aryl addends were thus formed and had good solubility in organic solvents [111]. Lu and Chen reported a GO-anthracene hybrid with blue-emission through the GO surface functionalization with aryl diazonium salts of 2-amino anthracene (Fig. 8.16). Different from the fluorescence quenching of GO to quantum dots in their hybrid materials, the anthryl-functionalized GO exhibit strong photoluminescence. This work presented a conception that graphene or GO could be made luminescent via surface covalent functionalization [112].

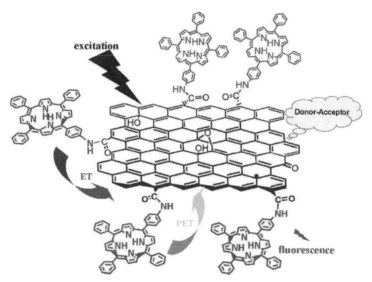

Figure 8.14 Scheme of the structure of the covalent binding porphyrin–graphene composite. Reprinted from reference [110] by permission of John Wiley & Sons Ltd.

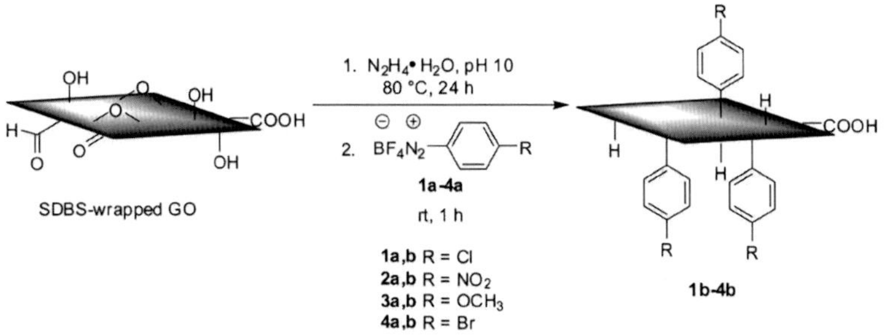

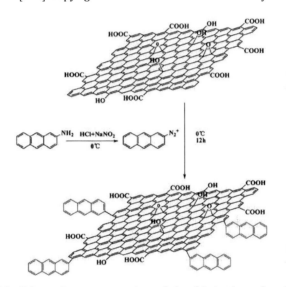

Figure 8.15 Starting with SDBS-wrapped GO, reduction and functionalization of intermediate SDBS-wrapped graphene with diazonium salts. Reprinted with permission from reference [111]. Copyright 2008 American Chemical Society.

Figure 8.16 Schematic representation of the fabrication of anthracene-functionalized GO through aryl diazonium reaction. Reprinted from reference [112] by permission of The Royal Society of Chemistry.

Some complicated graphene–organic molecule composite was also fabricated. Pyridine-functionalized graphene can be used as a building block in the assembly of metal organic framework (MOF). By reacting the pyridine covalently functionalized graphene with iron-porphyrin, a graphene–metalloporphyrin MOF with enhanced catalytic activity for oxygen reduction reactions (ORR) is synthesized. Results showed that the addition of pyridine-functionalized graphene changed the crystallization process of iron-porphyrin in the MOF, increased its porosity and enhanced the electrochemical charge transfer rate of iron-porphyrin. The chemical structures of the various subunits in the assembled MOF and the preparation process are illustrated in Fig. 8.17 [113].

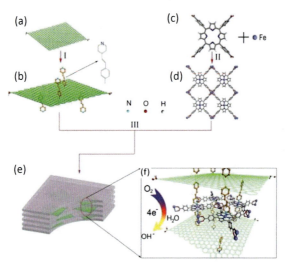

Figure 8.17 Schematic of the chemical structures of (a) reduced GO (r-GO), (b) G-dye, (c) TCPP, (d) (Fe-P)$_n$ MOF, (e) (G-dye-FeP)$_n$ MOF, and (f) magnified view of layers inside the framework of (G-dye-FeP)$_n$ MOF showing how graphene sheets intercalated between porphyrin networks. The synthetic process to form chemicals: (I) G-dye synthesized from r-GO sheets via diazotization with 4-(4-aminostyryl) pyridine, (II) (Fe-P)$_n$ MOF synthesized via reaction between TCPPs and Fe ions, (III) (G-dye-FeP)$_n$ MOF formed via reaction between (Fe-P)$_n$ MOF and G-dye. Reprinted with permission from reference [113]. Copyright 2012 American Chemical Society.

8.3 Conclusions and Future Outlook

In this chapter we discussed the functionalization and nanocomposite of chemically converted graphene and their applications. Since 2004, the initial report by Geim and Novoselov sent ripples of excitement through the scientific community. In the past years, the initial interest in the so-called "wonder" material has not faded. Functionalized graphene materials have shown wide applications in many fields such as supercapacitors, electrocatalysis, photocatalysts, electrochemical (bio-) sensors, transparent conductive film, electronics instruments, solar cells, and so on. Although numerous processes have been achieved, the research in this field is actually in its primary stages. Many challenges still remain in the functionalized graphene materials and their applications. First, the performance of these functionalized graphene materials is limited by their surface area and conductivity. Although the surface area and conductivity of ideal single-layer graphene are extraordinarily high, most practically functionalized graphene materials have structural defects and their surface cannot be fully exposed owing to the strong restacking effect of graphene sheets. Second, the specific structure of the functionalized graphene materials is usually indistinct and hard to control. Graphene-based composite materials with homogeneous distribution and structure are usually hard to obtain. Besides, the reproducibility of the composite structure during different synthesis is not satisfactory. Third, the chemical and physical stabilities of functionalized graphene materials still need improvements. Finally, the mechanisms of many applications of functionalized graphene materials are not completely clear. Researchers need to find solutions to these problems for the further applications of functionalized graphene materials.

Whatever the future holds for graphene-based materials research, we can be assured that there will continuously be interesting topics due to the unique structure and properties of graphene. Looking forward to the future of graphene-based materials research, it seems that devising the synthetic procedures to obtain high-quality graphene and in-depth exploration of the mechanism of the applications of functionalized graphene materials are the keys to unlock practical applications of graphene-based materials.

Acknowledgments

The authors are most grateful to the NSFC, China (no. 21175130, no. 21105096, no. 21225524, no. 21205112, and no. 21375124) and the Department of Science and Techniques of Jilin Province (no. 20120308 and no. 201215091) for their financial support.

References

1. Allen, M. J., Tung, V. C., and Kaner, R. B. (2009). Honeycomb carbon: A review of graphene. *Chem. Rev.*, **1**, pp. 132–145.
2. Geim, A. K., and Novoselov, K. S. (2007). The rise of graphene. *Nat. Mater.*, **3**, pp. 183–191.
3. Novoselov, K. S., McCann, E., Morozov, S. V., Fal'ko, V. I., Katsnelson, M. I., Zeitler,U., Jiang, D., Schedin, F., and Geim, A. K. (2006). Unconventional quantum Hall effect and Berry's phase of 2 pi in bilayer graphene. *Nat. Phys.*, **3**, pp. 177–180.
4. Novoselov, K. S., Geim, A. K., Morozov, S. V., Jiang, D., Zhang, Y., Dubonos, S. V., Grigorieva, I. V., and Firsov, A. A. (2004). Electric field effect in atomically thin carbon films. *Science*, **5696**, pp. 666–669.
5. Stoller, M. D., Park, S. J., Zhu, Y. W., An, J. H., and Ruoff, R. S. (2008). Graphene-based ultracapacitors. *Nano Lett.*, **10**, pp. 3498–3502.
6. Li, X. S., Cai, W. W., An, J. H., Kim, S., Nah, J., Yang, D. X., Piner, R., Velamakanni, A., Jung, I., Tutuc, E., Banerjee, S. K., Colombo, L., and Ruoff, R. S. (2009). Large-area synthesis of high-quality and uniform graphene films on copper foils. *Science*, **5932**, pp. 1312–1314.
7. Balandin, A. A., Ghosh, S., Bao, W., Calizo, I., Teweldebrhan, D., Miao, F., and Lau, C. N. (2008). Superior thermal conductivity of single-layer graphene. *Nano Lett.*, **3**, pp. 902–907.
8. Luo, B., Liu, S., and Zhi, L. (2012). Chemical approaches toward graphene-based nanomaterials and their applications in energy-related areas. *Small*, **8**, pp. 630–646.
9. Stankovich, S., Dikin, D. A., Piner, R. D., Kohlhaas, K. A., Kleinhammes, A., Jia, Y., Wu, Y., Nguyen, S. T., and Ruoff, R. S. (2007). Synthesis of graphene-based nanosheets via chemical reduction of exfoliated graphite oxide. *Carbon*, **7**, pp. 1558–1565.
10. Shao, Y. Y., Wang, J., Wu, H., Liu, J., Aksay, I. A., and Lin, Y. H. (2010). Graphene based electrochemical sensors and biosensors: A review. *Electroanalysis.*, **10**, pp. 1027–1036.
11. Li, D., Muller, M. B., Gilje, S., Kaner, R. B., and Wallace, G. G. (2008). Processable aqueous dispersions of graphene nanosheets. *Nat. Nanotech.*, **2**, pp. 101–105.
12. Georgakilas, V., Otyepka, M., Bourlinos, A. B., Chandra, V., Kim, N., Kemp, K. C., Hobza, P., Zboril, R., and Kim, K. S. (2012). Functionalization of graphene: Covalent and non-covalent approaches, derivatives and applications. *Chem. Rev.*, **112**, pp. 6156–6214.

13. Huang, X., Qi, X., Boey, F., and Zhang, H. (2012). Graphene-based composites. *Chem. Soc. Rev.*, **41**, pp. 666–686.
14. Moniruzzaman, M., and Winey, K. I. (2006). Polymer nanocomposites containing carbon nanotubes. *Macromolecules*, **16**, pp. 5194–5205.
15. Spitalsky, Z., Tasis, D., Papagelis, K., and Galiotis, C. (2010). Carbon nanotube-polymer composites: Chemistry, processing, mechanical and electrical properties. *Prog. Polym. Sci.*, **3**, pp. 357–401.
16. Ramanathan, T., Abdala, A. A., Stankovich, S., Dikin, D. A., Herrera-Alonso, M., Piner, R. D., Adamson, D. H., Schniepp, H. C., Chen, X., Ruoff, R. S., Nguyen, S. T., Aksay, I. A., Prud'homme, R. K., and Brinson, L. C. (2008). Functionalized graphene sheets for polymer nanocomposites. *Nat. Nanotech.*, **6**, pp. 327–331.
17. Liang, J., Huang, Y., Zhang, L., Wang, Y., Ma, Y., Guo, T., and Chen, Y. (2009). Molecular-level dispersion of graphene into polyvinyl alcohol) and effective reinforcement of their nanocomposites. *Adv. Funct. Mater.*, **14**, pp. 2297–2302.
18. Zhang, H. B., Zheng, W. G., Yan, Q., Yang, Y., Wang, J. W., Lu, Z. H., Ji, G. Y., and Yu, Z. Z. (2010). Electrically conductive polyethylene terephthalate/graphene nanocomposites prepared by melt compounding. *Polymer*, **5**, pp. 1191–1196.
19. Dasari, A., Yu, Z. Z., and Mai, Y. W. (2009). Electrically conductive and super-tough polyamide-based nanocomposites. *Polymer*, **16**, pp. 4112–4121.
20. Rafiee, M. A., Rafiee, J., Srivastava, I., Wang, Z., Song, H., Yu, Z. Z., and Koratkar, N. (2010). Fracture and fatigue in graphene nanocomposites. *Small*, **2**, pp. 179–183.
21. Zhang, K., Zhang, L. L., Zhao, X. S., and Wu, J. (2010). Graphene/polyaniline nanoriber composites as supercapacitor electrodes. *Chem. Mater.*, **4**, pp. 1392–1401.
22. Zhou, X., Wu, T., Hu, B., Yang, G., and Han, B. (2010). Synthesis of graphene/polyaniline composite nanosheets mediated by polymerized ionic liquid. *Chem. Commun.*, **21**, pp. 3663–3665.
23. Xu, Y., Hong, W., Bai, H., Li, C., and Shi, G. (2009). Strong and ductile polyvinyl alcohol)/graphene oxide composite films with a layered structure. *Carbon*, **15**, pp. 3538–3543.
24. Salavagione, H. J., Gomez, M. A., and Martinez, G. (2009). Polymeric modification of graphene through esterification of graphite oxide and polyvinyl alcohol. *Macromolecules*, **17**, pp. 6331–6334.

25. Ramanathan, T., Stankovich, S., Dikin, D. A., Liu, H., Shen, H., Nguyen, S. T., and Brinson, L. C. (2007). Graphitic nanofillers in PMMA nanocomposites-An investigation of particle size influence on nanocomposite and dispersion and their properties. *J. Polym. Sci. Polym. Phys.*, **15**, pp. 2097–2112.
26. Liang, J., Xu, Y., Huang, Y., Zhang, L., Wang, Y., Ma, Y., Li, F., Guo, T., and Chen, Y. (2009). Infrared-triggered actuators from graphene-based nanocomposites. *J. Phys. Chem. C*, **22**, pp. 9921–9927.
27. Wu, Q., Xu, Y., Yao, Z., Liu, A., and Shi, G. (2010). Supercapacitors based on flexible graphene/polyaniline nanofiber composite films. *ACS Nano*, **4**, pp. 1963–1970.
28. Stankovich, S., Dikin, D. A., Dommett, G. H. B., Kohlhaas, K. M., Zimney, E. J., Stach, E. A., Piner, R. D., Nguyen, S. T., and Ruoff, R. S. (2006). Graphene-based composite materials. *Nature*, **7100**, pp. 282–286.
29. Yang, H. F., Shan, C. S., Li, F. H., Zhang, Q. X., Han, D. X., and Niu, L. (2009). Convenient preparation of tunably loaded chemically converted graphene oxide/epoxy resin nanocomposites from graphene oxide sheets through two-phase extraction. *J. Mater. Chem.*, **46**, pp. 8856–8860.
30. Kim, H., and Macosko, C. W. (2009). Processing-property relationships of polycarbonate/graphene composites. *Polymer*, **15**, pp. 3797–3809.
31. Kim, I. H., and Jeong, Y. G. (2010). Polylactide/exfoliated graphite nanocomposites with enhanced thermal stability, mechanical modulus, and electrical conductivity. *J. Polym. Sci. Polym. Phys.*, **8**, pp. 850–858.
32. Kalaitzidou, K., Fukushima, H., and Drzal, L. T. (2007). A new compounding method for exfoliated graphite-polypropylene nanocomposites with enhanced flexural properties and lower percolation threshold. *Compos. Sci. Technol.*, **10**, pp. 2045–2051.
33. Yu, A., Ramesh, P., Itkis, M. E., Bekyarova, E., and Haddon, R. C. (2007). Graphite nanoplatelet-epoxy composite thermal interface materials. *J. Phys. Chem. C*, **21**, pp. 7565–7569.
34. Yan, J., Wei, T., Shao, B., Fan, Z., Qian, W., Zhang, M., and Wei, F. (2010). Preparation of a graphene nanosheet/polyaniline composite with high specific capacitance. *Carbon*, **2**, pp. 487–493.
35. Wang, D. W., Li, F., Zhao, J., Ren, W., Chen, Z. G., Tan, J., Wu, Z. S., Gentle, I., Lu, G. Q., and Cheng, H. M. (2009). Fabrication of graphene/polyaniline composite paper via in situ anodic electropolymerization for high-performance flexible electrode. *ACS Nano*, **7**, pp. 1745–1752.

36. Jang, J. Y., Kim, M. S., Jeong, H. M., and Shin, C. M. (2009). Graphite oxide/poly(methyl methacrylate), nanocomposites prepared by a novel method utilizing macroazoinitiator. *Compos. Sci. Technol.*, **2**, pp. 186–191.
37. Huang, Y., Qin, Y., Zhou, Y., Niu, H., Yu, Z. Z., and Dong, J. Y. (2010). Polypropylene/graphene oxide nanocomposites prepared by in situ Ziegler-Natta polymerization. *Chem. Mater.*, **13**, pp. 4096–4102.
38. Fim, F. D. C., Guterres, J. M., Basso, N. R. S., and Galland, G. B. (2010). Polyethylene/graphite nanocomposites obtained by in situ polymerization. *J. Polym. Sci. Polym. Chem.*, **3**, pp. 692–698.
39. Stankovich, S., Piner, R. D., Chen, X. Q., Wu, N. Q., Nguyen, S. T., and Ruoff, R. S. (2006). Stable aqueous dispersions of graphitic nanoplatelets via the reduction of exfoliated graphite oxide in the presence of poly(sodium 4-styrenesulfonate), *J. Mater. Chem.*, **2**, pp. 155–158.
40. Yang, H., Zhang, Q., Shan, C., Li, F., Han, D., and Niu, L. (2010). Stable, conductive supramolecular composite of graphene sheets with conjugated polyelectrolyte. *Langmuir*, **9**, pp. 6708–6712.
41. Shan, C., Yang, H., Han, D., Zhang, Q., Ivaska, A., and Niu, L. (2009). Water-soluble graphene covalently functionalized by biocompatible poly-L-lysine. *Langmuir*, **20**, pp. 12030–12033.
42. Liu, Z., Robinson, J. T., Sun, X., and Dai, H. (2008). PEGylated nanographene oxide for delivery of water-insoluble cancer drugs. *J. Am. Chem. Soc.*, **33**, pp. 10876–10881.
43. Muszynski, R., Seger, B., and Kamat, P. V. (2008). Decorating graphene sheets with gold nanoparticles. *J. Phys. Chem. C*, **14**, pp. 5263–5266.
44. Xu, C., Wang, X., and Zhu, J. (2008). Graphene-metal particle nanocomposites. *J. Phys. Chem. C*, **50**, pp. 19841–19845.
45. Li, F., Yang, H., Shan, C., Zhang, Q., Han, D., Ivaska, A., and Niu, L. (2009). The synthesis of perylene-coated graphene sheets decorated with Au nanoparticles and its electrocatalysis toward oxygen reduction. *J. Mater. Chem.*, **23**, pp. 4022–4025.
46. Kim, Y. K., Na, H. K., Lee, Y. W., Jang, H., Han, S. W., and Min, D. H. (2010). The direct growth of gold rods on graphene thin films. *Chem. Commun.*, **18**, pp. 3185–3187.
47. Huang, X., Li, S., Huang, Y., Wu, S., Zhou, X., Li, S., Gan, C. L., Boey, F., Mirkin, C. A., and Zhang, H. (2011). Synthesis of hexagonal close-packed gold nanostructures. *Nat. Commun.*, **2**.

48. Chen, H., Li, Y., Zhang, F., Zhang, G., and Fan, X. (2011). Graphene supported Au-Pd bimetallic nanoparticles with core-shell structures and superior peroxidase-like activities. *J. Mater. Chem.*, **44**, pp. 17658–17661.

49. Rao, C. V., Cabrera, C. R., and Ishikawa, Y. (2011). Graphene-Supported Pt-Au Alloy Nanoparticles: A highly efficient anode for direct formic acid fuel cells. *J. Phys. Chem. C*, **44**, pp. 21963–21970.

50. Guo, S., Dong, S., and Wang, E. (2010). Three-dimensional Pt-on-Pd bimetallic nanodendrites supported on graphene nanosheet: Facile synthesis and used as an advanced nanoelectrocatalyst for methanol oxidation. *ACS Nano*, **1**, pp. 547–555.

51. Lu, Y. Z., Jiang, Y. Y., Wu, H. B., and Chen, W. (2013). Nano-PtPd cubes on graphene exhibit enhanced activity and durability in methanol electrooxidation after CO stripping-cleaning. *J. Phys. Chem. C*, **6**, pp. 2926–2938.

52. Jiang, Y., Lu, Y., Li, F., Wu, T., Niu, L., and Chen, W. (2012). Facile electrochemical codeposition of "clean" graphene–Pd nanocomposite as an anode catalyst for formic acid electrooxidation. *Electrochem. Commun.*, **19**, pp. 21–24.

53. Jeong, G. H., Kim, S. H., Kim, M., Choi, D., Lee, J. H., Kim, J. H., and Kim, S. W. (2011). Direct synthesis of noble metal/graphene nanocomposites from graphite in water: Photo-synthesis. *Chem. Commun.*, **44**, pp. 12236–12238.

54. Kundu, P., Nethravathi, C., Deshpande, P. A., Rajamathi, M., Madras, G., and Ravishankar, N. (2011). Ultrafast microwave-assisted route to surfactant-free ultrafine Pt nanoparticles on graphene: Synergistic co-reduction mechanism and high catalytic activity. *Chem. Mater.*, **11**, pp. 2772–2780.

55. Lu, Y. Z., Jiang, Y. Y., and Chen, W. (2014). Graphene nanosheet-tailored PtPd concave nanocubes with enhanced electrocatalytic activity and durability for methanol oxidation, *Nanoscale*, **6**, pp. 3309–3315.

56. Zhang, S., Shao, Y. Y., Liao, H. G., Liu, J., Aksay, I. A., Yin, G. P., and Lin, Y. H. (2011). Graphene decorated with ptau alloy nanoparticles: Facile synthesis and promising application for formic acid oxidation. *Chem. Mater.*, **5**, pp. 1079–1081.

57. Liu, J., Fu, S., Yuan, B., Li, Y., and Deng, Z. (2010). Toward a universal "adhesive nanosheet" for the assembly of multiple nanoparticles based on a protein-induced reduction/decoration of graphene oxide. *J. Am. Chem. Soc.*, **21**, pp. 7279–7281.

58. Zhu, C. Z., Guo, S. J., Zhai, Y. M., and Dong, S. J. (2010). Layer-by-layer self-assembly for constructing a graphene/platinum nanoparticle three-dimensional hybrid nanostructure using ionic liquid as a linker. *Langmuir*, **10**, pp. 7614–7618.

59. Choi, Y., Gu, M., Park, J., Song, H. K., and Kim, B. S. (2010). Graphene multilayer supported gold nanoparticles for efficient electrocatalysts toward methanol oxidation. *Adv. Energ. Mater.*, **2**, pp. 1510–1518.

60. Lu, G. H., Mao, S., Park, S., Ruoff, R. S., and Chen, J. H. (2009). Facile, noncovalent decoration of graphene oxide sheets with nanocrystals. *Nano Res.*, **3**, pp. 192–200.

61. Kim, H., Kim, S. W., Park, Y. U., Gwon, H., Seo, D. H., Kim, Y., and Kang, K. (2010). SnO_2/graphene composite with high lithium storage capability for lithium rechargeable batteries. *Nano Res.*, **11**, pp. 813–821.

62. Liang, J. F., Wei, W., Zhong, D., Yang, Q. L., Li, L. D., and Guo, L. (2012). One-step in situ synthesis of SnO_2/graphene nanocomposites and its application as an anode material for Li-ion batteries. *ACS Appl. Mater. Inter.*, **1**, pp. 454–459.

63. Yoo, E., Kim, J., Hosono, E., Zhou, H. S., Kudo, T., and Honma, I. (2008). Large reversible Li storage of graphene nanosheet families for use in rechargeable lithium ion batteries. *Nano Lett.*, **8**, pp. 2277–2282.

64. Wang, D., Choi, D., Li, J., Yang, Z., Nie, Z., Kou, R., Hu, D., Wang, C., Saraf, L. V., Zhang, J., Aksay, I. A., and Liu, J. (2009). Self-assembled TiO_2-graphene hybrid nanostructures for enhanced li-ion insertion. *ACS Nano*, **4**, pp. 907–914.

65. Williams, G., Seger, B., and Kamat, P. V. (2008). TiO_2-graphene nanocomposites. UV-assisted photocatalytic reduction of graphene oxide. *ACS Nano*, **7**, pp. 1487–1491.

66. Zhang, Y., Tang, Z. R., Fu, X., and Xu, Y. J. (2010). TiO_2-graphene nanocomposites for gas-phase photocatalytic degradation of volatile aromatic pollutant: Is TiO_2-graphene truly different from other TiO_2-carbon composite materials? *ACS Nano*, **12**, pp. 7303–7314.

67. Zhang, H., Lv, X. J., Li, Y. M., Wang, Y., and Li, J. H. (2010). P25-graphene composite as a high performance photocatalyst. *ACS Nano*, **1**, pp. 380–386.

68. Lee, J. M., Pyun, Y. B., Yi, J., Choung, J. W., and Park, W. I. (2009). ZnO nanorod-graphene hybrid architectures for multifunctional conductors. *J. Phys. Chem. C*, **44**, pp. 19134–19138.

69. Lian, P., Zhu, X., Xiang, H., Li, Z., Yang, W., and Wang, H. (2010). Enhanced cycling performance of Fe_3O_4-graphene nanocomposite as an anode material for lithium-ion batteries. *Electrochim. Acta*, **2**, pp. 834–840.
70. Zhou, G., Wang, D. W., Li, F., Zhang, L., Li, N., Wu, Z. S., Wen, L., Lu, G. Q., and Cheng, H. M. (2010). Graphene-wrapped Fe_3O_4 anode material with improved reversible capacity and cyclic stability for lithium ion batteries. *Chem. Mater.*, **18**, pp. 5306–5313.
71. Wu, Z. S., Ren, W., Wen, L., Gao, L., Zhao, J., Chen, Z., Zhou, G., Li, F., and Cheng, H. M. (2010). Graphene anchored with Co_3O_4 nanoparticles as anode of lithium ion batteries with enhanced reversible capacity and cyclic performance. *ACS Nano*, **6**, pp. 3187–3194.
72. Liang, Y., Li, Y., Wang, H., Zhou, J., Wang, J., Regier, T., and Dai, H. (2011). Co_3O_4 nanocrystals on graphene as a synergistic catalyst for oxygen reduction reaction. *Nat. Mater.*, **10**, pp. 780–786.
73. Chen, S., Zhu, J., Wu, X., Han, Q., and Wang, X. (2010). Graphene oxide-MnO_2 nanocomposites for supercapacitors. *ACS Nano*, **5**, pp. 2822–2830.
74. Yan, J., Fan, Z., Wei, T., Qian, W., Zhang, M., and Wei, F. (2010). Fast and reversible surface redox reaction of graphene-MnO_2 composites as supercapacitor electrodes. *Carbon*, **13**, pp. 3825–3833.
75. Chen, Z. W., Higgins, D., Yu, A. P., Zhang, L., and Zhang, J. J. (2011). A review on non-precious metal electrocatalysts for PEM fuel cells. *Energ. Environ. Sci.*, **9**, pp. 3167–3192.
76. Xiao, J. and Yang, S. (2012). Nanocomposites of $Ni(OH)_2$/reduced graphene oxides with controllable composition, size, and morphology: Performance variations as pseudocapacitor electrodes. *Chempluschem*, **9**, pp. 807–816.
77. Wang, H., Casalongue, H. S., Liang, Y., and Dai, H. (2010). $Ni(OH)_2$ nanoplates grown on graphene as advanced electrochemical pseudocapacitor materials. *J. Am. Chem. Soc.*, **21**, pp. 7472–7477.
78. Wu, Z. S., Yang, S., Sun, Y., Parvez, K., Feng, X., and Muellen, K. (2012). 3D nitrogen-doped graphene aerogel-supported Fe_3O_4 nanoparticles as efficient eletrocatalysts for the oxygen reduction reaction. *J. Am. Chem. Soc.*, **22**, pp. 9082–9085.
79. Lee, J. S., You, K. H., and Park, C. B. (2012). Highly photoactive, low bandgap TiO_2 nanoparticles wrapped by graphene. *Adv. Mater.*, **8**, pp. 1084–1088.
80. Cao, A. N., Liu, Z., Chu, S. S., Wu, M. H., Ye, Z. M., Cai, Z. W., Chang, Y. L., Wang, S. F., Gong, Q. H., and Liu, Y. F. (2010). A facile one-step method to produce graphene-CdS quantum dot nanocomposites as promising optoelectronic materials. *Adv. Mater.*, **1**, pp. 103–106.

81. Huang, X., Zhou, X., Wu, S., Wei, Y., Qi, X., Zhang, J., Boey, F., and Zhang, H. (2010). Reduced graphene oxide-templated photochemical synthesis and in situ assembly of Au nanodots to orderly patterned Au nanodot chains. *Small*, **4**, pp. 513–516.

82. Xue, L., Shen, C., Zheng, M., Lu, H., Li, N., Ji, G., Pan, L., and Cao, J. (2011). Hydrothermal synthesis of graphene-ZnS quantum dot nanocomposites. *Mater. Lett.*, **2**, pp. 198–200.

83. Geng, X. M., Niu, L., Xing, Z. Y., Song, R. S., Liu, G. T., Sun, M. T., Cheng, G. S., Zhong, H. J., Liu, Z. H., Zhang, Z. J., Sun, L. F., Xu, H. X., Lu, L., and Liu, L. W. (2010). Aqueous-processable noncovalent chemically converted graphene-quantum dot composites for flexible and transparent optoelectronic films. *Adv. Mater.*, **5**, pp. 638–642.

84. Lin, Y., Zhang, K., Chen, W. F., Liu, Y. D., Geng, Z. G., Zeng, J., Pan, N., Yan, L. F., Wang, X. P., and Hou, J. G. (2010). Dramatically enhanced photoresponse of reduced graphene oxide with linker-free anchored CdSe nanoparticles. *ACS Nano*, **6**, pp. 3033–3038.

85. Watcharotone, S., Dikin, D. A., Stankovich, S., Piner, R., Jung, I., Dommett, G. H. B., Evmenenko, G., Wu, S. E., Chen, S. F., Liu, C. P., Nguyen, S. T., and Ruoff, R. S. (2007). Graphene-silica composite thin films as transparent conductors. *Nano Lett.*, **7**, pp. 1888–1892.

86. Yang, S., Feng, X., Wang, L., Tang, K., Maier, J., and Muellen, K. (2010). Graphene-based nanosheets with a sandwich structure. *Angew. Chem. Int. Ed.*, **28**, pp. 4795–4799.

87. Hao, L. Y., Song, H. J., Zhang, L. C., Wan, X. Y., Tang, Y. R., and Lv, Y. (2012). SiO_2/graphene composite for highly selective adsorption of Pb(II) ion. *J. Colloid Interface Sci.*, **369**, pp. 381–387.

88. Zhou, X., and Shi, T. J. (2012). One-pot hydrothermal synthesis of a mesoporous SiO_2-graphene hybrid with tunable surface area and pore size. *Appl. Surf. Sci.*, **259**, pp. 566–573.

89. Cao, L., Liu, Y., Zhang, B., and Lu, L. (2010). In situ controllable growth of prussian blue nanocubes on reduced graphene oxide: Facile synthesis and their application as enhanced nanoelectrocatalyst for H_2O_2 reduction. *ACS Appl. Mater. Interface*, **8**, pp. 2339–2346.

90. Jiang, Y. Y., Zhang, X. D., Shan, C. S., Hua, S. C., Zhang, Q. X., Bai, X. X., Dan, L., and Niu, L. (2011). Functionalization of graphene with electrodeposited Prussian blue towards amperometric sensing application. *Talanta*, **1**, pp. 76–81.

91. Lotya, M., Hernandez, Y., King, P. J., Smith, R. J., Nicolosi, V., Karlsson, L. S., Blighe, F. M., De, S., Wang, Z., McGovern, I. T., Duesberg, G. S., and Coleman, J. N. (2009). Liquid phase production of graphene

by exfoliation of graphite in surfactant/water solutions. *J. Am. Chem. Soc.*, **10**, pp. 3611–3620.

92. Xu, Y., Zhao, L., Bai, H., Hong, W., Li, C., and Shi, G. (2009). Chemically converted graphene induced molecular flattening of 5,10,15,20-tetrakis-1-methyl-4-pyridinio)porphyrin and its application for optical detection of cadmiumii) ions. *J. Am. Chem. Soc.*, **37**, pp. 13490–13497.

93. Tu, W. W., Lei, J. P., Zhang, S. Y., and Ju, H. X. (2010). Characterization, direct electrochemistry, and amperometric biosensing of graphene by noncovalent functionalization with picket-fence porphyrin. *Chem. Eur. J.*, **35**, pp. 10771–10777.

94. Huang, D. K., Lu, J. F., Li, S. H., Luo, Y. P., Zhao, C., Hu, B., Wang, M. K., and Shen, Y. (2014). Fabrication of cobalt porphyrin. electrochemically reduced graphene oxide hybrid films for electrocatalytic hydrogen evolution in aqueous solution. *Langmuir*, **23**, pp. 6990–6998.

95. Tang. H. J., Yin. H. J., Wang, J. Y., Yang, N. L., Wang, D., and Tang, Z. Y. (2013). Molecular architecture of cobalt porphyrin multilayers on reduced graphene oxide sheets for high-performance oxygen reduction reaction. *Angew. Chem. Int. Ed.*, **21**, pp. 5585–5589.

96. Wojcik, A., and Kamat, P. V. (2010). Reduced graphene oxide and porphyrin. an interactive affair in 2-d. *ACS nano*, **11**, pp. 6697–6706.

97. Zhang, X. F., and Xi, Q. (2011). A graphene sheet as an efficient electron acceptor and conductor for photoinduced charge separation. *Carbon*, **12**, pp. 3842–3850.

98. Zhang, C. Z., Hao, R., Yin, H., Liu, F., and Hou, Y. L. (2012). Iron phthalocyanine and nitrogen-doped graphene composite as a novel non-precious catalyst for the oxygen reduction reaction. *Nanoscale*, **23**, pp. 7326–7329.

99. Zhang, Q. A., Qiao, Y., Hao, F., Zhang, L., Wu, S. Y., Li, Y., Li, J. H., and Song, X. M. (2010). Fabrication of a biocompatible and conductive platform based on a single-stranded DNA/graphene nanocomposite for direct electrochemistry and electrocatalysis. *Chem. Eur. J.*, **27**, pp. 8133–8139.

100. Li, F., Bao, Y., Chai, J., Zhang, Q., Han, D., and Niu, L. (2010). Synthesis and application of widely soluble graphene sheets. *Langmuir*, **14**, pp. 12314–12320.

101. Jiang, Y. Y., Lu, Y. Z., Lv, X. Y., Han, D. X., Zhang, Q. X., Niu, L., and Chen, W. (2013). Enhanced catalytic performance of pt-free iron phthalocyanine by graphene support for efficient oxygen reduction reaction. *ACS Catalysis*, **6**, pp. 1263–1271.

102. Jiang, Y., Lu, Y., Wang, X., Bao, Y., Chen, W., and Niu, L. (2014). Cobalt-nitrogen complex on n-doped three-dimensional graphene framework as highly efficient electrocatalysts for oxygen reduction reaction. *Nanoscale*, **6**, pp. 15066–15072.

103. Yang, H. F., Shan, C. S., Li, F. H., Han, D. X., Zhang, Q. X., and Niu, L. (2009). Covalent functionalization of polydisperse chemically-converted graphene sheets with amine-terminated ionic liquid. *Chem. Commun.*, **26**, pp. 3880–3882.

104. Jiang, Y., Zhang, Q., Li, F., and Niu, L. (2012). Glucose oxidase and graphene bionanocomposite bridged by ionic liquid unit for glucose biosensing application. *Sens. Actuat. B Chem*, **1**, pp. 728–733.

105. Feng, Y. Y., Liu, H. P., Luo, W., Liu, E. Z., Zhao, N. Q., Yoshino, K., and Feng, W. (2013). Covalent functionalization of graphene by azobenzene with molecular hydrogen bonds for long-term solar thermal storage. *Sci. Rep.*, **3**, p. 3260.

106. Song, W. N., He, C. Y., Zhang, W., Gao, Y. C., Yang, Y. X., Wu, Y. Q., Chen, Z. M., Li, X. C., and Dong, Y. L. (2014). Synthesis and nonlinear optical properties of reduced graphene oxide hybrid material covalently functionalized with zinc phthalocyanine. *Carbon*, pp. 1020–1030.

107. Ragoussi, M. E., Katsukis, G., Roth, A., Malig, J., Torre, G. de la, Guldi, D. M., and Torres, T. (2014). Electron-donating behavior of few-layer graphene in covalent ensembles with electron-accepting phthalocyanines. *J. Am. Chem. Soc.*, **12**, pp. 4593–4598.

108. Tang, J. G., Niu, L., Liu, J. X., Wang, Y., Huang, Z., Xie, S. Q., Huang, L. J., Xu, Q. S., and Belfiore, L. A. (2014). Effect of photocurrent enhancement in porphyrin-graphene covalent hybrids. *Mater. Sci. Eng. C*, pp. 186–192.

109. Karousis, N., Sandanayaka, A. S. D., Hasobe, T., Economopoulos, S. P., Sarantopoulou, E., and Tagmatarchis, N. (2011). Graphene oxide with covalently linked porphyrin antennae: Synthesis, characterization and photophysical properties. *J. Mater. Chem.*, **1**, pp. 109–117.

110. Xu, Y., Liu, Z., Zhang, X., Wang, Y., Tian, J., Huang, Y., Ma, Y., Zhang, X., and Chen, Y. (2009). A graphene hybrid material covalently functionalized with porphyrin: Synthesis and optical limiting property. *Adv. Mater.*, **21**, pp. 1275–1279.

111. Lomeda, J. R., Doyle, C. D., Kosynkin, D. V., Hwang, W. F., and Tour, J. M. (2008). Diazonium functionalization of surfactant-wrapped chemically converted graphene sheets. *J. Am. Chem. Soc.*, **48**, pp. 16201–16206.

112. Lu, Y. Z., Jiang, Y. Y., Wei, W. T., Wu, H. B., Liu, M. M., Niu, L., and Chen, W. (2012). Novel blue light emitting graphene oxide nanosheets fabricated by surface functionalization. *J. Mater. Chem.*, **7**, pp. 2929–2934.
113. Jahan, M., Bao, Q. L., and Loh, K. P. (2012). Electrocatalytically Active Graphene-porphyrin MOF composite for oxygen reduction reaction. *J. Am. Chem. Soc.*, **15**, pp. 6707–6713.

Chapter 9

Development of Graphene-Based Nanostructures

Huaqiang Cao[a,b]

[a]*Department of Chemistry, Tsinghua University, Beijing 100084, People's Republic of China*
[b]*Department of Materials Science and Metallurgy, University of Cambridge, Cambridge CB3 0FS, United Kingdom*

hqcao@mail.tsinghua.edu.cn, hc444@cam.ac.uk

Graphene-based nanostructures are one of the most important classes of functional nanomaterials. Interest in graphene-based nanostructures received a great advance since the discovery of graphene in 2004. Since then there has been far-ranging research work on graphene-based nanomaterials. The synthesis of graphene-based nanostructures under soft chemical approach with control morphology, particle size remains a major challenge of nanochemistry. Also the applications of graphene-based nanostructures generated great interest due to their distinguished properties. In this chapter, we will review our recent research work on the synthesis and applications of graphene-based nanostructures, such as the synthesis of organic molecule–

Electrochemical Nanofabrication: Principles and Applications (Second Edition)
Edited by Di Wei
Copyright © 2016 Pan Stanford Publishing Pte. Ltd.
ISBN 978-981-4613-86-6 (Hardcover), 978-981-4613-87-3 (eBook)
www.panstanford.com

modified graphene, inorganic modified graphene. Some of these nanostructures have already been used in nanotechnological applications, for examples, lithium ion batteries, Ni-MH batteries, removal of dyes from water, and fluorescence quench.

9.1 Introduction

Graphene, being a single hexagonal carbon layer of sp^2 carbon atoms densely packed into a benzene-ring structure with two carbon atoms per unit cell, originally discovered by Novoselov and Geim and co-workers [1, 2], is attracting increasing interest as building blocks of novel materials and device structures, due to its room temperature quantum Hall effect (QHE) [3], the high electron mobility as high as 10,000 $cm^2 \cdot V^{-1} \cdot s^{-1}$ at room temperature, which is about 10 times higher than the mobility of commercial silicon wafers [4], good electrical conductivity of 7200 S/m [5], high room temperature thermal conductivity up to ~5300 $W \cdot m^{-1} \cdot K^{-1}$ [6], high surface area (calculated value, 2630 $m^2 \cdot g^{-1}$) [7], which make graphene alluring for certain applications. However, one possible route to obtain these unusual properties for applications would be to integrate graphene sheets in composite materials [8]. And the application of graphene or graphene-based materials is the driving force for intensively studying graphene [9]. Graphite oxide or graphene oxide (GO) paper [10], is a layered material produced by the oxidation of graphite, which owns hydroxyl and epoxide functional groups on their basal planes besides carbonyl and carboxyl groups located at the sheet edges [8]. So GO provide us a platform for synthesizing chemically modified graphene. Currently, the main interest in graphene are focused on two aspects: developing novel and simple scalable synthesis approaches for generating graphene nanosheets and finding novel distinguished physicochemical properties or graphene-based nanocomposites [11–14].

In this chapter, we review the progress in the applications of in graphene-based nanomaterials. Illustrative examples of the advances are provided, covering lithium ion battery (LIBs), NH-M battery, removal of dyes, and fluorescence quenching.

9.2 Fluorescence Quenching of Hybrid Graphene Material Covalently Functional with Indolizine

Graphene is a single atomic layer of carbon atoms bonded in a hexagonal lattice in perfect crystalline order, with two carbon atoms per unit cell. Each carbon atom is bonded to three other carbon atoms. Of the four valence states, three sp^2 orbitals form σ states with three neighboring carbon atoms and one p orbital develops into delocalized π and π^* states that form the highest occupied valence band (VB) and the lowest unoccupied conduction band (CB) [2]. Graphene has been regarded as a more promising candidate material than its predecessor carbon nanotubes [9], but the poor solubility and strong aggregation of graphene in water or some organic solvents restrains its processability and applications. Inspired by Coleman's work, i.e., single-walled carbon nanotubes (SWNTs) modified with pyridinium ylides via 1,3-dipolar cycloaddition reaction [15], we synthesize indolizine-modified graphene (termed as IMG) nanosheets and study its photoluminescence properties.

The synthesis process is composed of two steps (Scheme 9.1): In the first step, 30 mg reduced graphene oxide (RGO) synthesized by using a modified Hummer's method [16] was dissolved in 15 mL N,N-dimethylformamide (DMF) using sonication for 15 min, and then the mixture was heated to 140°C. In the second step, pyridunium salt reacts with RGO. Pyridine (0.05 mmol) was dissolved in 20 mL acetic ether, followed by adding ethyl 2-bromoacetate (BrCH$_2$COOC$_2$H$_5$, 0.05 mL) in an ice bath while stirring and reacting for 12 h at room temperature. The resulting white solid was washed with diethyl ether (3 × 20 mL) to remove excess pyridine and ethyl 2-bromoacetate to generate the pyridinium bromide salt (10.81 g, yield = 87.5%). Pyridinium salt (4.17 mmol) was then added to the above mixture followed by triethylamine (0.58 mL) after 30 min. The reaction mixture was refluxed for 5 days and the product was collected via centrifugation and washed by using acetonitrile and deionized water for three times, followed by washing by alcohol and acetone for three times. Then, the product was dispersed in alcohol followed by drying at 80°C for 5 h to afford 50.8 mg IMG.

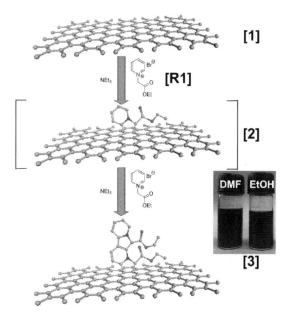

Scheme 9.1 Synthesis scheme of IMG. Inset: photographs of 1 mg/mL IMG in DMF and ethanol, after dispersion and standing for over 4 months. Copyright Institute of Physics Publishing Limited, reproduced with permission from Ref. 16.

The as-synthesized IMG was characterized by UV-vis spectroscopy, Fourier transform infrared (FT-IR) spectrum (obtained on a Nicolet 560 FT-IR spectrophotometer), Raman spectroscopy (obtained by a Renishaw, RM-1000 with excitation from the 514.5 nm line of an Ar-ion laser with a power of about 5 mW at room temperature), thermogravimetric analysis (TGA), X-ray photoelectron spectroscopy (XPS), transmission electron microscopy (TEM), atomic force microscopy (AFM), and Photoluminescence (PL) spectroscopy. Figure 9.1a presents the UV-vis absorption spectra of the IMG compared with RGO and pyridinium salt. IMG presents a weaker absorption peak centered at 261 nm in wavelength, which is different from the absorption peaks of RGO and pyridinium salt, suggesting that in the ground state attachment of organic groups has perturbed the electronic state of the graphene sheets.

Figure 9.1b is the FT-IR spectra of IMG as well as RGO and pyridinium salt. In the spectrum of IMG, the peak at 1703 cm^{-1}

can be attributed to the C=O stretch of the carboxylic group [17]. The peak at 1221 cm^{-1} is attributed to the C–O stretch vibration and 3381 cm^{-1} to the O–H stretch vibration [18]. The absorption peaks at 1627 and 1587 cm^{-1} are attributed to the characteristic peaks of indolizine, suggesting two characteristic bands of indolizine banded to the surface of RGO [15, 19]. The absorption peak at 1587 cm^{-1} is due to the presence of an aromatic ring on IMG sheets [18, 20]. The two peaks at 769 and 681 cm^{-1} are also characteristic peak of indolizine [21].

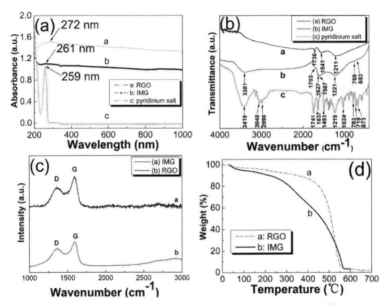

Figure 9.1 (a) UV-vis spectra of line a: RGO, line b: IMG, and line c: pyridinium salt dispersed in deionized water. (b) FT-IR spectra of line a: RGO, line b: IMG, and line c: pyridinium salt. (c) Raman spectra of line a: IMG and line b: RGO and (d) TGA pattern of line a: RGO and line b: IMG in air. Copyright Institute of Physics Publishing Limited, reproduced with permission from Ref. 16.

Raman spectra of IMG and RGO are shown in Fig. 9.1c. A weak peak at 1381 cm^{-1} (D band) and sharp peak at 1595 cm^{-1} (G band) for IMG, while 1360 cm^{-1} (D band) and 1584 cm^{-1} (G band) for RGO, were observed. The blueshifted G band of IMG compared with RGO is indicative of the severe disruption or disorder induced into the sp^2 carbon lattice of graphene [22]. This phenomenon demonstrated that graphene has been chemically modified after RGO is treated with pyridinium salt.

TGA analysis (Fig. 9.1d) indicates that there are three steps of weight change for IMG after heat treatment in air, i.e., the first two mass-loss knee points at about 142.1 and 275.4°C, respectively, attributed to the removal of the functional groups from IMG to yield CO, CO_2, and steam [15, 22, 23], and the third mass-loss knee point at around 482.0°C attributed to burning of the graphene skeleton. However, the TGA profile of RGO is quite different from that of IMG, which present the mass-loss knee point at 418.4°C with 14.7% weight loss. The TAG result indicates that there is one inodolizine group attached to RGO per 40 carbon atoms an IMG sheet.

To further evaluate the functional groups presenting in the IMG sample, X-ray photoelectron spectroscopy (XPS) data were investigated (Fig. 9.2). The elemental composition analysis shows the presence of 81.00 atomic percent (at.%) of C, 7.26 at.% of N, and 11.74 at.% of O in the IMG, respectively. It was estimated that there is about one indolizine group per 30 carbon atoms on the IMG sheets, which matches well with the TGA analysis considering the measurement error of XPS. The XPS spectra show that main peak C_{1s} locates at around 285 eV, which is attributed mainly to sp^2-hybridized graphitic carbon [23, 24]. The C_{1s} can be divided into four peaks: C_1, C_2, C_3, and C_4 (left inset in Fig. 9.2). The main peak C_1 is located at a binding energy of 284.4 eV with peak area proportion of 42.8%, which is attributed to the C–C bonding (sp^2 carbon) in defect-free graphite lattice. The peak C_2 located at 285.3 eV with peak area proportion of 35.1% is attributed to carbon in the C–C bonding in defect graphite lattice and C–N sp^2 bonding [25]. The large peak area of C_2 implies that the surface of graphene has been successfully functionalized with indolizine [26]. The C_3 peak at binding energy of 286.5 eV with peak area of 18.1% is assigned as carbon in C–O bond [27]. The C_4 peak at a binding energy of 288.0 eV with peak area proportion of 4.0% is attributed to carbon in C=O bond [25]. The right inset in Fig. 9.2 shows the presence of an N_{1s} peak at a binding energy of 399.8 eV, characteristic of sp^2 nitrogen [15]. The O_{1s} peak at 531.2 eV for IMG is due to a carbonyl group (C=O) and carboxylic group (COO) [28]. However, we cannot observe the Br peak, suggesting that there is no physical or chemical absorption of pyridinium salt on IMG. XPS data further demonstrated the successfully functionalization of RGO for generating IMG.

Fluorescence Quenching of Hybrid Graphene Material | 333

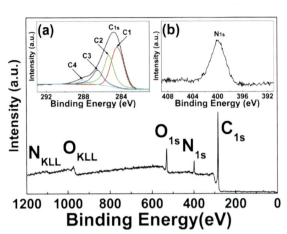

Figure 9.2 XPS Survey scan of IMG, inset (a) C_{1s} photoelectron spectrum, inset (b) N_{1s} photoelectron spectrum. Copyright Institute of Physics Publishing Limited, reproduced with permission from Ref. 16.

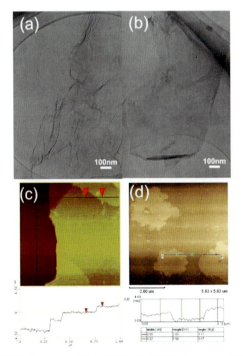

Figure 9.3 TEM images of (a) RGO, (b) IMG, tapping AFM images of (c) RGO, and (d) IMG on mica. Copyright Institute of Physics Publishing Limited, reproduced with permission from Ref. 16.

The morphology of RGO and IMG was characterized by TEM (Figs. 9.3a and 9.3b) and AFM (Figs. 9.3c and 9.3d). According to TEM observation, we can find large RGO and IMG sheets like crumpled silk veil waves, both of them are transparent and exhibit a stable nature under the electron beam. AFM analysis of RGO indicates that the average thickness of RGO is ~1.10 μm, suggesting the exfoliation of RGO down to one or two layers. Rougher surfaces of IMG can be observed by AFM (Fig. 9.3d), which is due to the covalent modification of RGO by indolizine groups.

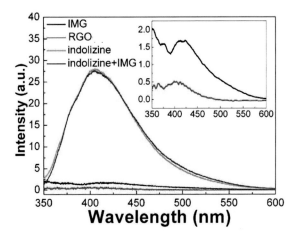

Figure 9.4 Fluorescence spectroscopic changes observed for black curve: 0.01 mg/mL of IMG; red curve: 0.01 mg/mL of RGO; green curve: 0.004 mg/mL of indolizine; and blue curve: simple mixture solution of 0.004 mg/mL of indolizine and 0.01 mg/mL of IMG. All of the samples were apart dispersed in DMF. Copyright Institute of Physics Publishing Limited, reproduced with permission from Ref. 16.

The fluorescence spectra of IMG, RGO, indolizine, and a mixed solution of indolizine and IMG were shown in Fig. 9.4. We can observe the photoluminescence intensity of IMG was only ~6.58% of that of indolizine under the same measurement condition, i.e., the photoluminescence intensity of IMG is quenched by a factor of ~93.4% on indolizine, showing that efficient energy transfer occurs along the indolizine/graphene interface, as demonstrated by Raman spectroscopy and density functional theory calculations [29, 30]. It is known that fluorescence quenching refers to any process that decreases the fluorescence intensity of a sample

[31]. The fluorescence quenching mechanisms include dynamic quenching and static quenching [32]. The detailed quenching mechanism of IMG need further study by measuring its other fluorescence parameters such as quantum yield and lifetime, beside intensity. The fluorescence property of IMG suggests that it may be of potential use in solar-harvesting applications.

9.3 Graphene-Based Materials Used as Electrodes in Ni-MH and Li-Ion Batteries

Our living standard greatly depends upon the amount of energy consumed per capital. However, most of our energy is currently produced from fossil fuels which lead to increase of greenhouse gases, including two main gases—water vapor and carbon dioxide, furthermore, to global warming [33]. Electrochemical energy production is under serious consideration as alternative energy sources, in particular, green secondary (rechargeable) battery is an efficient and cleanly new energy, which can be used repeatedly. It is regarded to provide an important technique approach for lightening energy, resource, and environment issues [34–48]. Battery devices are currently being developed to power an increasing diverse ranges of applications, including defense, spacecrafts, power tools, electric vehicles, computers, and other home appliances. A battery is essentially a series of cells, each containing two electrodes immersed in an electrically conducting medium called the electrolyte. Batteries are either non-rechargeable (primary) or rechargeable (secondary), but only rechargeable batteries are of interest for large-scale energy storage [33]. Especially, nickel metal hydride (Ni–MH) batteries and lithium ion batteries (LIBs) are most important devices among the secondary batteries. One of the key elements of these two batteries is that the same ion (H^+ for Ni–MH and Li^+ for LIBs) participates at both electrodes, being reversibly inserted and extracted from the electrode material, with the concomitant addition or removal of electrons [38].

Rechargeable Ni–MH and LIBs have become commercial reality for years [49–51]. Ni–MH battery, using hydrogen storage alloys as the negative electrode materials and $Ni(OH)_2$ as the positive electrode material, has been considered as one of the promising batteries, due to their advantages compared with traditional secondary batteries (such as Ni–Cd battery), such as high energy density both in terms of weight and volume, excellent cycle life, environmental friendliness, and non–memory effect, etc., which is able to reversibly deinsert/insert protons during discharge/charge [52–57]. For a long time, Ni–MH battery was the only battery available for power tools and one of the most promising choices for electric vehicles application, due to its excellent low-temperature and high-rate capabilities [58, 59].

However, the Ni–MH batteries also run under the strong competition by the rechargeable LIBs, which have a distinct advantage in the light of specific energy. The LIBs, with notably higher energy density and lighter weight, will replace Ni–MH batteries solve the problem of large-scale throughput.

There exist two polymorphs of $Ni(OH)_2$ at atmospheric pressure: metastable α-$Ni(OH)_2$ and stable β-$Ni(OH)_2$, all belonging to hexagonal brucite-type layered structure, which transforms into γ-NiOOH and β-NiOOH, respectively, after fully charging [60]. Although it is generally accepted that α-$Ni(OH)_2$ has higher theoretical electrochemical capacity (433 mAh·g^{-1}) due to an average oxidation state of 3.5 or higher for nickel in the γ-NiOOH, which can be easily transformed into β-$Ni(OH)_2$ after a few cycles [60]. In most case, the active material of positive electrode was the β-$Ni(OH)_2$ with its theoretical capacity of 289 mAh·g^{-1}. It is known that the extended or long life performance of Ni–MH batteries is usually restricted with the efficiency of the positive electrode, according to the principle of the rechargeable battery, which is controlled by electrochemical reversible reaction between $Ni(OH)_2$ and NiOOH, i.e., $Ni(OH)_2 + OH^- \leftrightarrow NiOOH + H_2O + e^-$ [58]. The $Ni(OH)_2$/NiOOH charge–discharge process is believed to be a solid-state, proton insertion/extraction reaction, whereby both protons and electrons are exchanged and the processes are regarded to be controlled by the bulk solid diffusion of protons.

As the microstructure of electrode materials greatly affected their electrochemical applications and practical capacities, great efforts have been devoted to control the size, morphology, additives and preparation routes of β-$Ni(OH)_2$. Nanotechnology is currently applied into improve the electrochemical performance of the rechargeable batteries, because nanomaterials not only change the electrode reaction pathway, which leads to high capacities and rechargeability, but also change the solid electrolytes' conductivity several-fold by adding nanograins in polymer-based electrolytes. Our previous work indicated that the highest discharge capacity of the β-Ni(OH)$_2$ nanoflowers with second-order structure nanoplate (20 nm in thickness) structure reached 249.3 mAh·g^{-1} at discharge current rate of 0.2 C (i.e., current density of 60 mA/g), which is attributed to the complex β-Ni(OH)$_2$ nanostructures [52]. He and co-workers found that granulated grains of nanoscale β-Ni(OH)$_2$ mixed with nanoscale Co(OH)$_2$ at Co/Ni = 1/20 used as positive electrode in Ni–MH battery presented the highest specific capacity reaching its theoretical value of 289 mAh·g^{-1} at 1 C [54]. This study indicates that the addition of Co(OH)$_2$ plays a crucial role in upswinging the electrochemical behavior of nanoscale Ni(OH)$_2$ because it enhance the conductivity of the active material when it transforms into CoO(OH), of which the conductivity is high, during charging. Kiani and co-workers reported a size effect study on between micro- and nanoparticles of β-Ni(OH)$_2$ as positive electrode material in Ni–MH batteries, which indicated that β-Ni(OH)$_2$ nanoparticles have a superior cycling reversibility and improved capacity of 270 mAh·g^{-1} compared with 241 mAh·g^{-1} of β-Ni(OH)$_2$ microparticles at a constant current density of 5 mA·g^{-1}, which is attribute to the increasing the number of available active sites in the process for β-Ni(OH)$_2$ nanoparticles [56]. Ni and co-workers also reported a β-Ni(OH)$_2$ nanoflowers with discharge capacity of 259 mAh·g^{-1} at a discharge rate of 16 mA for 2.5 h, which is also attributed to the special flowery nanostructure together with its high stability [61].

It is known that commercially available LIBs make use of layered LiCoO$_2$ in positive electrodes, though cobalt is expensive and poisonous [51, 62]. Hence, developing both environmentally friendly and cheap electrode materials in place of LiCoO$_2$ is an important issue in LIBs. The electrode reaction in LIBs depends upon the simultaneous intercalation/extraction of Li$^+$ ion and electron into the active electrode materials.

Without reference to the type of rechargeable battery, the battery performance, including discharge capacity, energy density, or cycle life, are related to the intrinsic property of the materials used as electrodes. So finding new materials as positive and negative electrodes, as well as right combination of electrode–electrolyte remains a great challenge. It has been demonstrated that carbon coating layer–modified electrode materials present improved rate performance attributed to carbon coating layer can significantly enhance the electronic conductivity of electrode materials, as well as good stability attributed to small volume change during Li^+ ion insertion/extraction [42, 44, 63–66].

Graphene is a flat monolayer of sp^2-bonded carbon atoms tightly packed into a two-dimensional honeycomb lattice, which is predicted to have a range of unusual properties, such as extraordinary electronic transport properties [67], elastic properties [68], etc. Graphene-based applications are regarded to come of age [69]. One possible route to make use of these properties for applications would be incorporate graphene sheets in a composite material [70, 71]. Another enthralling application is utilizing graphene in electric batteries [72, 73].

Here we introduce a simple synthesis of novel electrode material, i.e., the β-Ni(OH)$_2$@graphene composite (termed as NGC), and its use as a novel advanced positive electrode in rechargeable Ni–MH batteries and LIBs [74]. In a typical synthesis, aqueous solution of Ni(NO$_3$)$_2$ was added dropwise to the aqueous suspension of GO. After completing ion exchange, aqueous solution of equivalent NaOH was added dropwise to the above mixture. Then solid product was obtained after separation and drying, followed by hydrothermal treatment of the obtained solid product along with NaBH$_4$ at 80°C for 4 h, NGC was obtained. The TEM and AFM (Fig. 9.5) show that the sizes of β-Ni(OH)$_2$ nanoparticles on RGO sheets are in the range of 50–70 nm. The cross analysis of AFM indicates that the height of NGC is in the range of 1–16 nm and the RGO with mono- or bi-layer structure in NGC.

The electrochemical properties of as-synthesized NGC were first evaluated in Ni–MH battery. The charge–discharge curves of NGC as an active component of the positive electrode and hydrogen storage alloy as the negative electrode in Ni–MH at the current density of 60 mA/g are shown in Fig. 9.6a. After 25th cycles, the discharge curve is stable and remains constant up to 50 cycles with a discharge potential of 1.22 V. There is a long discharge platform, and the discharge specific capacity above 1.0 V is 283 mAh·g^{-1} (98% of the theoretical value of β-Ni(OH)$_2$) of 289 mAh·g^{-1} [75, 76]. Such improved discharge capacity value compared with other microscale structures can be attributed to the novel electrode material NGC, i.e., the introducing of graphene sheets into β-Ni(OH)$_2$ [51, 53, 57, 76], which can facilitate the transport and diffusion of both OH$^-$ ions and electrons in electrode, due to the good conductivity of graphene (Table 9.1).

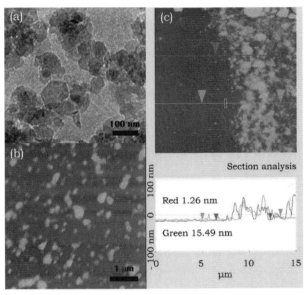

Figure 9.5 (a) TEM, (b, c) AFM images and corresponding height profile of NGC. Copyright the Royal Society of Chemistry, reproduced with permission from Ref. 74.

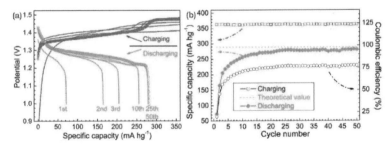

Figure 9.6 (a) The charge–discharge curves, and (b) cyclic performances of electrode fabricated with NGC at a current density of 60 mA g^{-1} in Ni–MH battery. Copyright the Royal Society of Chemistry, reproduced with permission from Ref. 74.

Table 9.1 The specific discharge capacities of electrodes fabricated with NGC, microflowers, microtubes, mesoporous β-Ni(OH)$_2$, nanotubes, and theoretical value

β-Ni(OH)$_2$ material	Current density /mA g^{-1}	Specific capacity /mA hg^{-1}	Reference
Theoretical value	~	289.1	51
NGC	60	282.5, 313.8[a]	This work
Microflowers	60	244.3	52
Microtubes	80	232.4	53
Mesoporous	70	241.5	57
Nanotubes	50	315.0[b]	76

Source: Copyright the Royal Society of Chemistry, reproduced with permission from Ref. 74.
[a]Calculated considering the weight ratio (90%) of β-Ni(OH)$_2$ in NGC.
[b]Larger than the theoretical value.

The cycle life of NGC electrode is shown in Fig. 9.6b. The curves of discharge capacity vs. cycle number present a significant increasing trend. In the total process, the charging capacities always stabilize at 361–365 mAh·g^{-1}, and the Coulombic efficiency gradually increases and keeps at 76–78%. The performances including specific capacity and stability of NGC are superior to those of most micro/nanostructure of β-Ni(OH)$_2$ [51–53, 56, 57, 76], and very close to nanotubes [76] (Table 9.1).

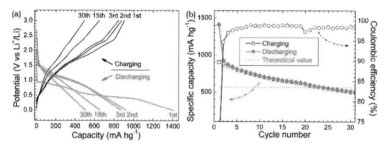

Figure 9.7 (a) The charge–discharge curves of 1–30 cycles, and (b) specific capacity retention of NGC with a current density of 200 mA g^{-1} in LIBs. Copyright the Royal Society of Chemistry, reproduced with permission from Ref. 74.

The as-synthesized NGC is used as the anode material for LIBs to study its electrochemcial properties (Fig. 9.7). Figure 9.7a shows the potential profiles for the 1st, 2nd, 3rd, 15th, and 30th cycles of the NGC/Li cell, which was charged and discharged at a constant current density of 200 mAh·g^{-1} and a fixed voltage window between 0.001–3.0 V (versus Li$^+$/Li). The first specific discharge capacity is as high as 1409 mAh·g^{-1} in the galvanostatic voltage profiles, due to the formation of a solid electrolyte interface (SEI) layer [77], by an irreversible insertion/deinsertion of Li into host structures or on Li alloying reaction (Eqs. 9.1–9.3) and possible interfacial Li storage [65, 78]. This initial irreversible capacity is more than 150% greater than that of the theoretical value of Ni(OH)$_2$ [the theoretical calculated specific capacity of Ni(OH)$_2$ is 563.3 mAh·g^{-1}, calculated through electrode reaction Eq. 9.1]. The extra capacity is probably contributed to the formation of surface polymeric layer due to the decomposition of the solvent in the electrolyte that caused the irreversible capacity loss [79, 80]. Similar phenomenon has also widely been observed in metal oxide electrodes, such as CuO nanoparticles [81], Co$_3$O$_4$ nanowires [37], Fe$_3$O$_4$ nanoparticles [82], and SnO$_2$ microspheres [83]. The load curves present continuous potential variation along with cycling. It is known that the electrode reactions in rechargeable LIBs depend on simultaneous intercalation of Li$^+$ and e$^-$ into the active intercalation host of a composite electrode. That means the NGC electrode reactions involves the formation and reduction of LiOH, as well as the redox of metal Ni nanoparticles. In the last step, a homogeneous distribution of Ni nanoparticles embedded in a LiOH matrix. The discharge capacity is found to fade to 927 mAh·g^{-1} for the 2nd cycle and 507 mAh·g^{-1} for the 30th cycle.

$$NGC + 2\ Li^+ + 2e^- \rightarrow Ni@graphene + 2\ LiOH \quad (9.1)$$

$$2\ Li^+ + 2\ e^- \rightarrow 2\ Li \quad (9.2)$$

$$NGC + 2\ Li \rightarrow 2\ LiOH + Ni@graphene \quad (9.3)$$

Figure 9.7b presents the cycling performance of the NGC. It has specific capacity 1409 mAh·g^{-1} with the Coulombic efficiency of 64.3% for the initial discharge, and 926.6 mAh·g^{-1} in capacity with 94.8% in Coulombic efficiency at the 2nd cycle, and 563.03 mAh·g^{-1} in capacity with 98.1% in Coulombic efficiency at the 5th cycle, and 507 mAh·g^{-1} in capacity with 98.5% in Coulobmic efficiency at 30th cycle. The Coulombic efficiency stays steadily above ~98% since 5th cycle.

We also synthesized Co_3O_4@RGO (CGC) [84], Fe_3O_4-C-RGO (GCF) [85], SnO_2-polyaniline-RGO (SPG) [86], and study their applications as anodes in LIBs.

In a typical synthesis of CGC, the aqueous solution of $CoCl_2$ was added into an aqueous suspension of GO under stirring. After completing ion exchange, an aqueous solution of NaOH (1.0 equivalent) was added dropwise into above mixture, followed by adding an aqueous solution of H_2O_2 with stirring. The as-obtained power was separated in deionized water for stirring and ultrasonic treatment, followed by adding $NaBH_4$ into the above mixture, and then heated. After washing the resulted black solid, CGC was obtained.

The morphology of CGC is shown in Fig. 9.8. We can find that the sizes of most Co_3O_4 nanoparticles anchored on graphene are in the range of 10–30 nm. AFM image provides an evidence of the dispersion of Co_3O_4 nanoparticles onto RGO sheets. The cross section analysis demonstrated that a height of ~1.5 nm for the graphene and a height of 10–50 nm for Co_3O_4 nanoparticles, respectively.

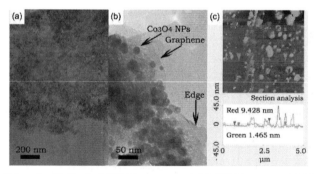

Figure 9.8 (a, b) Typical TEM images, and (c) AFM images of CGC. Inset in (c) is cross section analysis of CGC. Copyright American Chemical Society, reproduced with permission from Ref. 84.

Figure 9.9a shows charge/discharger curves of CGC at a current density of 200 mAh·g^{-1}. The first discharge specific capacity of CGC is about 1276 mAh·g^{-1}, larger than that of its theoretical capacity of 838 mAh·g^{-1} [84]. This phenomenon is attributed to the formation of SEI layer. Figure 9.9b shows the cycling performance of CGC at a current density of 0.2 C (1 C = 1000 mA·g^{-1}). It has specific capacity of 941 mAh·g^{-1} at the 2nd cycle fading from 1276 mAh·g^{-1} at 1st cycle, and the capacity remains at 740 mAh·g^{-1} after 60 cycles, corresponding to 83.1%

of the theoretical prediction capacity of CGC. Figure 9.9c shows the cycling performance of CGC electrode at different charge/discharge rates after a slow charge and hold at 3 V to fully charge the material. At a 0.2 C rate (corresponding to a time of 5 h to fully discharge the capacity), the CGC discharge to an average capacity of 909.518 mAh·g^{-1}, while the CGC reaches about 216.788 mAh·g^{-1} at the highest rate tested (5 C), corresponding to a time 720 s to fully discharge the capacity. The capacity decreased stepwise accompanying with the rate increase. When the rate returned to 0.2 C, the material discharged to an average capacity of 653.446 mAh·g^{-1}. A control experiment was carried out with Co_3O_4@graphite composites as anode material in LIBs at a rate of 0.2 C, which exhibits a much lower discharge capacity than those of CGC (Fig. 9.9d), suggesting that the superior performance of CGC over Co_3O_4@graphite due to the graphene in CGC.

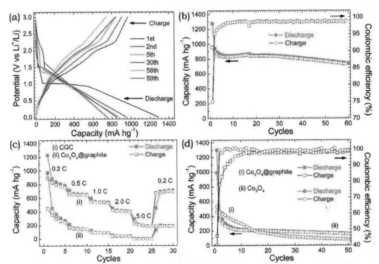

Figure 9.9 (a) Charge–discharge curves of CGC, (b) cyclic performances of electrodes fabricated by CGC at current rate of 0.2 C (1 C = 1000 mA·g^{-1}, corresponding to the full discharge in 5 h, a rate of n C corresponds to the full discharge in 1/n h), (c) the charge-discharge performances of CGC and Co_3O_4@graphite at various current rates, and (d) cyclic performances of electrodes fabricated by Co_3O_4@graphite and Co_3O_4 at current rate of 0.2 C. The weight of CGC in the working electrode was used to estimate the specific discharge capacity of the battery. Copyright American Chemical Society, reproduced with permission from Ref. 84.

Developing novel electrode materials to obtain high-performance in LIBs is a great challenge. Obviously, it regarded as an approach to synthesize hybrid materials composed of conductive matters and/or wrapped within high conductive matters [40, 87–89]. If the active materials were wrapped in conductive matters such as carbon shell, the carbon shell can effectively avoid the morphological changes of electrode during the charging/discharging process, further avoid the pulverization problem. On the other hand, the carbon shell wrapped electrode materials can prevent the agglomeration of graphene. The approach to synthesize three-dimensional (3D) composite electrode materials by encapsulating the active electrode materials in carbon shell and anchoring on conductive graphene sheets could be expected to enhance the electrochemical performance in LIBs. Fe_3O_4 is a promising candidate for electrode material, due to its high theoretical capacity and low expense [90, 91]. So we introduce a facile method to construct a 3D composite composed of RGO sheets anchored with carbon shells wrapped within Fe_3O_4 nanoparticles (termed as GCF) (Scheme 9.2).

Scheme 9.2 The three-dimensional structure of the GCF. Copyright the Royal Society of Chemistry, reproduced with permission from Ref. 85.

The morphology of GFC was characterized by TEM, HRTEM, and AFM (Fig. 9.10). The sizes of Fe_3O_4 particles wrapped in carbon shells (FOG) on RGO are in the range of 10–20 nm (Figs. 9.10a–c). The thickness of carbon shell is 2–3 nm (Figs. 9.10c,d). The lattice fringes with a period of ∼0.297 nm corresponding to (220) lattice planes of Fe_3O_4 were observed [92, 93]. AFM analysis demonstrated that the heights of FOG are in the range of 10–20 nm, corresponding to the diameters of FOG, which are homogeneously anchored on the RGO sheets (Figs. 9.10e,f).

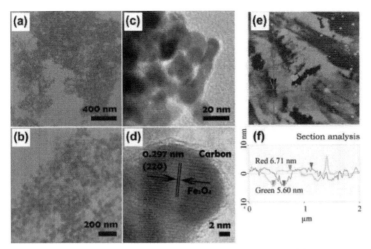

Figure 9.10 (a, b) TEM images, (c, d) HRTEM images, (e) AFM image of the GCF, and (f) its section analysis. Copyright the Royal Society of Chemistry, reproduced with permission from Ref. 85.

The as-synthesized GFC are used as anode material for LIBs to study the electrochemical properties (Fig. 9.11). Figure 9.11a shows the potential profiles for the 1st, 2nd, 5th, 10th, 50th, and 100th cycles of the GFC/Li cell at 0.2 C (1 C = 1000 mA·g^{-1}). The first specific discharge of GFC is as high as 1426 mAh·g^{-1}, which exceeds the theoretical capacity of Fe$_3$O$_4$ (926 mAh·g^{-1}) [94]. And the specific discharge of GFC at 2nd, 3rd, and 4th cycles of GFC are 952, 899.4, and 874.5 mAh·g^{-1}, respectively, while the specific capacity of 100th cycle is 842.7 mAh·g^{-1} which is 96.4% of the discharge capacity in the 4th cycle (Fig. 9.11b). The cycling response at various rates is shown in Fig. 9.11c, which shows the discharge of the material at various rates, ranging from 0.2 C to 5 C, then back to 0.2 C, after a slow charge and hold at 3 V to fully charge the material, which presents a decrease trend with increasing rate from 0.2 C to 5 C (Fig. 9.11c). A control experiment was carried out by using commercial Fe$_3$O$_4$ as anode material in LIBs, which presents much substantially lower capacities than those of GFC (Fig. 9.11d). The enhanced electrochemical performance of GFC to commercial Fe$_3$O$_4$ is attributed to the unique 3D structure of GFC, which can restrain the pulverization problem, as well as the excellent conductivity of carbon shell and RGO, especially, the stable close structure of carbon shells.

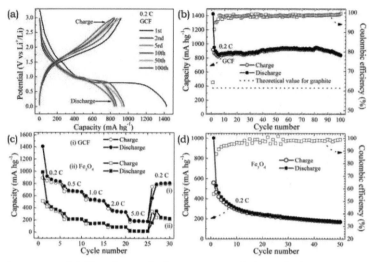

Figure 9.11 (a) The charge-discharge curves of electrode fabricated with the GCF at current rate of 0.2 C (1 C = 1000 mA·g^{-1}, corresponding to the full discharge in 1 h, a rate of n C corresponds to the full discharge in $1/n$ h), (b) Cyclic performances of the GCF at current rate of 0.2 C (c) the charge-discharge performances of the GCF at various current rates, and (d) Fe$_3$O$_4$ at current rate of 0.2 C. The weight of the GCF in the working electrode was used to estimate the specific discharge capacity of the battery. Copyright the Royal Society of Chemistry, reproduced with permission from Ref. 85.

SnO$_2$-based nanostructured materials are important anode materials in LIBs, due to their high theoretical capacity (782 mAh·g^{-1}), low discharge potential, and low toxicity [42, 66, 87, 95–99]. However, there is a severe problem that should be overcome for their practical applications, i.e., pulverization problem generated by great strain in the host materials in LIBs [95]. It is known that there is a molar volume change of 311% caused by the alloying reaction between Sn with Li$^+$ during Li$^+$ insertion process to form Li$_{4.4}$Sn [87], which can lead to a rapid reduction in capacity [34]. In order to overcome this problem, we design SnO$_2$-polyaniline-reduced graphene oxide (SPG, Scheme 9.3) nanostructured material used as anode in LIBs. Polyaniline (PANI), with theoretical capacity ranging from approximately 100 to 140 mAh·g^{-1}, is an excellent electronically conducting polymer material that has been used in batteries, because it can be oxidized and reduced at very high rates [100] and offers excellent mechanical stability, good electrical conductivity, and ease of processing [101]. The structure of SPG can stabilize the electrochemical performance besides its excellent electronic conductivity.

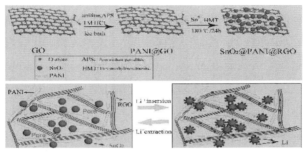

Scheme 9.3 Synthesis and structure of SnO$_2$@PANI@RGO (SPG) hybrid nanostructures. Enhanced cyclability via SnO$_2$@PANI@RGO 3D flexible structure with various pores, and Li$^+$ can be intercalated into both SnO$_2$ and PANI@RGO. Copyright the Royal Society of Chemistry, reproduced with permission from Ref. 86.

The structure and morphology of SPG was characterized by TEM, HRTEM, and AFM (Fig. 9.12), which show irregularly shaped individual sheets larger than 1μm in size (Figs. 9.12a–c) and averaging 4.66 nm in height (Figs. 9.12e,f). The PANI nanoplates are ~12 nm in size (Figs. 9.12a,b). Figures 9.12c and 9.12e can demonstrate that the SnO$_2$ nanoparticles are anchored on the RGO sheets.

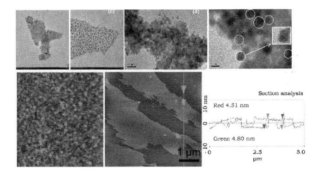

Figure 9.12 (a) and (b) TEM images of PANI@GO with different magnifying power, (c) TEM image of SPG, (d) HRTEM image of SPG, (e) AFM image of SPG, (f) the cross section analysis of SPG. Copyright the Royal Society of Chemistry, reproduced with permission from Ref. 86.

The as-synthesized SPG is used as anode material for LIBs (Fig. 9.13). Figure 9.13a shows the potential profiles for the 1st, 2nd, 5th, 25th, and 50th cycles of the SPG/Li cell at 0.2 C

(1 C = 1000 mA·g^{-1}), which exhibits the specific capacities to be 1211.4, 845.9, 798.2, 611.9, and 577.7 mAh·g^{-1}, respectively (Fig. 9.13a). The initial irreversible capacity is larger than the theoretical capacity of SPG (778.4 mAh·g^{-1}), which is attributed to the formation of SEI. We can observe three quasi-plateaus appearing at approximately 1.22, 0.55, and 0.3 V in the first discharge curve and might be ascribed to the formation of the SEI, Li$_2$O, and Li$_x$Sn (0 ≤ x ≤ 4.4) alloy, respectively [98]. The SnO$_2$ electrodes react with electrolyte in the following process: [99]

$$Li + e^- + electrolyte \leftrightarrow SEI (Li) \tag{9.4}$$

$$SnO_2 + 4Li^+ + 4e^- \leftrightarrow Sn + 2Li_2O \tag{9.5}$$

$$Sn + xLi^+ + xe^- \leftrightarrow Li_xSn\ (0 \leq x \leq 4.4) \tag{9.6}$$

The introduction of PANI@RGO nanostructure in the SPG plays an important role; i.e., it can increase the electrode conductivity and function as a buffer layer to hold stress caused by the volume change.

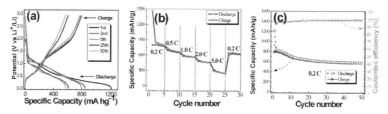

Figure 9.13 (a) Charge/discharge profiles for the SPG between 1 mV and 3 V at a rate of 0.2 C, (b) Cycle performance of the SPG electrode at various charge/discharge rates, (c) Cyclic performances for the SPG at 0.2 C: charge and discharge capacities as well as Coulombic efficiency as a function of cycle number. Copyright the Royal Society of Chemistry, reproduced with permission from Ref. 86.

The cycling response at various rates is present in Fig. 9.13b, which shows the discharge of the material at various rate after a slow charge and hold at 3 V to fully charge the material, which presents a decrease trend with increasing rate from 0.2 C to 5 C (Fig. 9.13b). When the rate is returned back to 0.2 C, the SPG discharges a capacity of 624.04 mAh·g^{-1} (i.e., 80.2% of the theoretical capacity of SPG). This result suggests that the structure of SPG remains stable even under high rate cycling. The cycling performance of SPG is shown in Fig. 9.13c, which measured the long-cycle characteristics up to 50 cycles at a current rate of 0.2 C. We can find that the SPG discharges 573.6 mAh·g^{-1} after 50 cycles, which amounts to ~73.7% of its theoretical value with Coulombic efficiency of 99.26%.

The SPG exhibits excellent electrochemical performance as anode in LIBs, due to its structures. Firstly, the Sn nanoparticles can reduce the pulverization during the charge/discharge processes, because the Sn nanoparticles are separated by the PANI@RGO layers. Secondly, the Sn nanoparticles will attach in the surface of the PANI@RGO layers, which is helpful for electron and lithium ion conduction. Thirdly, the specific structure of SPG could provide more reactive sites on the surface, and the SnO$_2$ nanoparticles provide a short diffusion length for Li$^+$ insertion, which could enhance the charge transfer and electron conduction.

A control experiment was carried out with RGO as the anode material in LIBs at a rate of 0.2 C under the same conditions as those of SPG, which shows a specific capacity of 147.2 mAh·g^{-1} at the 50th cycle, only amounting to 25.7% of that of SPG, suggesting that the capacity contribution of RGO is low (Fig. 9.14).

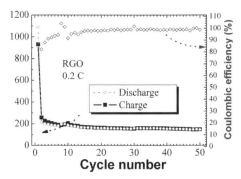

Figure 9.14 Cyclic performances for RGO at 0.2 C: charge and discharge capacitances as well as Coulombic efficiency as a function of cycle number. Copyright the Royal Society of Chemistry, reproduced with permission from Ref. 86.

9.4 Removal of Dye from Water by Cu$_2$O@Graphene

It is known that graphene owns large specific surface area (SSA) 2630 m^2·g^{-1} (calculated value) [7], and excellent elastic properties (second-and third-order elastic stiffness of 340 Newton per meter ($N \cdot m^{-1}$) and -690 $N \cdot m^{-1}$, respectively,) and intrinsic strength (breaking strength of 42 $N \cdot m^{-1}$) [68]. Here we introduced graphene-based hybrids, i.e., Cu_2O nanoparticles anchored on reduced graphene oxide (termed as CGC), afforded a high adsorption capacity for dyes such as rhodamine B (RhB) and methylene blue (MB). A reactive filtration film was assembled with the CGC and applied to remove dye contaminants from waste water [102].

The morphology of the CGC was studied by TEM and AFM (Fig. 9.15). We can observe the sizes of Cu_2O nanoparticles anchored on RGO are in the range of 5–8 nm (Figs. 9.15a,b). AFM analysis indicates the RGO height is 1–2 nm, and the Cu_2O height (i.e., its diameter) is 5–8 nm, which matches well with TEM observation (Figs. 9.15c,d). The N_2 adsorption–desorption isotherms and corresponding Barrett–Joyner–Halenda (BJH) pore-size distribution plot of CGC suggest the SSA of CGC is 90.2 $m^2 \cdot g^{-1}$ with the average pore width of 7.3 nm (Fig. 9.16). The total pore volumes with pore size from 1.7 to 300 nm are 0.25 $cm^3 \cdot g^{-1}$ for CGC (Fig. 9.17). The lower SSA than that of graphene (2630 $m^2 \cdot g^{-1}$, calculated value) [7], but larger than that of pure Cu_2O (SSA = 24.6 $m^2 \cdot g^{-1}$), is attributed to the composite of CGC (10 wt.% in RGO and 90 wt.% in Cu_2O).

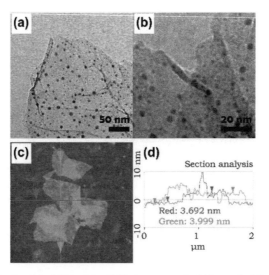

Figure 9.15 (a, b) TEM, (c) AFM images of CGC, and (d) the cross section analysis. Copyright the Royal Society of Chemistry, reproduced with permission from Ref. 102.

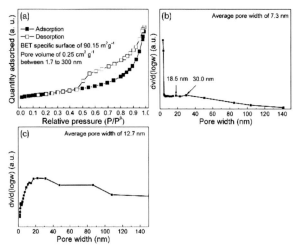

Figure 9.16 (a) N_2 adsorption–desorption isotherms at 77 K, and the pore width distribution of CGC calculated from (b) desorption branch and (c) adsorption branch. Copyright the Royal Society of Chemistry, reproduced with permission from Ref. 102.

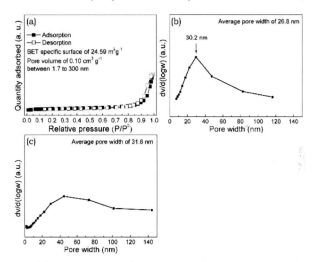

Figure 9.17 (a) N_2 adsorption–desorption isotherms at 77 K, and the pore width distribution of Cu_2O calculated from (b) desorption branch and (c) adsorption branch. Copyright the Royal Society of Chemistry, reproduced with permission from Ref. 102.

The adsorption ability of CGC was studied by measuring the UV-vis spectra after RhB and MB solution was mixed with CGC, which clearly demonstrates the CGC owns the significant adsorption ability for organic dyes RhB (Fig. 9.18) compared with GO and Cu_2O (Fig. 9.19), and MO (Fig. 9.20) from aqueous solutions. The filtering of RhB may result from physical adsorption on the RGO surface [103, 104], due to larger SSA of CGC.

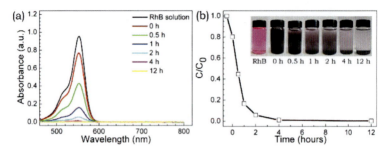

Figure 9.18 (a) VU–Vis spectra, of the original (1.0 × 10^{-5} M, 20 mL) and after treatment in RhB solutions with CGC (100 mg) for various times, and (b) C/C_0 curve of RhB concentration versus time. The insets in (b) are photo images of corresponding solution. Copyright the Royal Society of Chemistry, reproduced with permission from Ref. 102.

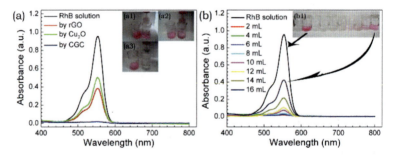

Figure 9.19 VU–Vis spectra and pictures of (a) the original and filtered RhB solutions with RGO (Inset a1), Cu_2O (Inset a2), and CGC (Insets a3), and (b) filtrated collected at different filter amounts of RhB aqueous solution. The amounts of adsorbents are 100 mg. Copyright the Royal Society of Chemistry, reproduced with permission from Ref. 102.

Also we studied the electrochemical properties of CGC used as electrode in supercapacitors (Fig. 9.21), due to its large SSA and high conductivity. As shown in Fig. 9.21a, the average specific capacitance values are 31.0, 26.0, and 24.0 F/g, corresponding to the discharge currents of 100, 200, and 400 mA/g, respectively. The CGC exhibits good cycling stability without obvious fading even after 5000 cycles (24.0 F/g at 5000th cycle). In a control experiment, we measured the galvanostatic discharge capacity of

Cu$_2$O at a discharge current of 200 mA/g condition, which presents a specific capacity of 8.7 F/g (Fig. 9.21b). These results demonstrate that the interaction between RGO sheets and Cu$_2$O nanoparticles.

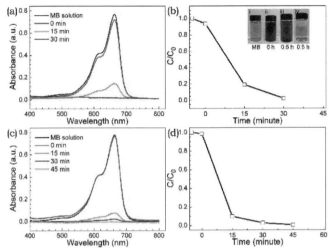

Figure 9.20 (a) VU–Vis spectra, of the original (2.0 × 10^{-5} M, 20 mL) and after treatment in MB solutions with CGC (100 mg) for various times, and (b) C/C_0 curve of MB concentration versus time. The insets in (b) are photo images of corresponding solution, and (iv) Water after adsorbents removed. (c) VU–Vis spectra, of the original (2.0 × 10^{-5} M, 20 mL) and after treatment in MB solutions with reused CGC (100 mg) for various times, and (d) C/C_0 curve of MB concentration versus time. The reused CGC was obtained from the used CGC by washing with ethanol (10 mL). Copyright the Royal Society of Chemistry, reproduced with permission from Ref. 102.

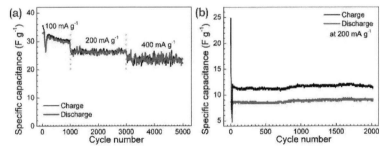

Figure 9.21 (a) The discharge curves of CGC electrodes at various current rates, and (b) the cycle performances of Cu$_2$O electrode at current rate of 200 mA·g^{-1}. Copyright the Royal Society of Chemistry, reproduced with permission from Ref. 102.

9.5 Summary

The past several years have witnessed significant advances in the synthesis and applications of graphene-based nanocomposites. Organic- and inorganic structure–modified graphene have been synthesized by covalent reaction or chemical deposition after charge-exchange processes. Several applications relying on the unique properties of graphene-based nanocomposites have been reviewed here, such as fluorescence quenching, electrochemical properties as well as applied in energy storage devices, and adsorption properties for applying in dealing with waste water. It clearly is an exciting age to research graphene-based materials.

Acknowledgments

Financial support from 973 Program (2013CB933800) and the National Natural Science Foundation of China (21271112, 21231005) are acknowledged. I also thank my co-workers Prof. G. Yin, Mr. X. Wu, Dr. B. Li, Prof. M. Qu, Mr. J. Shao, Mr. H. Zheng, Mr. Y. Lu, Mr. J. Yin, Mr. G. Li, Prof. D. Qian, Mr. R. Liang, and Miss J. Zhang.

References

1. Novoselov, K. S., Geim, A. K., Morozov, S. V., Jiang, D., Zhang, Y., Dubonos, S. V., Grigorieva, I. V., and Firsov, A. A. (2004). Electric field effect in atomically thin carbon films, *Science*, **306**, pp. 666–669.

2. Ohta, T., Bostwick, A., Seyller, T., Horn, K., and Rotenberg, E. (2006). Controlling the electronic structure of bilayer graphene, *Science*, **313**, pp. 951–954.

3. Novoselov, K. S., Jiang, Z., Zhang, Y., Morozov, S. V., Stormer, H. L., Zeitler, U., Maan, J. C., Boebinger, G. S., Kim, P., and Geim, A. K. (2007). Room-temperature quantum hall effect in graphene, *Science*, **315**, p. 1379.

4. van den Brink, J. (2007). Graphene: From strength to strength, *Nat. Nanotechnol.*, **2**, pp. 199–201.

5. Li, D., Müller, M. B., Gilje, S., Kaner, R. B., and Wallace, G. G. (2008). Processable aqueous dispersions of graphene nanosheets, *Nat. Nanotechnol.*, **3**, pp. 101–105.

6. Balandin, A. A., Ghosh, S., Bao, W., Calizo, I., Teweldebrhan, D., Miao, F., and Lau, C. N. (2008). Superior thermal conductivity of single-layer graphene., *Nano Lett.*, **8**, pp. 902–907.

7. Stoller, M. D., Park, S., Zhu, Y., An, J., and Ruoff, R. S. (2008). Graphene-based ultracapacitors, *Nano Lett.*, **8**, pp. 3498–3502.

8. Stankovich, S., Dikin, D. A., Dommett, G. H. B., Kohlhaas, K. M., Zimney, E. J., Stach, E. A., Piner, R. D., Nguyen, S. T., and Ruoff, R. S. (2006). Graphene-based composite materials, *Nature*, **442**, pp. 282–286.

9. Brumfiel, G. (2009). Graphene gets ready for the big time, *Nature*, **458**, pp. 390–391.

10. Dikin, D. A., Stankovich, S., Zimney, E. J., Piner, R. D., Dommett, G. H. B., Evmenenko, G., Nguyen, S. T., and Ruoff, R. S. (2007). Preparation and characterization of graphene oxide paper, *Nature*, **448**, pp. 457–460.

11. Meyer, J. C., Geim, A. K., Katsnelson, M. I., Novoselov, K. S., Booth, T. J., and Roth, S. (2007). The structure of suspended graphene sheets, *Nature*, **446**, pp. 60–63.

12. Räder, H. J., Rouhanipour, A., Talarico, A. M., Palermo, V., Samorì, P., and Müllen, K. (2006). Processing of giant graphene molecules by soft-landing mass spectrometry, *Nat. Mater.*, **5**, pp. 276–280.

13. Jiao, L., Wang, X., Diankov, G., Wang, H., and Dai, H. (2010). Facile synthesis of high-quality graphene nanoribbons, *Nat. Nanotechnol.*, **5**, pp. 321–325.

14. Malig, J., Jux, N., Kiessling, D., Cid, J.-J., Vázquez, P., Torres, T., and Guldi, D. M. (2011). Towards tunable graphene/phthalocyanine–PPV hybrid systems, *Angwe. Chem. Int. Ed.*, **50**, pp. 3561–3565.

15. Bayazit, M. K., and Coleman, K. (2009). Fluorescent single-walled carbon nanotubes following the 1,3-dipolar cycloaddition of pyridinium ylides, *J. Am. Chem. Soc.*, **131**, pp. 10670–10676.

16. Wu, X., Cao, H., Li, B., Yin, G. (2011). The synthesis and fluorescence quenching properties of well soluble hybrid graphene material covalently functionalized with indolizine, *Nanotechnology*, **22**, pp. 075202 (1–8).

17. Liang, Y. Y., Wu, D. Q., Feng, X. L., Müllen, K. (2009). Dispersion of graphene sheets in organic solvent supported by ionic interactions, *Adv. Mater.*, **21**, pp. 1679–1683.

18. Lomeda, J. R., Doyle, C. D., Kosynkin, D. V., Hwang, W. F., and Tour, J. M. (2008). Diazonium functionalization of surfactant-wrapped chemically converted graphene sheets, *J. Am. Chem. Soc.*, **130**, pp. 16201–16206.

19. Padwa, A., Austin, D. J., Precedo, L., and Zhi, L. (1993). Cycloaddition reactions of pyridinium and related azomethine ylides, *J. Org. Chem.*, **58**, pp. 1144–1150.

20. Allongue, P., Delamar, M., Desbat, B., Fagebaume, O., Hitmi, R., Pinson, J., and Savéant, J. M. (1997). Covalent modification of carbon surfaces by aryl radicals generated from the electrochemical reduction of diazonium salts, *J. Am. Chem. Soc.*, **119**, pp. 201–207.
21. Liu, Y., Hu, H. Y., Liu, Q. J., Hu, H. W., and Xu, J. H. (2007). Synthesis of polycyclic indolizine derivatives via one-pot tandem reactions of N-ylides with dichloro substituted α,β-unsaturated carbonyl compounds, *Tetrahedron*, **63**, pp. 2024–2031.
22. Wilson, N. R., Pandey, P. A., Beanland, R., Young, R. J., Kinloch, I. A., Gong, L., Liu, Z., Suenaga, K., Rourke, J. P., York, S. J., and Sloan, J. (2009). Graphene oxide: structural analysis and application as a highly transparent support for electron microscopy, *ACS Nano*, **3**, pp. 2547–2556.
23. Jeong, H. K., Lee, Y. P., Lahaye, R. J. W. E., Park, M. H., An, K. H., Kim, I. J., Yang, C. W., Park, C. Y., Ruoff, R. S., and Lee, Y. H. (2008). Evidence of graphitic AB stacking order of graphite oxides, *J. Am. Chem. Soc.*, **130**, pp. 1362–1366.
24. Wei, D. C., Liu, Y. Q., Zhang, H. L., Huang, L. P., Wu, B., Chen, J. Y., and Yu, G. (2009). Scalable synthesis of few-layer graphene ribbons with controlled morphologies by a template method and their applications in nanoelectromechanical switches, *J. Am. Chem. Soc.*, **131**, pp. 11147–11154.
25. Zhang, W. X., Cui, J. C., Tao, C. A., Wu, Y. G., Li, Z. P., Ma, L., Wen, Y. Q., and Li, G. T. (2009). A strategy for producing pure single-layer graphene sheets based on a confined self-assembly approach, *Angew. Chem. Int. Ed.*, **48**, pp. 5864–5868.
26. Bekyarova, E., Itkis, M. E., Ramesh, P., Berger, C., Sprinkle, M., de Heer, W. A., and Haddon, R. C. (2009). Chemical modification of epitaxial graphene: Spontaneous grafting of aryl groups, *J. Am. Chem. Soc.*, **131**, pp. 1336–1337.
27. Park, S., An, J. H., Piner, R. D., Jung, I., Yang, D. X., Velamakanni, A., Nguyen, S. T., and Ruoff, R. S. (2008). Aqueous suspension and characterization of chemically modified graphene sheets, *Chem. Mater.*, **20**, pp. 6592–6594.
28. Campos-Delgado, J., Romo-Herrera, J. M., Jia, X. T., Cullen, D. A., Muramatsu, H., Kim, Y. A., Hayashi, T., Ren, Z. F., Smith, D. J., Okuno, Y., Ohba, T., Kanoh, H., Kaneko, K., Endo, M., Terrones, H., Dresselhaus, M. S., and Terrones, M. (2008). Bulk production of a new form of sp^2 carbon: crystalline graphene nanoribbons, *Nano Lett.*, **8**, pp. 2773–2778.

29. Rao, C. N. R., Sood, A. K., Subrahmanyam, K. S., and Govindaraj, A. (2009). Graphene: the new two-dimensional nanomaterial, *Angew. Chem. Int. Ed.*, **48**, pp. 7752–7777.
30. Treossi, E., Melucci, M., Liscio, A., Gazzano, M., Samorì, P., and Palermo, V. (2009). High-contrast visualization of graphene oxide on dye-sensitized glass, quartz, and silicon by fluorescence quenching, *J. Am. Chem. Soc.*, **131**, pp. 15576–15577.
31. Lakovicz, J. R. (2006). *Principles of Fluorescence Spectroscopy*, 3rd ed. (Springer-Verlag Berlin, Germany).
32. Albani, J. R. (2007). *Principles and Applications of Fluorescence Spectroscopy*, (Blackwell Publishing, UK).
33. Andrews, J., and Jelley, N. (2007). *Energy Science: Principles, Technologies, and Impacts* (Oxford University Press, UK).
34. Tarascon, J. M., and Armand, M. (2001). Issues and challenges facing rechargeable lithium batteries, *Nature*, **414**, pp. 359–367.
35. Wagemaker, M., Kentgens, A. P. M., and Mulder, F. M. (2002). Equilibrium lithium transport between nanocrystalline phases in intercalated TiO_2 anatase, *Nature*, **418**, pp. 397–399.
36. Kang, K., Meng, Y. S., Bréger, J., Grey, C. P., and Ceder, G. (2006). Electrodes with high power and high capacity for rechargeable lithium batteries, *Science*, **311**, pp. 977–980.
37. Nam, K. T., Kim, D. W., Yoo, P. J., Chiang, C. Y., Meethong, N., Hammond, P. T., Chiang, Y. M., and Belcher, A. M. (2006). Virus-enabled synthesis and assembly of nanowires for lithium ion battery electrodes, *Science*, **312**, pp. 885–888.
38. Armand, M., and Tarascon, J. M. (2008). Building better batteries, *Nature*, **451**, pp. 652–657.
39. Kang, B., and Ceder, G. (2009). Battery materials for ultrafast charging and discharging, *Nature*, **458**, pp. 190–193.
40. Magasinki, A., Dixon, P., Hertzberg, B., Kvit, A., Ayala, J., and Yushin, G. (2010). High-performance lithium-ion anodes using a hierarchical bottom-up approach, *Nature Mater.*, **9**, pp. 353–358.
41. Wu, X. L., Jiang, L. Y., Cao, F. F., Guo, Y. G., and Wan, L. J. (2009). $LiFePO_4$ nanoparticles embedded in a nanoporous carbon matrix: superior cathode material for electrochemical energy-storage devices, *Adv. Mater.*, **21**, pp. 2710–2714.
42. Lou, X. W., Li, C. M., and Archer, L. A. (2009). Designed synthesis of coaxial SnO_2@carbon hollow nanospheres for highly reversible lithium storage, *Adv. Mater.*, **21**, pp. 2536–2539.

43. Zhang, H. X., Feng, C., Zhai, Y. C., Jiang, K. L., Li, Q. Q., and Fan, S. S. (2009). Cross-stacked carbon nanotube sheets uniformly loaded with SnO_2 nanoparticles: a novel binder-free and high-capacity anode material for lithium-ion batteries, *Adv. Mater.*, **21**, pp. 2299–2304.

44. Martin, C., Alias, M., Christien, F., Crosnier, O., Bélanger, D., and Brousse, T. (2009). Graphite-grafted silicon nanocomposite as a negative electrode for lithium-ion batteries, *Adv. Mater.*, **21**, 4735–4741.

45. Park, M. H., Kim, K., Kim, J., and Cho, J. (2010). Flexible dimensional control of high-capacity Li-ion-battery anodes: from 0D hollow to 3D porous germanium nanoparticle assemblies, *Adv. Mater.*, **22**, pp. 415–418.

46. Martha, S. K., Grinblat, J., Haik, O., Zinigrad, E., Drezen, T., Miners, J. H., Exnar, I., Kay, A., Markovsky, B., and Aurbach, D. (2009). $LiMn_{0.8}Fe_{0.2}PO_4$: an advanced cathode material for rechargeable lithium batteries, *Angew. Chem. Int. Ed.*, **48**, pp. 8559–8563.

47. Yu, Y., Gu, L., Wang, C., Dhanabalan, A., van Aken, P. A., and Maier, J. (2009). Encapsulation of Sn@carbon nanoparticles in bamboo-like hollow carbon nanofibers as an anode material in lithium-based batteries, *Angew. Chem. Int. Ed.*, **48**, pp. 6485–6489.

48. Li, H. Q., Wang, Y. G., Na, H. T., Liu, H. M., and Zhou, H. S. (2009). Rechargeable Ni-Li battery integrated aqueous/nonaqueous system, *J. Am. Chem. Soc.*, **131**, pp. 15098–15099.

49. Tang, Y., and Li, W. (2007). *Nickel Metal Hydride (Ni–MH) Battery* (Chemical Industry Press, Beijing, China).

50. Huang, K., Wang, Z., and Liu, S. (2007). *Principle and Key Technique of Li–Ion Battery* (Chemical Industry Press, Beijing, China).

51. Linden, D., and Reddy, T. B. (2002). *Handbook of Batteries*, 3rd ed. (The McGraw–Hill Companies, Inc., USA).

52. Cao, H. Q., Zheng, H., Liu, K. Y., and Warner, J. H. (2010). Bioinspired peony-Like β-$Ni(OH)_2$ nanostructures with enhanced electrochemical activity and superhydrophobicity, *ChemPhysChem*, **11**, pp. 489–494.

53. Tao, F. F., Guan, M. Y., Zhou, Y. M., Zhang, L., Xu, Z., and Chen, J. (2008). Fabrication of nickel hydroxide microtubes with micro- and nano-scale composite structure and improving electrochemical performance, *Cryst. Growth Des.*, **8**, pp. 2157–2162.

54. He, X. M., Li, J. J., Cheng, H. W., Jiang, C. Y., and Wan, C. R. (2005). Controlled crystallization and granulation of nano-scale β-$Ni(OH)_2$ cathode materials for high power Ni-MH batteries, *J. Power Sources*, **152**, pp. 285–290.

55. Snook, G. A., Duffy, N. W., and Pandolfo, A. G. (2007). Assessment of coal and graphite electrolysis on carbon fiber electrodes, *J. Power Sources*, **168**, pp. 513–523.

56. Kiani, M. A., Mousavi, M. F., and Ghasemi, S. (2010). Size effect investigation on battery performance: Comparison between micro- and nano-particles of β-Ni(OH)$_2$ as nickel battery cathode material, *J. Power Sources*, **195**, pp. 5794–5800.

57. Li, B. J., Ai, M., and Xu, Z. (2010). Mesoporous β-Ni(OH)$_2$: synthesis and enhanced electrochemical performance, *Chem. Commun.*, **46**, pp. 6267–6269.

58. Ovshinsky, S. R., Fetcenko, M. A., and Ross, J. (1993). A Nickel metal hydride battery for electric vehicles, *Science*, **260**, pp. 176–181.

59. Winter, M., and Brodd, R. J. (2004). What are batteries, fuel cells, and supercapacitors?, *Chem. Rev.*, **104**, pp. 4245–4270.

60. Wu, F., and Yang, H. (2009). *Green Secondary Batteries: New Systems and Research Techniques* (Science Publish Press, Beijing, China).

61. Ni, X. M., Zhao, Q. B., Zhang, Y. F., Song, J. M., Zheng, H. G., and Yang, K. (2006). Large scale synthesis and electrochemical characterization of hierarchical β-Ni(OH)$_2$ flowers, *Solid State Sci.*, **8**, pp. 1312–1316.

62. Kim, J. K., and Manthiram, A. (1997). A manganese oxyiodide cathode for rechargeable lithium batteries, *Nature*, **390**, pp. 265–267.

63. Bonino, F., Brutti, S., Reale, P., Scrosati, B., Gherghel, L., Wu, J., and Müllen, K. (2005). A disordered carbon as a novel anode material in lithium-ion cells, *Adv. Mater.*, **17**, pp. 743–746.

64. Wang, H., Abe, T., Maruyama, S., Iriyama, Y., Ogumi, Z., and Yoshikawa, K. (2005). A disordered carbon as a novel anode material in lithium-ion cells, Ordered two- and three-dimensional arrays self-assembled from water-soluble nanocrystal–micelles, *Adv. Mater.*, **17**, pp. 2857–2590.

65. Du, N., Zhang, H., Chen, B., Wu, J., Ma, X., Liu, Z., Zhang, Y., Yang, D., Huang, X., and Tu, J. (2007). Porous Co$_3$O$_4$ nanotubes derived from Co$_4$(CO)$_{12}$ clusters on carbon nanotube templates: A highly efficient material for Li-battery applications, *Adv. Mater.*, **19**, pp. 4505–4509.

66. Zhang, W. M., Hu, J. S., Guo, Y. G., Zheng, S. F., Zhong, L. S., Song, W. G., and Wan, L. J. (2008). Tin-nanoparticles encapsulated in elastic hollow carbon spheres for high-performance anode material in lithium-ion batteries, *Adv. Mater.*, **20**, pp. 1160–1165.

67. Novoselov, K. S., Geim, A. K., Morozov, S. V., Jiang, D., Katsnelson, M. I., Grigorieva, I. V., Dubonos, S. V., and Firsov, A. A. (2005). Two-

dimensional gas of massless Dirac fermions in graphene, *Nature*, **438**, pp. 197–200.

68. Lee, C., Wei, X., Kysar, J. W., and Hone, J. (2008). Measurement of the elastic properties and intrinsic strength of monolayer graphene, *Science*, **321**, pp. 385–388.

69. Geim, A. K., and Novoselov, K. S. (2007). The rise of graphene, *Nature Mater.*, **6**, pp. 183–191.

70. Ruoff, R. (2008). Graphene: calling all chemists, *Nat. Nanotechnol.*, **3**, pp. 10–11.

71. Zhu, J. (2008). Graphene production: New solutions to a new problem, *Nature Nanotechnol.*, **3**, pp. 528–529.

72. Yang, S. B., Feng, X. L., Wang, L., Tang, K., Maier, J., and Müllen, K. (2010). Graphene-based nanosheets with a sandwich structure, *Angew. Chem. Int. Ed.*, **49**, pp. 4795–4799.

73. Kim, K. S., Zhao, Y., Jang, H., Lee, S. Y., Kim, J. M., Kim, K. S., Ahn, J. H., Kim, P., Choi, J. Y., and Hong, B. H. (2009). Large-scale pattern growth of graphene films for stretchable transparent electrodes, *Nature*, **457**, pp. 706–710.

74. Li, B., Cao, H., Shao, J., Zheng, H., Lu, Y., and Qu, M. (2011). Improved performances of β-Ni(OH)$_2$@reduced-graphene-oxide in Ni-MH and Li-ion batteries, *Chem. Commun.*, **47**, pp. 3159–3161.

75. Linden, D., and Reddy, T. B. (2002). *Handbook of Batteries*, 3rd ed. (McGraw-Hill, USA).

76. Cai, F. S., Zhang, G. Y., Chen, J., Gou, X. L., Liu, H. K., and Dou, S. X. (2004). Ni(OH)$_2$ tubes with mesoscale dimensions as positive-electrode materials of alkaline rechargeable batteries, *Angew. Chem. Int. Ed.*, **43**, pp. 4212–4216.

77. Wu, F. (2007). *Green Rechargeable Batteries and New System Research Process* (Science Publisher, Beijing China).

78. Poizot, P., Larunelle, S., Grugeon, S., and Tarascon, M. (2002). Rationalization of the low-potential reactivity of 3d-metal-based inorganic compounds toward Li, *J. Electrochem. Soc.*, **149**, pp. A1212–A1217.

79. Laruelle, S., Grugeon, S., Poizot, P., Dolle, M., Dupont, L., and Tarascon, J. M. (2002). On the origin of the extra electrochemical capacity displayed by MO/Li cells at low potential, *J. Electrochem. Soc.*, **149**, pp. A627–A634.

80. Binotto, G., Larcher, D., Prakash, A. S., Urbina, R. H., Hegde, M. S., and Tarascon, J. M. (2007). Synthesis, characterization, and Li-

electrochemical performance of highly porous Co_3O_4 powders, *Chem. Mater.*, **19**, pp. 3032–3040.

81. Poizot, P., Laruelle, S., Grugeon, S., Dupont, L., and Tarascon, J. M. (2000). Nano-sized transition-metal oxides as negative-electrode materials for lithium-ion batteries, *Nature*, **407**, pp. 496–499.

82. Ni, S., Wang, X., Zhou, G., Yang, F., Wang, J., Wang, Q., and He, D. (2009). Hydrothermal synthesis of Fe_3O_4 nanoparticles and its application in lithium ion battery, *Mater. Lett.*, **63**, pp. 2701–2703.

83. Yu, Y., Chen, C. H., and Shi, Y. (2007). A tin-based amorphous oxide composite with a porous, spherical, multideck-cage morphology as a highly reversible anode material for lithium-ion batteries, *Adv. Mater.*, **19**, pp. 993–997.

84. Li, B., Cao, H., Shao, J., Li, G., Qu, M., and Yin, G. (2011). Co_3O_4@graphene composites as anode materials for high-performance lithium ion batteries, *Inorg. Chem.*, **50**, pp. 1628–1632.

85. Li, B., Cao, H., Shao, J., and Qu, M. (2011). Enhanced anode performances of the Fe_3O_4-carbon-rGO three dimensional composite in lithium ion batteries, *Chem. Commun.*, **47**, pp. 10374–10376.

86. Liang, R., Cao, H., Qian, D., Zhang, J., and Qu, M. (2011). Designed synthesis of SnO_2-polyaniline-reduced graphene oxide nanocomposites as an andoe material for lithium-ion batteries, *J. Mater. Chem.*, **21**, pp. 17654–17657.

87. Chiang, Y. M. (2010). Building a better battery, *Science*, **330**, pp. 1485–1486.

88. Lima, M. D., Fang, S. L., Lepró, X., Lewis, C., Ovalle-Robles, R., Carretero-González, J., Castillo-Martínez, E., Kozlov, M. E., Oh, J., Rawat, N., Haines, C. S., Haque, M. H., Aare, V., Stoughton, S., Zakhidov, A. A., and Baughman, R. H. (2011). Biscrolling nanotube sheets and functional guests into yarns, *Science*, **331**, pp. 51–55.

89. Arstrong, A. R., Lyness, C., Panchmatia, P. M., Islam, M. S., and Bruce, P. G. (2011). The lithium intercalation process in the low-voltage lithium battery anode $Li_{1+x}V_{1-x}O_2$, *Nat. Mater.*, **10**, pp. 223–229.

90. Taberna, P. L., Mitra, S., Poizot, P., Simon, P., and Tarascon, J. M. (2006). High rate capabilities Fe_3O_4-based Cu nano-architectured electrodes for lithium-ion battery applications, *Nat. Mater.*, **5**, pp. 567–573.

91. Ban, C. M., Wu, Z. C., Gillaspie, D. T., Chen, L., Yan, Y. F., Blackburn, J. L., and Dillon, A. C. (2010). Nanostructured Fe_3O_4/SWNT electrode: Binder-free and high-rate Li-ion anode, *Adv. Mater.*, **22**, pp. E145–E149.

92. Wang, J. Z., Zhong, C., Wexler, D., Idris, N. H., Wang, Z. X., Chen, L. Q., and Liu, H. K. (2011). Graphene-encapsulated Fe_3O_4 nanoparticles with 3D laminated structure as superior anode in lithium ion batteries, *Chem. Eur. J.*, **17**, pp. 661–667.

93. Shi, W. H., Zhu, J. X., Sim, D. H., Tay, Y. Y., Lu, Z. Y., Zhang, X. J., Sharma, Y., Srinivasan, M., Zhang, H., Hng, H. H., and Yan, Q. Y. (2011). Achieving high specific charge capacitances in Fe_3O_4/reduced graphene oxidenanocomposites, *J. Mater. Chem.*, **21**, pp. 3422–3427.

94. Paek, S. M., Yoo, E. J., and Honma, I. (2009). Enhanced cyclic performance and lithium storage capacity of SnO_2/graphene nanoporous electrodes with three-dimensionally delaminated flexible structure, *Nano Lett.*, **9**, pp. 72–75.

95. Huang, J. Y., Zhong, L., Wang, C. M., Sullivan, J. P., Xu, W., Zhang, L. Q., Mao, S. X., Hudak, N. S., Liu, X. H., Subramanian, A., Fan, H., Qi, L., Kushima, A., and Li, J. (2010). In situ observation of the electrochemical lithiation of a single SnO_2 nanowire electrode, *Science*, **330**, pp. 1515–1520.

96. Yu, Y., Chen, C. H., and Shi, Y. (2007). A tin-based amorphous oxide composite with a porous, spherical, multideck-cage morphology as a highly reversible anode material for lithium-ion batteries, *Adv. Mater.*, **19**, pp. 993–997.

97. Park, M. S., Wang, G. X., Kang, Y. M., Wexler, D., Dou, S. X., and Liu, H. K. (2007). Preparation and electrochemical properties of SnO_2 nanowires for application in lithium-ion batteries, *Angew. Chem. Int. Ed.*, **46**, pp. 750–753.

98. Wang, C., Zhou, Y., Ge, M., Xu, X., Zhang, Z., and Jiang, J. Z. (2010). Large-scale synthesis of SnO_2 nanosheets with high lithium storage capacity, *J. Am. Chem. Soc.*, **132**, pp. 46–47.

99. Li, J., Zhao, Y., Wang, N., and Guan, L. (2011). Frontal polymerizations carried out in deep-eutectic mixtures providing both the monomers and the polymerization medium, *Chem. Commun.*, **47**, pp. 5328–5330.

100. Nyholm, L., Nyström, G., Mihranyan, A., and Strømme, M. (2011). Toward flexible polymer and paper-based energy storage devices, *Adv. Mater.*, **23**, pp. 3751–3769.

101. Liu, Y., Matsumura, T., Imanishi, N., Hirano, A., Ichikawa, T., and Takeda, Y. (2005). Preparation and characterization of Si/C composite coated with polyaniline as novel anodes for Li-ion batteries, *Electrochem. Solid St.*, **8**, pp. A599–A602.

102. Li, B., Cao, H., Yin, G., Lu, Y., and Yin, J. (2011). Cu_2O@reduced graphene oxide composite for removal of contaminants from water and supercapacitors, *J. Mater. Chem.*, **21**, pp. 10645–10648.
103. Li, B., and Cao, H. (2011). ZnO@graphene composite with enhanced performance for the removal of dye from water, *J. Mater. Chem.*, **21**, pp. 3346–3349.
104. Li, H. B., Gui, X. C., Zhang, L. H., Wang, S. S., Ji, C. Y., Wei, J. Q., Wang, K. L., Zhu, H. W., Wu, D. H., and Cao, A. Y. (2010). Carbon nanotube sponge filters for trapping nanoparticles and dye molecules from water, *Chem. Comm.*, **46**, pp. 7966–7968.

Chapter 10

Recent Advances in Multidimensional Electrode Nanoarchitecturing for Lithium-Ion and Sodium-Ion Batteries

Gregorio Ortiz, Pedro Lavela, Ricardo Alcántara, and José L. Tirado

Inorganic Chemistry Laboratory, University of Córdoba,
Bldg. Marie Curie, Rabanales Campus, 14071 Córdoba, Spain

iq1ticoj@uco.es

This chapter presents a critical review on the newly developed procedures for multidimensional electrode nanoarchitecturing for Li- and Na-ion batteries. Starting from nt-TiO_2 utilization, first-row transition metal oxide nanocomposites are examined. Metal foams for 2D and 3D battery architectures and graphene–transition metal oxide heterostructures with unusual performance for battery applications are discussed.

10.1 Introduction

Mass and volume reduction to increase capacity and energy density is the key objective in current battery performance, particularly for transport applications. In addition to the increasingly difficult

Electrochemical Nanofabrication: Principles and Applications (Second Edition)
Edited by Di Wei
Copyright © 2016 Pan Stanford Publishing Pte. Ltd.
ISBN 978-981-4613-86-6 (Hardcover), 978-981-4613-87-3 (eBook)
www.panstanford.com

finding of new, high-capacity electrodes, the avoidance of binding and conductive additive weights could improve the specific capacity of lithium-ion batteries. Enhanced contact between components and reduction in diffusion lengths facilitate improved rate performances and allow fast charge and power release. Cost reduction for large-scale stationary applications in renewable energy management and distribution networks calls for lithium replacement by less expensive and more abundant sodium, with the subsequent tackling of costly copper collectors. Other needs include battery miniaturization and integration in electronic circuits, flexible batteries and origami battery concepts. All these research fronts can get substantial help from multidimensional electrode nanoarchitecturing techniques. From self-organized nanostructures to 3D printers, the possibilities of generating these nanostructures are many, and recent developments have given a clear confirmation of their diversity. For these reasons, an organized revision of the most attractive possibilities with a critical discussion on their possibilities is necessary, this being the main goal of this chapter.

10.2   Newly Developed Procedures for nt-TiO$_2$ Utilization

Since TiO$_2$ in bulk state exhibits poor electrochemical behavior, the nanoparticles are preferred for electrode materials, particularly in the form of nanotubes. The first method for synthesizing TiO$_2$ nanotubes (nt-TiO$_2$) was reported by Kasuga et al. following a chemical procedure [1]. The template method can also be used to prepare nt-TiO$_2$ [2]. Later, the electrochemical anodic oxidation method was used to obtain self-organized nt-TiO$_2$ where the nanotubes are aligned in a direction and in a compact form [3–5]. Self-organized nt-TiO$_2$ prepared by anodization is particularly useful for batteries [6].

One of the advantages of the preparation of self-organized nt-TiO$_2$ by anodization of Ti is the easy control of the nanotube microstructure, length, diameter and aspect ratio. Thus, it was found that the nanotube rate follows a parabolic-type law:

$$L^2 = kt \qquad (10.1)$$

where L is the average nanotube length, t is time and k is the rate constant [7]. The value of the rate constant increases with the anodization voltage. In principle, the specific capacity (Q) measured per area unit (mA h cm^{-2}) increases with L. However, the relationship between Q and L is not linear; also the gravimetric capacity can decrease for longer nanotube and, apparently, the extent of the reaction between the lithium ions and nt-TiO$_2$ is relatively hindered for the longer nanotubes.

The rate of nanotube growth can be increased by high-intensity ultrasonication during the anodization process [8]. Under ultrasonication, the transient curve (current density plotted vs. anodization time) is modified probably due to more rapid dissolution of the titanium foil.

The array of nt-TiO$_2$ obtained by anodization exhibits amorphous character. After annealing at around 300–550°C, amorphous titania can be transformed to polycrystalline anatase-type structure while the nanotube morphology remains nearly unchanged [9, 10]. Annealing at higher temperatures (typically about 600°C) drives to formation of rutile phase. However, some authors have found that the treatment at 450°C leads to cracks and to the subtle formation of a small amount of rutile [11]. In TEM observations the structure transformation of TiO$_2$ also can be induced by electron beam. It is worth to note that nt-TiO$_2$ obtained by hydrothermal process can collapse to other morphologies during heating [12]. It has been reported that amorphous nt-TiO$_2$ has better rate capability than anatase [13]. It seems that the surface of amorphous TiO$_2$ plays an important role for accommodation of lithium and sodium (pseudocapacity effect) [14, 15]. However, the question of what is the most suitable phase of nt-TiO$_2$ for using in batteries is still not completely elucidated. In addition, initially amorphous nt-TiO$_2$ can be in situ transformed into a cubic phase (Li$_2$Ti$_2$O$_4$) after electrochemical reaction with lithium trough an irreversible pseudoplateau at ca. 1.1 V in the first discharge [16].

On the other hand, the film of titanium used for anodization to prepare nt-TiO$_2$ can be previously deposited on a substrate such as silicon [17]. This method allows that after a complete anodization of Ti to form nt-TiO$_2$, the nt-TiO$_2$ layer still has good electrical contact with the substrate which acts like current collector.

In the literature the capacity values for nt-TiO$_2$ electrode are given using several units: mA h cm^{-2} (areal capacity), mA h g^{-1} (gravimetric capacity) or mA h cm^{-3} (volumetric capacity). For giving areal and volumetric capacities, geometrical parameters (i.e., fingerprint area and film thickness) are taken into account. The determination of the weight of the nt-TiO$_2$ amount which contains the battery electrode is usually solved by following several procedures. A very rough estimation can be to assume that the mass of active material (m) is equals to $m = dv$, where d is the density of TiO$_2$ (for anatase d = 3.9 g cm^{-3}) and v is the volume of the nanotube array film (taken from the SEM-observed thickness and area of the electrode) [18, 19]. In addition, the mass of the nanotubes can be estimated by weighing the electrode (titanium substrate and nanotubes) before and after removing the nanotubes film, for example by peeling off the nanotube film from Ti substrate using adhesives [6, 16, 20]. This last procedure can be used for foil-type electrodes but it cannot be used for non-planar geometries where it is impossible to fully remove the nanotubes from the substrate (e.g., a foam-type electrode).

Wei et al. used two-step anodization to prepare nt-TiO$_2$ for lithium ion batteries and they found that the tube walls of the nanotubes annealed at 350°C (anatase-type) contained two types of crystalline layers: the inner wall layer and the outer wall layer [21]. In addition, the same authors revealed that nt-TiO$_2$ with thinner walls give lower areal capacities. González et al. used anodization of Ti by applying a ramp of voltage (from 20 to 120 V) as a way to improve the electrochemical behavior in sodium cell [15]. The application of a ramping of voltage favors the formation of double-walled nt-TiO$_2$ (Fig. 10.1).

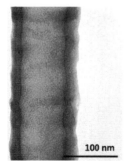

Figure 10.1 TEM of single-walled (left) and double-walled (right) nt-TiO$_2$.

A drawback of TiO$_2$ in comparison with graphite is its lower gravimetric capacity. The capacity may be increased by using composite materials that combine the stable structure of TiO$_2$ with an element or compound that exhibits higher gravimetric capacity, such as a Li-alloying element. In this sense, Tian et al. obtained the SnO$_2$@TiO$_2$ composite by the encapsulation of SnO$_2$ nanoparticles into hollow TiO$_2$ nanowires [22]. Composite electrodes of self-organized, conductive titania (TiO$_{2-x}$)-C nanotubes coated with silicon via plasma-enhanced chemical vapor deposition also have been reported [18].

Another challenge of the use of nt-TiO$_2$ in batteries is its effective application in all-solid-state microbatteries. Organic polymers such as PAN [23] and PMMA-PEO copolymer [24] have been used to cover the surface of nt-TiO$_2$. The total and homogeneous filling of the nanotube with solid electrolyte is particularly challenger. One approach may be the electrodeposition of the polymer in the nanotube. It is not easy to avoid the accumulation of polymer (or inorganic salt) on the top of the nanotube openings avoiding its effective filling.

The preparation by hydrothermal method and properties of nanotubes of Na$_2$Ti$_3$O$_7$ have been reported [25]. In contrast to the Na$_2$Ti$_3$O$_7$ bulk powder which exhibits a two-phase plateau at 0.3 V [26], the nanotube exhibits a sloping voltage–capacity curve, which suggests a pseudocapacitive mechanism. In addition, the nanotubes exhibit better rate capability than the bulk powder.

10.3 First-Row Transition Metal Oxide Nanocomposites with Unusual Performance

10.3.1 Conversion Electrodes

The optimization of lithium-ion batteries requires large efforts to overcome the limited capacity of the graphitic electrodes. Although lithium–metal alloys are able to deliver large capacities, the drastic change in volume occurring upon cycling is an important drawback limiting their extended use in commercial batteries [27]. With these precedents, Li-driven conversion reactions were regarded as a breakthrough in the research of anode materials for Li-ion batteries at the beginning of this century [28, 29]. In these reactions, transition metal oxides (TMO) could be fully reduced to their metallic state delivering high capacity values and the reaction could also be reversed at room temperature, according to the following reaction:

$$2y\text{Li}^+ + 2y\text{e}^- + \text{M}_x\text{O}_y \leftrightarrows y\text{Li}_2\text{O} + x\text{M} \qquad (10.2)$$

The product consists of nanometric metallic particles homogeneously dispersed in a Li_2O matrix. Thus, the alkaline oxide acts as an oxygen reservoir, which can be readily turned back to the metal state during the oxidation process. It is feasible because of the large surface of contact between Li_2O and M, facilitating the oxidation of M to M_xO_y on charging though in an amorphous state [30]. Due to the exchange of several electrons, the specific capacities are much higher than the theoretical capacity of graphite and similar to those of Li alloying metals.

Nevertheless, some important drawbacks hinder their extended use in commercial batteries. On the one hand, the large and flat plateau observed during the first discharge is not reproducible in successive discharges. It becomes a sloping curve shifted to a significantly higher voltage. Also, a non-negligible irreversible effect appears in the first discharge. On the other hand, a huge voltage hysteresis is observed between the discharge and charge branches resulting in poor energy efficiency. The irreversible amorphization of the initial phase during the first discharge is an inherent effect of the lithium-driven conversion reactions which induces a large difference in free energy and consequently in the equilibrium reaction potential [31].

Several alternatives have been proposed to figure it out. Thus, oxygen can be substituted by distinct counter-ions such as sulfide, phosphide, and fluoride [32–34]. It can be an interesting tool to diminish the voltage hysteresis [35]. Also, electrode materials with nanostructured design are more flexible and may accommodate the strains exerted by the drastic volume changes occurring upon cycling. A number of methods of synthesis favoring the production of nanometric particles of these TMO have been essayed. Thus notorious improvements in reversibility and diminution of the voltage hysteresis have been achieved by electrodeposition [36–38] and the use of templates [39].

A different approach resides in the advantageous combination of transition metal oxides with other materials being either electroactive or not. This second phase provides desirable properties that compensate the inherent limitations of the transition metal oxide resulting in a synergistic effect that enhances the electrochemical behavior of the composite. In this section, we will focus our attention in the validity of this approach to manufacture new electrode materials with excellent performances when undergoing lithium-driven conversion reactions.

10.3.2 Composites with Carbon Materials

The use of a carbon phase intimately mixed with the electroactive material to prepare TMO/C composites has been extensively essayed. This approach considers that the poor electronic conductivity of the Li_2O/M matrix seriously limits their electrochemical performance. The better electronic conductivity of carbon materials along with the possibility of being prepared in a large variety of particle sizes and shapes would ensure a suitable dispersion of carbon in the eventual composite. Also, the carbon matrix may suppress the aggregation of TMO particles during the synthesis and buffer the large volume changes during charge and discharge.

A single way to prepare these composites consists of milling the electroactive material with a carbon precursor as sugar, citric acid, oleic acid, etc. [40, 41]. This carbon-containing precursor is converted to a carbon matrix upon thermal treatment in inert conditions. This is a cost-effective procedure to produce MnO/C composites with a good cycling performance. It is due to the decreased particle size and carbon coating, though the initial columbic efficiency and large polarization need to be clearly improved. Nevertheless, it is suggested that reductive carbon monoxide can be evolved during the carbonization process promoting metal reduction and the undesirable appearance of impurities [41]. In the way of producing more homogeneous TMO/C composites, the dispersion of the precursors of the components before precipitation has been essayed. Thus, citric acid is a good chelating agent that ensures an efficient dispersion of the transition metal in sol-gel precursors, but also leaves a carbon residue after the thermal treatment. Yao et al. proposed the hydrothermal synthesis of $ZnFe_2O_4$/C from the transition metal chlorides resulting in mesoporous microspheres and outstanding capacity, rate capability and cycling stability. A specific reversible capacity of 1100 mA h g^{-1} after 100 cycles at the specific current of 0.05 A g^{-1} was reported [42].

An alternative approach claims that hierarchical heterostructures may overcome problems as the inevitable enhanced volume change occurring upon the conversion reaction. Thus, coating processes yield core–shell structures composed by a carbon shell surrounding the oxide. It provides a residual void space buffering the large volume change and restraining secondary agglomeration of the active material. Coating procedures commonly employ TMO particles and carbon-containing molecules as reagents. During the process, these molecules attach to the surface of the oxide. Further thermal treatment will leave a carbon shell surrounding the active material. Pyrrole easily implements the chemical oxidation polymerization for polypyrrole by means of the inducement of surfactants resulting in a homogeneous polypyrrole layer coating [43]. Alternatively, the composites can be also prepared from powdered carbon and TMO solution precursor. The coprecipitation of the preexisting carbon and the newly formed TMO yields homogeneously dispersed composites. This preparation route has the advantage of controlling the final morphology of the carbon phase. Lee et al. discovered remarkable differences between composites using carbons of distinct dimensionalities, as carbon black nanoparticles and micron-sized graphitic flakes. The latter option leads to the so-called "balls-on-plate" oxide-C configuration [44]. This morphology favors a lower temperature for the MnO formation, leading to significantly enhanced cycling stability and rate performance. This result opens new opportunities of performance improvement by considering the wide range of carbon morphologies that can be achieved by dealing with nanotubes, nanowires, or graphene among others. Thus, carbon nanotubes provide an interconnected network in which TMO particles can be embedded by several means as the decomposition of a carbonate precursor [45], electrical wire pulse method [46], etc. This carbon phase produces unique electrical and mechanical properties along with good chemical stability, which is reflected in an improved performance.

10.3.3 Composites with Other Metal Oxides

An emerging attractive approach is to directly grow integrated heterostructures containing two or more metal oxides on the surface of conducting substrates. Most of these nanoarchitectures can be directly used as anodes without the necessity of adding any binder or conductive substance. Guo et al. proposed MnO_2 electrodeposited onto ZnO nanorod array grown on Ni foil. MnO_2 burr tube shell structure leads to high electrochemical activity while the internal ZnO core support ensures good structural stability. This synergistic effect provides 1137 mA h g^{-1} after 40 cycles at a current density of 120 mA g^{-1} [47]. Ultra-thin NiO nanosheets can be uniformly grew on porous $ZnCo_2O_4$ nanowires, resulting in the formation of hierarchically architectured core/shell 1D nanoarrays. The capacities recorded after 50 cycles were 357 mA h g^{-1}, while only 152 mA h g^{-1} were determined for non-composited $ZnCo_2O_4$ [48]. Our group has recently prepared interesting vertical Fe_2O_3 nanowires grown onto self-organized TiO_2 nanotubes. Both hematite and titania demonstrated to be electrochemically active versus lithium contributing to an overall capacity of 468 µA h cm^{-2} at a rate of 25 mA cm^{-2}. The ordered arrangements of 1D nanostructures exhibited good capacity retention. Likely, this nanoarchitecturing may alleviate the stress caused by volume change during continuous discharge–charge cycles, and suppress the degradation of the material when increasing the number of cycles [49]. A suitable control of the electrodeposition parameters evidenced the possibility of achieving different morphologies as microballs. Nevertheless, a poor performance was detected for the latter samples, likely as a consequence of the lower surface of contact with the electrolyte. In a different approach, thin film electrodes consisting of co-deposited of iron and nickel oxides (NiO and Fe_2O_3) can be easily prepared by electrochemical deposition on a titanium substrate and further thermal treatment at low temperature. Cross-sectional views of the film from electron microscopy images revealed a thickness of ca. 5 mm (Fig. 10.2a). The lithium cells assembled with these free additives electrodes featured a meaningful diminution of the discharge polarization and improved coulombic efficiency as a result of the combination of nanocrystalline and thin film character of the starting material. A maximum first-discharge capacity close to 1325 µA h cm^{-2} was achieved after the first cycle (Fig. 10.2b) [37, 38].

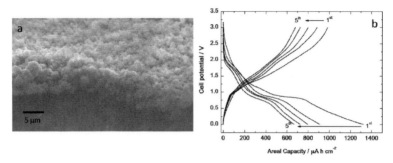

Figure 10.2 (a) Scanning electron microscopy picture of the thin film. (b) Voltage versus areal capacity plot for the first five cycles.

Interestingly, TMO may also create composites with tin compounds whose lithium reaction mechanism differs substantially because is based on the formation of reversible Li–Sn alloys. Li et al. evidenced a good electrochemical performance on CuO/SnO$_2$ composites. Tin oxide nanoparticles are coating CuO nanotubes. After discharging, the conversion reaction induces the reduction of CuO to metallic copper while retaining the tubular morphology. It helps to alleviate the volume changes of Sn during cycling. Moreover, the tubular metallic Cu improves the electrical conductivity of the electrodes [50]. More recently, ternary composites have been proposed including a carbon phase to promote the electrical conductivity and two active materials contributing positively to the overall capacity. Zhang et al. revealed that the unique structure of SnO$_2$-Fe$_2$O$_3$@C and the synergistic effect between SnO$_2$ and Fe$_2$O$_3$, are responsible for the excellent capacity over 1000 mA h g^{-1} which is maintained after 380 cycles at current density of 400 mA g^{-1} [51].

Composites containing both TMO and Li$_2$O may provide significant benefits to the electrochemical performance of lithium test cells. Chen et al. prepared CoO-Li$_2$O composite films by the electrostatic spray deposition on a conducting nickel-foam substrate. The reticular film revealed an excellent rate capability with specific capacity of 650 mA h g^{-1} when cycled at a 5 C rate. The contribution of the alkaline oxide is multiple. Thus, it hinders the particle growth of CoO, hence, maximizing the surface area of the active material and the interface with the polymeric gel-like film. Otherwise, Li_2O may act as an oxidant converting CoO into Co_3O_4 or even Co_2O_3 during charging [52].

10.3.4 Composites with Metals

An alternative approach to improve the electrochemical performance of a TMO anode is the preparation of nanostructures consisting of an inert but conducting metallic core surrounded by a shell of a selected TMO. For this purpose, special synthetic routes allow the hierarchical separation of phases. Thus, the use of electrostatic spray deposition led to the formation CoO/Co composite thin films consisted of hollow spherical particles delivering a high reversible capacity of 1182.1 mA h g^{-1} at 70th cycle and retaining 69.5% of the initial capacity at 10 C [53]. Gold nanoparticles have also been chosen due to their high electronic conductivity, their thermodynamically stable interface with Co_3O_4, and the potential to catalyze electrochemical reactions at the nanoscale. Belcher et al. propose hybrid gold–cobalt oxide wires prepared by a virus-templated synthesis. The cell delivered 94% of its theoretical capacity at a rate of 1.12 C and 65% at a rate of 5.19 C. Thus, the authors evidenced that nanoscale battery components can be designed and assembled on the basis of biological principles [54].

10.3.5 Composites with Polymers

Polymers are soft conducting materials, which may act as effective protective matrices of TMO active materials. These composites have been recently envisaged because of their flexibility and single preparation routes. Polypyrrole (PPy) performs suitable mechanical flexibility and chemical stability during the electrochemical reaction. Uniform Fe_3O_4/PPy composite nanospheres have been prepared by a hydrothermal process followed by an ultrasonication-assisted polymerization. Lithium cells assembled with this material reveal a high reversible capacity of 544 mA h g^{-1} after 300 cycles [55]. Zheng et al. also evidenced the beneficial effect of polypyrrole in CuO/PPy obtained by a surfactant-assisted reaction. A good reversible value of 613 mA h g^{-1} was recorded after 80 cycles [56]. These improvements in cycling performance and rate capability are explained in terms of the capability of the polymeric shell to prevent the dissolution and aggregation of nanomaterials.

10.4 Metal Foams for 2D and 3D Battery Architectures

Designing special 2D or 3D (Fig. 10.3) architectures is one of the most promising routes to solve several problems for many electrode materials and to improve their electrochemical behavior. Certain applications demand thin electrodes with high areal capacity values (more than 3 mA h cm^{-2} units referred to as the fingerprint area) and being able to deliver high power. Unfortunately, very thick electrodes use to give poor electrochemical performance, particularly when the electrode is prepared by using conventional slurry based processing. The optimization of the mass loading, and the thickness of the active material film, can be critical to achieve good electrochemical behavior. In order to achieve this goal and to prepare negative electrodes of rocking-chair batteries, the deposition of Li-alloying elements or compounds on a metallic substrate may be suitable. Thus, the use of substrates such as copper [57, 58], titanium [59, 60] and nickel [61] has been reported. The use of aluminum substrate is usually reserved for positive electrode. By convention, the metal substrate, which acts like current collector and structural backbone, is not included in the calculation of mass to give the gravimetric capacity of the electrode (mA h g^{-1} units). In addition, these electrodes are self-supported and do not need the use of binder additives, which usually are non-conductive polymers (PVDF, PTFE, etc.). If the relative amount of active material is very small with respect to the metal substrate, the relative cost of the substrate would be increased. The adherence and cycling performance also can be affected by the mass loading. Very stable electrochemical performance of materials cast to very thin electrodes with a capacity loading of about 0.1 mA h cm^{-2} does not guarantee their stability when tested at higher loadings [62]. In fact, in many papers reported in the literature that show excellent cycling behavior for many cycles frequently the reported capacity values are in the order of μA h cm^{-2}. For example, the use of vertically aligned nanotubes of very long length (hundreds of micrometers) deposited on a flat metal substrate may be a way to achieve high areal capacities and good cycling behavior [63].

Figure 10.3 Schematic representations of 2D (left) and 3D (right) lithium-ion batteries.

The method most commonly reported for the deposition of Li-alloying elements on metallic substrate is electroplating. Electrodeposited tin films consist of individual tin grains separated by grain boundaries and almost all tin films have columnar grain microstructure, although more work should be carried out to study the texture of the films using electron back scatter diffraction (EBSD) [64]. The preparation of binary and ternary alloys [65] or compounds is more challenger than the simple electroplating of a single element and the undesired occurrence of impurities is very common. The properties of the resulting film are strongly affected by many experimental conditions (bath composition, temperature, stirring, current density, etc.) of the electroplating method and, consequently, the procedure must be very carefully controlled and monitored. One of the problems of this method is to control the homogeneity of the electrodeposited film. Several additives can be used to improve the homogeneity of the film, such as gelatin, pyrophosphite ions, tartaric acid and glycine. In addition, the electrode can be pressed after electrodeposition to improve the adherence of the active material to the substrate.

Vapor deposition [66, 67] and sputtering [68] are alternative methods for electrodeposition on a metallic substrate. However, these methods are less common than the electrochemical ones employing flat substrates, as derived from the number of reports using the latter technique.

Metal foils, meshes and foams of controlled porosity are commercialized, for example by Goodfellow[TM]. Because of its cellular structure containing a large fraction of pores, the metal foam can be a good substrate. 3D electrodes consist of interconnected porous networks (Fig. 10.4). In some reported procedures, a certain roughness or porosity is induced on the surface of flat metallic substrates before electroplating the active material, for example by treatment with acid solution, or with more elaborated procedures such as electroless plating [69]. The pore sizes and wall structures of the metal foams can be attuned by adjusting the experimental conditions during metal deposition, in particular, the pore size is controlled by the generated hydrogen bubbles [70, 71]. In this method, the bubbles act like a sort of dynamic template. In addition, metal foam produced by chemical vapor deposition (CVD) of nickel from $Ni(CO)_4$ onto an open cell polyurethane foam has been reported [72]. The use of non-conductive substrate for electrodeposition of a metallic foam has been also reported [73].

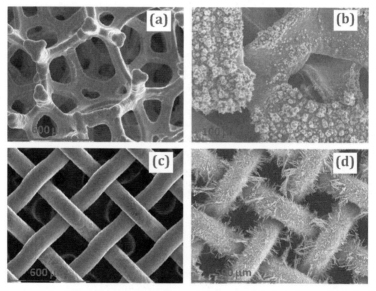

Figure 10.4 SEM micrographs of (a) metal foam, (b) $CoSn_2$ electrodeposited on metal foam, (c) wire mesh and (d) Sn electrodeposited on wire mesh.

For 2D electrodes, the optimal thickness of the active material film is limited by the resulting electrochemical behavior, mainly because of the poor response of thick electrodes at high rates.

The use of 3D substrates may be more advantageous than 2D [74]. One reason is the larger surface area of active material that is exposed to the electrolyte solution (for a fixed fingerprint area). The larger electrode/electrolyte surface contact can improve the utilization of the active material, allow a good kinetic behavior and high areal capacity. Unfortunately, the enhanced electrode/electrolyte surface contact also may increase the irreversible decomposition of the electrolyte solution and decrease the coulombic efficiency. In addition, the free space in the electrode can buffer the volume change of the active material and to improve the cycling stability. The three-dimensional character allows increasing the material loading per unit of fingerprint area. Another advantage of the 3D electrode is the reduced internal impedance. On the other hand, like the metal foam has low density, the resulting volumetric capacity of the electrode (mA h cm^{-3} units) may be decreased with respect to 2D thin electrode architecture.

Porous and rough metals substrates are being extensively studied in the last few years like electrodes for lithium and sodium batteries [65, 69, 75].

Hertzberg et al. used a Cu foil like substrate of a thin Si tube confined by a carbon shell layer [76]. This special architecture can be viewed like 1D hybrid particle on a 2D metallic substrate forming a 3D electrode. Interesting, the mass loading for this electrode was relatively high (5–8 mg cm^{-2}) and the observed gravimetric capacity was high (about 800 mA h g^{-1}).

González et al. first reported $CoSn_2$ on metal foam obtained by electrodeposition of amorphous Co–Sn alloy and subsequent annealing [77]. It was observed that low current densities and short electrodeposition times favor the formation of discrete submicrometric particles (Volmer–Weber model) and the electrochemical behavior in lithium cell was very good, while the capacity in sodium cell was lower.

On the other hand, although tin electrodeposition on metallic substrate deserved significant interest for developing batteries, it is worth to note that the anodic oxidation of a tin foil can produce porous electrodes [78]. Similarly, electrodeposited transition metals can be oxidized to form porous oxide films on flat metal substrate [38]. These porous transition metal oxides—although they are based on conversion reactions—exhibit a considerable pseudocapacitance.

Besides metals and alloys, other elements and compounds can be deposited on foams. Thus, NiO nanostructures can be directly grown on Ni foam by heating in air, and then the resulting NiO foam may perform as a conversion electrode [79]. Cheng et al. prepared a CuS 3D network starting from a copper foam and using a facile in situ melt-diffusion strategy [80]. Nickel foam has been reported as a template for deposition of few-layer graphene by CVD followed by Ni etching [81]. Then, the graphene network foam obtained can be used for the deposition of sulfur. The replacement of Ni by C decreases the mass of the electrode.

Within the field of rechargeable non-aqueous lithium-air batteries, the use of metallic foam also has been proposed. A sponge-like epsilon-MnO_2 nanostructure was obtained by direct growth of epsilon-MnO_2 on Ni foam through an electrodeposition method [82]. The observed improvement of the electrochemistry was associated with the 3D nanoporous structures, oxygen defects and the absence of side reactions related to conductive additives and binders.

One can easily visualize 1D, 2D, and 3D electrodes. However, the use of 3D microelectrodes in actual 3D batteries still remains a great challenge. To the best of our knowledge, the effective fabrication of a 3D battery using electrodes based on metal foam, or 1D electrodes, has not been yet reported, and the use of the techniques reported above for the deposition of active electrode materials and electrolyte seems to be not very suitable for this purpose. A possibility may be the use of 3D printers [83] to make these batteries.

10.5 Graphene–Transition Metal Oxide Heterostructures for Battery Applications

The graphene era has brought extraordinary findings in different fields of application, as a result of the extensive research focused in this novel material [84, 85]. Lithium-ion battery researchers also tested graphene as active electrode material and found exciting results, as reported in the literature [86, 87]. Theoretically, monolayer graphene could store lithium twice as much as graphite. The intercalation of lithium is not allowed in a monolayer; instead, graphene is capable to accommodate lithium ions by a surface mechanism without intra-lattice diffusion. Consequently graphene behaves more closely to a supercapacitor in a way that provides higher power densities than conventional Li-ion batteries, while enhancing cell capacity [88]. Regarding the working voltage of graphene electrodes versus lithium, the material behaves as a typical anode, and may form composite electrodes with other active anode materials. In addition, the conductive properties of graphene can be exploited in the composites where it can provide electrical wiring between less-conductive solids, including active cathode or anode materials. The different chemical and conductive properties of the components in most graphene-based composite electrodes allow us to describe them as heterostructures. The layered structure of graphene in conjunction with other nanostructured components may also provide interesting nanoarchitectured materials with unusual properties (Fig. 10.5), which require the development of adept preparation procedures.

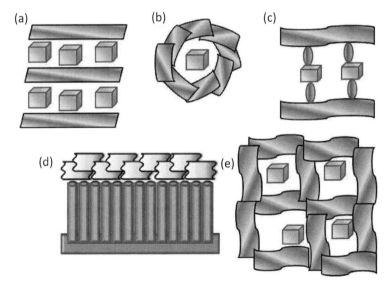

Figure 10.5 Artist's impression of different metal oxide–graphene heterostructures: (a) Self-assembled 2D heterostructure, (b) graphene wrapped core–shell MO_x@G structure, (c) surfactant and/or functionalized graphene self-organized heterostructure, (d) graphene coverage of self-assembled metal oxide nanostructures, and (e) metal oxide nanoparticles trapped inside porous, three-dimensional graphene networks.

The aim of this section is to provide a systematic analysis of selected heterostructures so far considered in the literature for energy storage applications with special emphasis on lithium batteries. Irrespective of the short time period in which this field of research has been developed, the number of possibilities already tested in the literature is unexpectedly large. Moreover, the chemical nature of the materials that have been tested in graphene-based composites for energy storage applications in the literature is highly diversified and requires previous classification. Two different sequences are followed here. The first is based on synthetic route and the other on material application: Li-ion (intercalation anodes: TiO_2, conversion anodes, cathodes), and supercapacitors.

10.5.1 Synthetic Route

The search for a convenient preparative route of graphene-based nanoarchitectured composite electrodes should first consider the synthesis procedure of the graphene-related material to be incorporated. Reduced graphene oxide (rGO) is probably the most frequently used material, although CVD graphene and mechanically cleaved exfoliated graphite has also been tested [86, 89]. Thermally and chemically exfoliated graphene by the Hummers or Hummers-derived methods is commonly used [90]. The methods to form the composite expand from purely physical methods, such as electrophoretic deposition, to real chemical reactions. Other factors such as the choice of pre- or post-graphenization can affect significantly the performance of the resulting graphene-transition metal oxide hybrid electrodes in electrochemical energy storage devices [91].

Electrophoresis—the motion of charged particles dispersed in a fluid under the influence of an applied electric field—is a well-documented procedure for particle sorting and separation. The method is particularly useful for the deposition of particles with size ranging from the nanoscale. Also, the deposition may take place from gas (e.g., in particle precipitation CVD) or liquid (colloids) media. Either positively charged particles moving towards the negative electrode, cataphoresis, or negatively charged particles to the positive, anaphoresis, can be used.

GO and rGO nanolayers suspended in aqueous solutions have negative zeta potential in a wide pH interval [92]. Negatively charged graphene suspensions can be used to deposit layers of this carbon form on a convenient substrate. Moreover, nanostructured deposits can be developed by co-deposition of different negatively charged particles. In this way, anaphoresis of rGO in combination with platinum nanoparticles was used to obtain graphene/platinum composite electrodes on ITO-coated glass at 5 V and used as electrocatalytic electrodes [93].

Regarding battery applications, self-organized TiO_2 nanotubes (nt-TiO_2) prepared by titanium anodization were found to be interesting candidates for the negative electrode of lithium ion microbatteries, due to their safe voltage vs. Li and environmentally benign nature [6]. Combined electrophoretic deposition-anodization methods have been used to fabricate reduced graphene oxide–TiO_2 nanotube films [94]. More recently, self-organized amorphous titania nanotubes with electrophoretically deposited graphene film were also obtained [95].

A borderline method involving both simple electrostatic interactions and chemical reactions is the co-assembly method of fabrication of graphene-encapsulated metal oxide nanoparticles developed by Yang et al. [96]. The co-assembly process is driven by the mutual electrostatic interactions between negatively charged graphene oxide and positively charged oxide nanoparticles. However, the procedure is followed by a chemical reduction reaction. Similarly, the electrostatic interactions between two oppositely charged components may yield graphene@metal oxide core–shell nanostructures. Thus, a ring-opening reaction between epoxy groups and amine groups in GO platelets yields product that interact with amine-modified MO nanostructures [97].

Electrochemical processes leading to the reduction of GO can be coupled with transition metal oxide deposition. In this way, a facile one-step electrochemical preparation of graphene-based heterostructures for both electrodes of Li-ion batteries was reported and tested for Co_3O_4 and Fe_2O_3 conversion anodes and V_2O_5 cathode [98]. The metal oxides/graphene hybrids were electrochemically synthesized by simultaneous deposition of metal oxides and exfoliation of graphene sheets. The hybrid electrodes showed high specific capacities with excellent cycling stability, as discussed below.

Preparation of graphene films coated on both sides with NiO nanoparticles was achieved by co-electrodeposition [99]. Polyethyleneimine –$(CH_2CH_2NH)n$– in aqueous solutions is a polycationic polymer that can be used to stabilize graphene oxide suspensions. Simultaneously, the multiple nitrogen atoms in the polymer may act as a polydentate ligand of transition metal ions such as Ni^{2+}. In this way, positively charged complexes of Ni^{2+} and GO and polyethyleneimine were prepared and electrodeposited on an stainless steel substrate, resulting in the electrochemical deposition of PEI-modified GO/$Ni(OH)_2$ at the electrode surface under an applied elec. field. The as-synthesized film is then converted to graphene/NiO after annealing at 350°C [99].

Soft templates such as swollen micelles formed by surfactant molecules are a common tool to carry out chemical reactions in a restricted volume and thus obtain nanocrystalline products. On the other hand, one of the major problems in the preparation of orderly stacked graphene-MO composites is the hydrophobic nature of graphene surfaces. One possible way to circumvent this obstacle is the functionalization of graphene surfaces and/or the use of surfactant molecules that link together the metal oxide particles with hydrophobic pristine graphene. Sodium dodecyl sulfate surfactant has been used to assist the stabilization of graphene in aqueous solutions and facilitate the self-assembly of different forms of titanium dioxide with graphene. The resulting nanoarchitectures contained alternating layers of the nanocrystalline metal oxide prepared in situ with graphene or graphene stacks. Alternatively, the graphene or graphene stacks can be incorporated into liquid crystal-templated nanoporous structures to form high surface area, conductive networks [100].

On the other hand, a ternary self-assembly approach using functionalized graphene as fundamental building blocks to construct stable, ordered oxide–graphene nanocomposites has been reported [101]. The method uses functionalized graphene sheets prepared by the thermal expansion of graphite oxide, which contain ca. 80% single-sheet graphene along with stacked graphene. The versatility of the method was demonstrated by the preparation of group 4 oxides (SiO_2, SnO_2) as well as transition metal oxides (MnO_2, NiO). 2D core–shell ternary structures were also prepared from poly-(dimethyldiallyl ammonium) chloride functionalized graphene via an in situ hydrolysis of corresponding metal salt [102]. In situ polymerization of phenol and formaldehyde in the presence of G@MO led to the formation of a phenol-formaldehyde resol encapsulated product, namely G@MO@PF. Finally, thermal treatment of G@MO@PF at 500°C in N_2 led to the formation of a carbon shell coated over G@MO. The method was successfully applied to post transition (G@SnO_2@C) and transition metal oxides (G@Fe_3O_4@C), could be fabricated, indicating that the 2D core–shell architecture is constructed by confining the well-defined graphene based metal oxides nanosheets (G@MO) within carbon layers. The resulting 2D carbon-coated graphene/metal oxides nanosheets (G@MO@C) combine the electroactivity of the oxides with the good electrical conductivity of graphene. The problems of scalability in the preparation procedures commonly found in the synthesis of nanoarchitectured materials and the drawbacks associated to the use of binder additives in the preparation of electrodes were simultaneously circumvented in the porous graphene–metal oxide@C electrodes reported by Zhu et al. [103]. Related with the latter, graphene-based metal oxides are further protected by graphene nanosheets through a stepwise heterocoagulation method [104]. The product shows a layered sandwich-like, graphene-based composite structure.

Simple sol-gel methods have been extensively used in the preparation of nanodispersed materials, although room-temperature aqueous procedures may yield inhomogeneities in particle size and shape. Non-aqueous and thermally assisted variants—such as the thermal decompositions of oleate complexes in high boiling point organic solvents [105]—have proven more effective to achieve monodispersed materials. Returning to graphene/metal oxide nanoarchitectures, a fast one-pot method, using benzyl alcohol and microwave-assisted thermal activation has been suggested to obtain SnO_2- or Fe_3O_4-based graphene heterostructures [106]. The microwave heating–assisted the growth of metal oxide nanoparticles, which, in combination with the reduction of GO, and allowed a selective MO-crystallization on the surface of the graphene layers.

Sub-critical and supercritical solvothermal routes are known to be very useful in the preparation of nanomaterials, including battery materials [107, 108]. By a careful selection of thermodynamic conditions—temperature, pressure and composition—control over the processes involved in solvothermal methods—nucleation, growth, and agglomeration phenomena—can be provided, and finally crystal structure, as well as particle size and shape can be tailored.

Zhu et al. [109] proposed an environmentally innocuous method of preparation by using a single-step solvothermal route in ethanol solutions. The procedure leads to simultaneous rGO reduction and iron or cobalt oxide precipitation due to the fact that the GO/rGO layers act as heterogeneous nucleation seeds during the precipitation of the metal oxide nanocrystals. In a related approach, Han et al. [110] were able to obtain $Li_4Ti_5O_{12}$ particles anchored to rGO by solvothermal treatment of H_2O/EtOH-based suspensions of graphite oxide and the oxide powder. The process involves reduction of GO and attachment of the mixed oxide nanoparticles within a single step.

The most commonly used solvothermal routes that use water as the solvent are usually named by using the geological term introduced by Sir Roderick Murchison: hydrothermal. Purely hydrothermal procedures have also been successfully used to prepare transition metal oxide nanoparticles-reduced graphene oxide composites. Park et al. [111] developed an in situ hydrothermal synthesis of Mn_3O_4 nanoparticles on nitrogen-doped graphene. The resulting nanocomposite showed promising results as high-performance anode materials for Li-ion batteries.

Using a dynamic collection of droplets in a gas medium is also a well-known procedure to restrict the reaction volumes and obtain nanoparticulate materials by a thermally promoted reaction (spray pyrolysis). Chidembo et al. [112] used an in situ spray pyrolysis approach to fabricate metal oxide–graphene composites with highly porous morphologies. The materials exhibited unique globular structures comprising metal oxide nanoparticles intercalated between graphene sheets.

Metal-organic frameworks (MOFs) were used as precursors of metal oxides embedded inside porous, three-dimensional graphene networks (3DGNs) [89]. The 3DGNs served as a template for the preparation of MOF/3DGN composites. Subsequently, the desired metal oxide/3DGN composites were obtained by a two-step annealing process: first under argon and then in air. The successful preparation of two example composites, ZnO/3DGN and Fe_2O_3/3DGN, was reported.

10.5.2 Metal Oxides Involved in Energy Storage System

Titanium dioxide is a well-known transition metal oxide that displays reversible lithium intercalation reactions in all its crystalline forms at potential ranges that are compatible with the anode of Li-ion batteries. In addition, titanium is abundant in nature and titania is cheap, non-toxic, and environmentally friendly. Moreover, nanodispersed and nanoarchitectured titania materials show an extraordinary improvement in electrode performance, and capacities close to the theoretical $Li_{0.5}TiO_2$. Nanostructured anatase electrodes were early shown to increase lithium insertion rates [113]. More recently, self-organized TiO_2 nanotube layers prepared by titanium anodization have been proposed as an alternative electrode for lithium-ion microbatteries [17]. The powder-free nanofabrication of the active material directly on the current collector makes unnecessary the use of additives to improve the mechanical (polymer binders) or conductive (carbon black) properties. Regarding $TiO_2(B)$, the use of this solid in the form of nanowires provides a maximum capacity of 305 mA hg^{-1}, and excellent cyclability and rate performance [107]. Even the initially less attractive rutile gets closer to anatase in the form of nanorods although the reaction with lithium involves two consecutive structural phase transformations [114].

Concerning graphene/nanoparticle-TiO_2 hybrid nanostructures, anatase and rutile polymorphs have been tested [100]. The gravimetric capacity of the G/anatase hybrid was 96 mA hg^{-1} at 30 C, i.e., almost four times the observed value of nanocrystalline anatase TiO_2. The specific capacity of the rutile hybrid nanostructure was 87 mA hg^{-1} at the same rate, a value twice as much as nanodispersed rutile. Both rutile and anatase hybrid nanostructures showed Coulombic efficiencies greater than 98% and good capacity retention over 100 cycles at C rat. Later, Yang et al. [115] fabricated sandwich-like, graphene-based titania nanosheets as an anode for lithium-ion batteries by a nanocasting method. On the other hand, TiO_2-graphene nanocomposite synthesized by a gas/liquid interface reaction were shown to store lithium in an enlarged potential window of 0.01–3.0 V [116]. Also, Zhang et al. studied polyaniline-TiO_2-rGO nanocomposites, which provided a first discharge capacity of 496.3 mA h g^{-1} and delivered a capacity of 149.8 mA h g^{-1} at 5 C rate [117]. Lee et al. used graphene nanosheets as a platform for the 2D ordering of mesoporous 2D aggregates of anatase TiO_2 nanoparticles [118]. Finally, Wang et al. used hydrogenated TiO_2-reduced-graphene oxide nanostructured composites and found a high rate capability (166.3 mA h g^{-1} at 5 C rate) and improved cycling performance (97.6% capacity retention after 100 cycles) [119].

Concerning self-organized titania nanotubes/rGO heterostructures [95], the capacities achieved in the 1.0–2.6 V range were found to be higher than in any other TiO_2-based electrode previously reported. Moreover, effective energy densities were particularly high, due to the absence of binding additives. An unusual rate performance was also found: Even for extremely high rates close to 300 C, significant capacities above 50 mA h g^{-1} were measured, which means that the cell could be charged in a few seconds [95]. The improved behavior results from a reduction in charge transfer resistance within the heterostructure.

Regarding the $Li_4Ti_5O_{12}$ mixed oxide, with spinel–related structure and $(Li)_{8a}[Li_{1/3}Ti_{5/3}]_{16d}O_4$ cation distribution, pointed out by Ohzuku et al. [120] as a zero strain intercalation electrode, it is also benefited by forming nanocomposites with graphene. $Li_4Ti_5O_{12}$ nanoparticles were anchored to rGO nanosheets by using the solvothermal method discussed above [110]. The surface modified rGO nanosheets have an unusual performance in lithium test cells.

As mentioned above, conversion reactions generated great expectations in the field of electrode materials for Li ion batteries after their first definition by Tarascon's group [28, 30]. In order to minimize the negative effects of the volume changes upon cycling, the high electrical conductivity of different carbon-based materials was discussed. Now we extend the discussion to graphene, which has been recently studied as conductive matrix to improve the cycle performance of conversion electrodes. Despite the recent graphene advent, many transition metal oxides have been tested so far.

Among them, cobalt oxides provide a good number of examples, such as the co-assembly fabrication of graphene-encapsulated oxide nanoparticles previously discussed [96]. This process yields a G/Co_3O_4 composite electrode with unusually high initial reversible capacity of about 1100 mA hg^{-1} and retaining around 1000 mA h g^{-1} after 130 cycles. Other examples of the use of cobalt oxides/graphene heterostructures include the combination of Co_3O_4 nanorods with graphene nanosheets [121]. By using the electrochemical synthesis, the Co_3O_4/graphene depicted a discharge capacity of 880 mA h g^{-1} during the 40th cycle at 0.3 C [98]; the (CuO, Co_3O_4)/rGO composites [122], and $Co(OH)_2$ hexagonal rings anchored to graphene oxide layers [123]. Wang et al. [124] reported a CuO/graphene nanocomposite as a high-performance anode material for lithium-ion batteries.

Regarding iron oxides, a first report on graphene-wrapped Fe_3O_4 electrodes showed interesting reversible capacities and capacity retention on prolonged cycling vs. lithium [125]. More recently, graphene/Fe_3O_4 nanrods provided almost tenfold higher capacity than bare magnetite [126]. Moreover, a high rate capacity up to ca. 670 mA h g^{-1} was achieved when the current density recovered to 0.1 C, due to the enhanced electronic conductivity of the overall electrode and stabilization of the nanoarchitectured electrode [121]. Concerning ferric oxide, Zhou et al. [97] reported graphene/Fe_2O_3 core–shell composite material. In order to evaluate the potential interest of this material as the anode of Li-ion batteries, the material was tested in lithium cells, which also showed a high reversible capacity, improved cycling stability, and excellent rate capability as compared with the well-known electrochemical performance of hematite nanoparticles. Later, the Fe_2O_3/graphene anode prepared by using the electrochemical preparation, was able to deliver a discharge capacity of 894 mA h g^{-1} during the 50th cycle at 0.3 C [98]. More recently, Huang et al. prepared two-dimensional nanosheets, which include C-Fe_2O_3-rGO [127]. Tannic acid, a natural polyphenol, was used as a dispersing agent to introduce a metal precursor on the surface of rGO, and the metal precursor was subsequently converted to the corresponding metal oxide nanoparticles by thermal annealing in a vacuum. Also, by using multiwalled carbon nanotubes (MWCNT), a three-dimensional rGO/MWCNTs/Fe_2O_3 ternary composite has been tested [128]. Finally, composites including 3D graphene networks (3DGN) and Fe_2O_3 were obtained from metal organic frameworks (MOF) and used in a lithium-ion battery. At a current density of 0.2 A g^{-1}, the Fe_2O_3/3DGN composite provided a capacity per gram of the electrode of 864 mA h g^{-1} after 50 cycles, whereas only 24.3 mA h g^{-1} was observed for the Fe_2O_3 electrode [89].

Graphene-encapsulated mesoporous metal oxides (e.g., Co_3O_4, Cr_2O_3, and NiO) can be obtained by adjusting the pH of two suspensions: dispersed mesoporous metal oxides to the 5–7 range and graphene oxide to 7–8. Then the resulting suspensions are mixed in the presence of reducing agents to give the heterostructured materials, which were tested as anode materials for lithium-ion batteries [129]. Also regarding nickel oxides, NiO-graphene hybrid nanoarchitectures showed enhanced capacity and rate performance, which were associated with a decrease in voltage polarization, particularly when cycled at high rates [130]. In this way, when the hybrid was cycled at a current density of 100 mA g^{-1}, it delivered a specific capacity of ca. 650 mA h g^{-1} after 35 cycles, with 86% retention. The capacity decreased to still interesting values of ca. 510 and 370 mA h g^{-1} when the current density was increased 4 and 8 times, respectively.

Nickel hydroxide was studied by [131] who obtained β-Ni(OH)$_2$@reduced-graphene-oxide composites with improved performances in both Ni-MH and Li-ion batteries. Preparation via an electrochemical method of graphene films coated on both sides with NiO nanoparticles has also been reported [99]. Finally, Tao et al. [132] also examined graphene-encapsulated NiO prepared by the heterocoagulation method and found improved electrochemical performances. Thus, the discharge capacity of the composite containing 51.8 wt.% NiO was ca. 570 mA h g^{-1} after 100 cycles at 0.1 C, more than twice the capacity of the bare NiO graphite electrode. In addition, ca. 310 mA h g^{-1} was retained at 1 C for the composite. The improvement in conductivity by graphene encapsulation was evidenced from electrochemical impedance spectroscopy experiments that showed a charge transfer resistance of the composite practically half of the value than in the absence of graphene.

Manganese oxides were first considered less interesting for conversion reactions due to their highly negative free energy of formation. However, Wang et al. [133] described a Mn_3O_4-rGO hybrid delivering a specific capacity of ca. 810 mA h g^{-1} when cycled at 40 mA g^{-1} and based on the mass of the Mn_3O_4 nanoparticles in the hybrid, exclusively. Such value is close to the theoretical capacity for Mn_3O_4 (936 mA h g^{-1}) in the full conversion reaction to Mn(0) + Li_2O, and ca. 300 mA h g^{-1} higher than for bare Mn_3O_4 electrodes. In addition, the specific capacity was preserved to almost 780 and

390 mA h g^{-1} after increasing current density by 10 and 40 times, respectively. Related with this, Latorre et al. [134] prepared a hybrid graphene–nickel/manganese mixed oxide with a maximum capacity value of 1030 mA h g^{-1} during the first discharge, and capacity values higher than 400 mA h g^{-1} were still achieved after 10 cycles. Finally, Park et al. [111] used Mn_3O_4 nanoparticles on nitrogen-doped graphene as high-performance anode materials for lithium ion batteries. Although penalized by a high irreversible capacity in the first cycle, commonly found in conversion electrodes (see previous sections), the composite delivered ca. 800 mA h g^{-1} after 40 cycles at 200 mA g^{-1}.

The oxides of heavier, second-row transition metal elements are commonly discarded for their use in high gravimetric capacity applications. Nevertheless, an enhanced nanoscale conduction capability of a MoO_2/Graphene composite for high performance anodes in lithium-ion batteries should be highlighted. The composite electrode showed a reversible capacity of 605 mA h g^{-1} in the initial cycle at current density of 540 mA g^{-1} and upon increasing the current density to 2045 mA g^{-1} the electrode shows a reversible capacity of 300 mA h g^{-1} [135].

A recent review on graphene in lithium ion battery cathode materials [136], evidenced that most of the literature on this topic is connected to $LiFePO_4$. In addition, $Li_3V_2(PO_4)_3$/graphene composites were considered in the review. As a result of the poor electronic conductivity of the olivine, graphene can be considered as a conducting additive [137]. The composites with outstanding rate performance in lithium ion batteries can be easily prepared, e.g., by co-precipitation method [138]. Moreover, the degree of interaction and electrical wiring may vary from simple mixtures to stronger anchoring.

Concerning transition metal oxide cathodes, earlier works were mostly limited to $LiMn_2O_4$ [136]. However, recent studies have found that vanadium oxides and the 4.8 V spinel $LiNi_{0.5}Mn_{1.5}O_4$ also improve their performance in graphene composites. The V_2O_5/graphene cathode prepared by electrochemical method delivered a discharge capacity of 208 mA h g^{-1} after 100 cycles at 6.8 C [98]. Later, Zhao et al. [139] showed the versatility of vanadium oxides/rGO composites for lithium-ion batteries and supercapacitors with improved electrochemical performance. Regarding the high voltage spinel cathodes, Prabakar et al. [140]

prepared graphene-sandwiched $LiNi_{0.5}Mn_{1.5}O_4$ cathode composites for enhanced high voltage performance. Also, Fang et al. used graphene oxide-coated $LiNi_{0.5}Mn_{1.5}O_4$ as high voltage cathode for lithium ion batteries with high energy density and long cycle life [141].

Even more, for layered second-row transition metal oxides it was possible to obtain binder-free, nanobelt-MoO_3/G film electrodes by using microwave-assisted hydrothermal methods. Irrespective of the large molecular weight of the oxide, the gravimetric capacity could reach 291 mA h g^{-1} at 100 mA g^{-1}, with retention of ca. 60% after 100 cycles [142]. An exciting achievement was also the flexible properties of the electrode that could finally lead to new applications in bendable batteries or the origami concept [143].

Porous $NiCo_2O_4$ hexagonal ring-graphene hybrids have been prepared by dehydrating β-$Ni_{0.33}Co_{0.67}(OH)_2$ platelets, and tested as electrode materials for both supercapacitors and lithium-ion batteries. These materials exhibit a large capacity, high rate capability, and excellent cycling stability [123]. Also, the vanadium oxides/rGO composites developed by Zhao et al. were shown to versatile for the cathodes of lithium-ion batteries as well as in supercapacitors with improved performance [139].

By using aqueous electrolyte, Chang et al. [144] could broaden the operation potential, reaching values close to 2.0 V, together with 307 F g^{-1} gravimetric capacitance, leading to 42.6 Wh kg^{-1} energy density and 276 W kg^{-1} power density. Moreover, a supercapacitor with rGO/MnO_2 and rGO/MoO_3 electrodes could be cycled for 1000 cycles due to the improved ion transfer, as a result formation of a microporous system during cycling. Finally, a manganosite (MnO)/microwave-exfoliated GO composite material displayed areal capacitances of 0.107 F cm^{-2}, and 82% capacity retention at 0.5 A g^{-1} after 15,000 cycles [145].

10.6 Surface Modification of Nanostructrures for Improved Battery Performance

Surface modification treatments have been largely explored to improve the ionic and electronic properties of a huge variety of electrode materials. In this section some of the most important coating treatments on cathodes such as spinel-type ($LiMn_2O_4$)

and the LiFePO$_4$ compounds will be highlighted, because of their relevance as positive materials for large-scale battery applications such as electric vehicles (EVs) and hybrid electric vehicles (HEVs) [146]. Particularly, LiMn$_2$O$_4$ and related compositions exhibit severe capacity fading especially at elevated temperature during cycling due to the Mn^{2+} dissolution [147], Jahn–Teller distortion [148], and decomposition of electrolyte solution on the electrode [149]. On the other hand, the use of LiFePO$_4$ as battery cathodes is hindered by its poor rate capability principally due to its intrinsic low electronic conductivity (10^{-9} S cm^{-1} at room temperature [150]) and slow diffusion rate of lithium ions [151]. Finally, the surface treatment of TiO$_2$ as anodes will be discussed. Titania is one of the best candidates to replace graphite due to its inherent safety, relatively high potential, low self-discharge, low cost and environmental friendliness [6, 17]. The surface treatment for layer of self-organized TiO$_2$ nanotubes have been proportionally less considered as compared to cathode materials.

Modification by coating has proven to be an important method to achieve optimal electrochemical performances for cathodes and anodes. Therefore, much attention is now focused on improving the electronic and ionic conductivity. Among the different surface modification treatments, the most remarkable are the following:

(a) Carbon coating

A typical example where the graphene nanosheets (GN) are homogenously incorporated into the porous hierarchical network of LiFePO$_4$ (LFP) was described by Yang et al. [152]. The steps of this synthesis are schemed in Fig. 10.6a. The procedure starts with the dispersion of GN in deionized water, then proceeds with the self-assembly of graphene with the LFP precursor, and ends with the crystallization of the LFP/G precursor. The CO and CO$_2$ are released from the degradation of these precursors during annealing, resulting in the formation of a porous 3D network. This particular LFP/GN composite exhibited a capacity of 146 mA h g^{-1} at 17 mA g^{-1} over 100 cycles, which is about 1.4 times higher than that of porous LFP (104 mA h g^{-1}). Further studies demonstrated that carbon coating plays an exceptional role for LiFePO$_4$ cathode. A comparison of the effects of using carbon nanotubes (CNT) or carbon black (CB) as additive in LiFePO$_4$ was performed by Liu et al. [153]. They found that those electrodes with CNT additive show better electrochemical performances in terms of capacity retention > 99%. With CNT additive, the polarization voltage was decreased from 0.3 to 0.2 V, and the impedance was decreased from 423.2 to 36.88 ohm. Better crystal structure retention was observed when using CNT. In summary, Carbon is an outstanding electronic conductor and can form a continuous transport path in the whole electrode, but it is not a good ionic conductor [154].

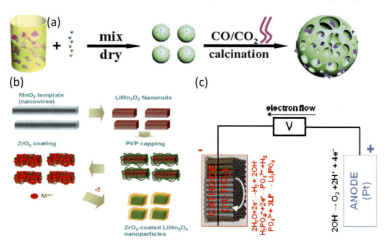

Figure 10.6 Scheme representing the preparation of some the most outstanding surface treatment for: (a) 3D porous networks of LiFePO$_4$/graphene (with permission from Elsevier adapted from ref. [152], (b) PVP capping onto the spinel nanorods (adapted from [155]), and (c) Li$_3$PO$_4$ electrodeposition on self-organized TiO$_2$ nanotubes.

(b) Metal and oxide coating

Modification by metal or oxide coating has proven to be an important method to enhance the electrochemical performance of cathode materials. For example, a number of materials such as ZrO$_2$, Co$_3$O$_4$, SiO$_2$, Li$_2$O-B$_2$O$_3$, and silver coating [156, 157], were applied to cover the surface of the particles in order to reduce the contact between the active material and the electrolyte and hence is possible to prevent the dissolution of manganese. For instance, the silver coating can be a promising tool to preserve the capacity even at high current densities. LiMn$_2$O$_4$ spinel nanorods can be sheathed with polyvinyl pyrrolidone (PVP) and coated with

ZrC_2O_4 precursors in aqueous solution as shown schematically in Fig. 10.6b [155]. A TEM image of the 1 wt.% ZrO_2-coated spinel nanoparticles after thermal treatment at 600°C is shown in Fig. 10.7a. The coating layers are uniformly distributed on the particle surface exhibiting a coating thickness about 6 nm (Fig. 10.7a) as a result of the PVP burn-off at high temperature in air. Also, it can be discerned the distances of the neighboring lattice fringes with a d-spacing value of 4.7 Å, corresponding to de (111) plane of the spinel phase. The thickness can be attuned up to 12 nm and the coating layer is amorphous if the concentration increases to 2 wt.% [155]. The electrochemical cycling results clearly showed that nanoscale ZrO_2 coating significantly improved the rate capability and cycle life at 65°C in spite of very high surface area of the spinel nanoparticles. The reason for this observed improvement was associated with the fact that the ZrO_2 coating layer minimized Mn dissolution from the spinel lattice.

(c) Conducting polymer coating

During the past decade, it has been shown that conducting polymers, including redox polymers can have positive effect on the performance of $LiFePO_4$ [158] and other cathode materials such as $LiCoO_2$ [159]. A recent methodology [160] reported an improved way for the fabrication and use of conducting polymer/$LiFePO_4$, which relies on the intrinsic oxidation power of $Li_{1-x}FePO_4$ rather than on an external oxidant as the driving force of the polymerization process. Figure 10.7b shows that the thickness of the polymer coating can be about 2 nm for isolated particles. Therefore, the risk of residual oxidant or oxidant by-products leaching from the polymer into the battery electrolyte, which would cause destruction on the anode electrode process, is eliminated. Moreover, the polymerization propagation requires the reinsertion of lithium into the partially de-lithiated lithium iron phosphate, as well as the transport of lithium ions and electrons through the deposited polymer coating. These are the functionality characteristics of an effective conducting coating for $LiFePO_4$. In turn, the deposition of polymers onto nanostructures by electrochemical techniques is a convenient way to ensure the optimum filling of the nanostructures. Indeed, electropolymerization is particularly powerful to control the deposition of different polymers into various porous materials to serve as a way to get electrolyte interface [161]. Very recently, the use of electrodeposition to fill TiO$_2$ nanotubes with a layer of poly(methyl methacrylate)-polyethylene oxide, i.e., PMMA-(PEO) 475 has been reported [24]. The resulting electrochemical behavior can be affected in a positive way as a consequence of such coating onto 3D nanostructures. Moreover, the surface of the TiO$_2$ nanotubes is crucial to reach high specific capacity, rate performance and cycling stability. A different alternative to improve the stability of nanotubular titanium dioxide electrodes in lithium batteries has been proposed using polyacrylonitrile (PAN) by electropolymerization [23]. The effects on PAN on self-organized TiO$_2$ nanotubes fabricated with different aspect ratios were evaluated in depth. The resulting electrode material showed an areal capacity value of 500 µA h cm^{-2} when using 0.0–3.0 V as a potential window in Li-cells at slow rates. An areal capacity value of 260 µAh cm^{-2} is delivered at 75 C rate evidencing that the ion-conducting PAN layer ensures lithium ion access to the nanotubes, provides enhanced mechanical stability to the electrode and lower charge transfer resistance. Most probably, this PAN coating protects the open end surface from undesirable reactions with the electrolyte.

(d) Li$_3$PO$_4$ surface coating

Li$_3$PO$_4$ is a fast solid lithium ionic conductor that can be envisaged as an outstanding coating material [162]. Li$_3$PO$_4$-coated LiCoO$_2$ particles provide good capacity retention in a wide range voltage window. Moreover, the power density of the lithium cell increased up to 50% when disordered Li$_3$PO$_4$ films coated LiCoO$_2$ as compared to powders [163]. The cycling performance of LiMn$_2$O$_4$ by coating with Li$_3$PO$_4$ via ball milling was improved mostly because of the decrease of the cell impedance [156]. These results are supported on the basis that the coating of the film is quite uniform and the average thickness of the shell is around 7 nm (Fig. 10.7c). In this context, the group of Prof. Yang has studied the effect of the confinement of lithium phosphate with double-layer surface coating on Li$_2$CoPO$_4$F materials for high voltage cathodes (~5 V vs. Li/Li$^+$) [164]. The Li$_3$PO$_4$-coated Li$_2$CoPO$_4$F shows a high reversible capacity of 154 mA h g^{-1} at 1 C current rate, and excellent rate capability (141 mA h g^{-1}) at 20 C, exhibiting energy density up to 700 W h kg^{-1}.

More critical for battery applications is the fact that an irregular coating of cathode and anode materials may lead to poor connectivity of the particles and hence loss of performance. This possibility has been less explored in anode materials. It would be an important improvement to the conventional TiO$_2$ system whether low temperature methods could be found. The design of novel nanoarchitectured anodes with a tailored morphology (nanotubes, nanowires, etc.) and crystallinity is a hot topic studied by many groups. We will focus on the improvement of a self-organized anatase nanotube (nt-TiO$_2$) electrode by protecting the surface layer with a deposit of lithium phosphate ionic conductor. Figure 10.7d shows how the electrolytic Li$_3$PO$_4$ films can be successfully deposited on the nanotube array by an electrochemical procedure consisting of proton reduction with subsequent increase in pH, hydrogen phosphate dissociation and Li$_3$PO$_4$ deposition on the surface of the cathode. The Li$_3$PO$_4$ polymorph (γ or β) in the deposit could be tailored by modifying the electrodeposition parameters, such as time, current density or temperature [165]. The TEM micrographs of the nt-TiO$_2$/γ-Li$_3$PO$_4$ powder scraped from the Ti sheet reveal the occurrence of lithium phosphate inside the nt-TiO$_2$ and in the external wall of the nanotubes (Fig. 10.7d).

The effect of different Li$_3$PO$_4$ film thickness on the nt-TiO$_2$ affected drastically the electrochemical performance as explained in detail in Ref. [165]. The results therein aim to show a way of preparing low crystalline and amorphous materials into the negative electrode to increase the interphase area between the active materials and the electrolyte. An example of an all-solid-state Li/Li$_3$PO$_4$/LiCoO$_2$ battery obtained by using pulsed laser deposition (PLD) grown-up Li$_3$PO$_4$ thin films showed excellent excellent intercalation properties and electrochemical stability in the operating voltage range from 0 to 4.7 V vs. Li$^+$/Li^0 [166]. PLD technique is still too expensive from the industrial point of view to develop such battery technology worldwide.

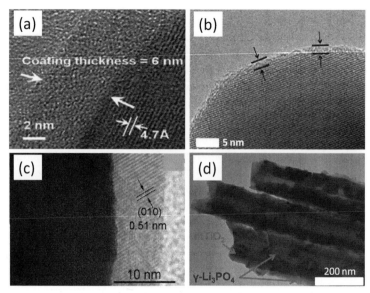

Figure 10.7 TEM images of some spectacular surface coating: (a) of the ZrO$_2$-coated with spinel nanoparticle after firing at 600°C (coating concentration of about 1 wt.%) (adapted from [155] with permission from Elsevier), (b) PEDOT-LiFePO$_4$ for isolated particles [160], (c) Li$_3$PO$_4$-coated LiMn$_2$O$_4$ (adapted from [156]) and (d) titania nanotubes with electrodeposited Li$_3$PO$_4$ using current density of −3.7 mAcm^{-2} during 60 s.

Considering electrochemical fabrication routes as a cheap and simple way to prepare electrodes, a revolutionary concept of small scale battery design based on TiO$_2$/Li$_3$PO$_4$ (LiPF$_6$ in EC: DEC)/LiFePO$_4$ is shown here. As expected from their respective negative (1.7–1.9 V vs. lithium) and positive (3.4–3.5 V vs. lithium) potential electrodes, the resulting operational voltage of the battery is about 1.5–2.0 V in accordance to that observed experimentally. Lopez et al. [165] showed the galvanostatic charge/discharge curves where lithium ions are shuttling from LiFePO$_4$ to nt-TiO$_2$/γ-Li$_3$PO$_4$ and going back from nt-Li$_x$TiO$_2$/γ–Li$_3$PO$_4$ to Li$_{1-x}$FePO$_4$, respectively.

A common characteristic is the occurrence of a very flat voltage profile centered about 1.7 V. The small hysteresis between charge and discharge is a clear evidence of high reversibility and relatively fast kinetic. Indeed, this Li-ion battery was tested using a wide windows rate (C/2, C, 2 C and 5 C) exhibiting an excellent rate capability with capacities ranging from 120 to 100 mA h g^{-1} from C/2 to 5 C, respectively. Therefore, one of the most representative results is that recorded after cell cycling at a C rate and at 25 or 55°C, exhibiting a maximum reversible capacity of 133.7 and 159 mA h g^{-1} in charge, respectively. This significant change of capacity can be attributed to a lower polarization (0.1 V less) observed on charge/discharge for the experiment recorded at 55°C. However, these capacity values are very close to the theoretical values, having in mind that the battery is cathode limited.

Additional experiments were also performed for a large number of cycles, which showed different capacity tendency as compared to the rate. An unusual behavior was the best capacity retention over 160 cycles is observed at high rates. For instance, the full cell cycled at C/2, 1 C and 2 C exhibited 20, 30 and 56 mA h g^{-1}, while the capacity was 82 mA h g^{-1} when cycled at 5 C. A lithium-ion cell based on titanium oxide nanotubes and bare LiFePO$_4$ exhibited only 70 mA h g^{-1} at C/10 at the first reversible cycle [167]. The increased capacity in batteries using coated TiO$_2$ can be ascribed to an improvement of the Li ion diffusion when Li$_3$PO$_4$ layer is applied. Moreover, a reduced contact area between the electrodes and the electrolyte can be achieved.

Acknowledgements

The authors are grateful to MEC (MAT2014-56470-R) and Junta de Andalucía (FQM288, FQM7206, and FQM6017) for financial support. GOJ is grateful to the *Ramón y Cajal* programme.

References

1. Kasuga, T., Hiramatsu, M., Hoson, A., Sekino, T., and Niihara, K. (1998). Formation of titanium oxide nanotube, *Langmuir*, **14**, pp. 3160–3163.
2. Michailowski, A., Al-Mawlawi, D., Cheng, G. S., and Moskovits, M. (2001). Highly regular anatase nanotubule arrays fabricated in porous anodic templates, *Chem. Phys. Lett.*, **349**, pp. 1–5.
3. Gong, D., Grimes, C. A., Varghese, O. K., Hu, W., Singh, R. S., Chen, Z., and Dickey, E. C. (2001). Titanium oxide nanotube arrays prepared by anodic oxidation, *J. Mater. Res.*, **16**, pp. 3331–3334.

4. Kowalski, D., Kimb, D., and Schmuki, P. (2013). TiO_2 nanotubes, nanochannels and mesosponge: Self-organized formation and applications, *Nano Today*, **8**, pp. 235–264.
5. Regonini, D., Bowen, C. R., Jaroenworaluck, A., and Stevens, R. (2013). A review of growth mechanism, structure and crystallinity of anodized TiO_2 nanotubes, *Mater. Sci. Eng. R*, **74**, pp. 377–406.
6. Ortiz, G. F., Hanzu, I., Djenizian, T., Lavela, P., Tirado, J. L., and Knauth, P. (2009a). Alternative Li-ion battery electrode based on self-organized titania nanotubes, *Chem. Mater.*, **21**, pp. 63–67.
7. González, J. R., Alcántara, R., Ortiz, G. F., Nacimiento, F., and Tirado, J. L. (2013). Controlled growth and application in lithium and sodium batteries of high-aspect-ratio, self-organized titanium nanotubes, *J. Electrochem. Soc.*, **160**, pp. A1390–A1398.
8. González, J. R., Alcántara, R., Nacimiento, F., Ortiz, G. F., Tirado, J. L., Zhecheva, E., and Stoyanova, R. (2012). Long-length titania nanotubes obtained by high-voltage anodization and high-intensity ultrasonication for superior capacity electrode, *J. Phys. Chem. C*, **115**, pp. 20182–20190.
9. Varghese, O. K., Gong, D. W., Paulose, M., Grimes, C. A., and Dickey, E. C. (2003). Crystallization and high-temperature structural stability of titanium oxide nanotube arrays, *J. Mater. Res.*, **18**, pp. 156–165.
10. Perathoner, S., Passalacqua, R., Centi, G., Su, D. S., and Weinberg, G. (2007). Photoactive titania nanostructured thin films synthesis and characteristics of ordered helical nanocoil array, *Catal. Today*, **122**, pp. 3–13.
11. Albu, S. P., Tsuchiya, H., Fujimoto, S., and Schmuki, P. (2010). TiO_2 Nanotubes–annealing effects on detailed morphology and structure, *Eur. J. Inorg. Chem.*, **27**, pp. 4351–4356.
12. Yan, J., Song, H., Yang, S., and Chen, X. (2009). Effect of heat treatment on the morphology and electrochemical performance of TiO_2 nanotubes as anode materials for lithium-ion batteries, *Mater. Chem. Phys.*, **118**, pp. 367–370.
13. Fang, H. T., Liu, M., Wang, D. W., Sun, T., Guan, D. S., Li, F., Zhou, J., Sham, T. K., and Cheng, H. M. (2009). Comparison of the rate capability of nanostructured amorphous and anatase TiO_2 for lithium insertion using anodic TiO_2 nanotube arrays, *Nanotechnology*, **20**, pp. 225701–22507.
14. Borghols, W. J. H., Lützenkirchen-Hecht, D., Haake, U., Chan, W., Lafont, U., Kelder, E. M., van Eck, E. R. H., Kentgens, A. P. M., Mulder, F. M., and Wagemaker, M. (2010). Lithium storage in amorphous TiO_2 nanoparticles, *J. Electrochem. Soc.*, **157**, pp. A582–A588.

15. González, J. R., Alcántara, R., Nacimiento, F., Ortiz, G. F., and Tirado, J. L. (2014). Microstructure of the epitaxial film of anatase nanotubes obtained at high voltage and the mechanism of its electrochemical reaction with sodium, *CrystEngComm.*, **16**, pp. 4602–4609.

16. Xiong, H., Ildirim, H., Shevchenko, E. V., Prakapenka, V. B., Koo, B., Slater, M. D., Balasubramanian, M., Sankaranarayanan, S. K. R. S., Greeley, J. P., Tepavcevic, S., Dimitrijevic, N. M., Podsiadlo, P., Johnson, C. S., and Rajh, T. J. (2012). Self-improving anode for lithium-ion batteries based on amorphous to cubic phase transition in TiO_2 nanotubes, *J. Phys. Chem. C*, **116**, pp. 3181–3187.

17. Ortiz, G. F., Hanzu, I., Knauth, P., Lavela, P., Tirado, J. L., and Djenizian, T. (2009b). TiO_2 nanotubes manufactured by anodization of Ti thin films for on-chip Li-ion 2D microbatteries, *Electrochim. Acta*, **54**, pp. 4262–4268.

18. Brumbarov, J., and Kunze-Liebhäuser, J. (2014). Silicon on conductive self-organized TiO_2 nanotubes: A high capacity anode material for Li-ion batteries, *J. Power Sources*, **258**, pp. 129–133.

19. Lamberti, A., Garino, N., Sacco, A., Bianco, S., Manfredi, D., and Gerbaldi, C. (2013). Vertically aligned TiO_2 nanotube array for high rate Li-based micro-battery anodes with improved durability, *Electrochim. Acta*, **102**, pp. 233–239.

20. Bi, Z., Paranthaman, M. P., Menchhofer, P. A., Dehoff, R. R., Bridges, C. A., Chi, M., Guo, B., Sun, X. G., and Dai, S. (2013). Self-organized amorphous TiO_2 nanotube arrays on porous Ti foam for rechargeable lithium and sodium ion batteries, *J. Power Sources*, **222**, pp. 461–466.

21. Wei, W., Oltean, G., Tai, C. W., Edström, K., Björefors, F., and Nyholm, L. (2013). High energy and power density TiO_2 nanotube electrodes for 3D Li-ion microbatteries, *J. Mater. Chem. A*, **1**, pp. 8160–8169.

22. Tian, Q., Zhang, Z., Yang, L., and Hirano, S. (2014). Encapsulation of SnO_2 nanoparticles into hollow TiO_2 nanowires as high performance anode materials for lithium ion batteries, *J. Power Sources*, **253**, pp. 9–16.

23. Nacimiento, F., González, J. R., Alcántara, R., Ortiz, G. F., and Tirado, J. L. (2013). Improving the electrochemical properties of self-organized titanium dioxide nanotubes in lithium batteries by surface polyacrylonitrile electropolymerization, *J. Electrochem. Soc.*, **160**, pp. A3026–A3035.

24. Kyeremateng, N. A., Dumur, F., Knauth, P., Pecquenard, B., and Djenizian, T. (2011). Electropolymerization of copolymer electrolyte into titania nanotube electrodes for high-performance 3D microbatteries, *Electrochem. Commun.*, **13**, pp. 894–897.
25. Zhao, L., Qi, L., and Wang, H. (2013). Sodium titanate nanotube/graphite, an electric energy storage device using Na^+-based organic electrolytes, *J. Power Sources*, **242**, pp. 597–603.
26. Senguttuvan, P., Rousse, G., Seznec, V., Tarascon, J. M., and Palacín, R. M. (2011). $Na_2Ti_3O_7$: Lowest voltage ever reported oxide insertion electrode for sodium ion batteries, *Chem. Mater.*, **23**, pp. 4109–4111.
27. Tirado, J. L. (2003). Inorganic materials for the negative electrode of lithium-ion batteries: State of the art and future prospects, *Mater. Sci. Eng. R*, **40**, pp. 103–136.
28. Poizot, P., Laruelle, P. S., Grugeon, S., Dupont, S. L., and Tarascon, J. M. (2000). Nano-sized transition-metal oxides as negative-electrode materials for lithium-ion batteries, *Nature*, **407**, p. 496.
29. Tarascon, J. M., Grugeon, S., Morcrette, M., Laruelle, S., Rozier, P., and Poizot, P. (2005). New concepts for the search of better electrode materials for rechargeable lithium batteries, *C. R. Chim.*, **8**, pp. 9–15.
30. Poizot, P., Laruelle, S., Grugeon, S., and Tarascon, J. M. (2002). Rationalization of the low-potential reactivity of 3D-metal-based inorganic compounds toward Li, *J. Electrochem. Soc.*, **149**, pp. A1212–A1217.
31. Cabana, J., Monconduit, L., Larcher, D., and Palacín, M. R., (2010). Beyond intercalation-based Li-ion batteries: The state of the art and challenges of electrode materials reacting through conversion reactions, *Adv. Mater.*, **22**, pp. E170–E192.
32. Boyanov, S., Womes, M., Monconduit, L., and Zitoun, D. (2009). Mössbauer spectroscopy and magnetic measurements as complementary techniques for the phase analysis of FeP electrodes cycling in Li-ion batteries, *Chem. Mater.*, **21**, pp. 3684–3692.
33. Wang, Q., Gao, R., and Li, J. H. (2007). A Porous, self-supported Ni_3S_2/Ni nanoarchitectured electrode operating through efficient lithium-driven conversion reactions, *Appl. Phys. Lett.*, **90**, p. 143107.
34. Badway, F., Pereira, N., Cosandey, F., and Amatucci, G. G. (2003). Carbon-metal fluoride nanocomposites structure and electrochemistry of FeF_3:C, *J. Electrochem. Soc.*, **150**, pp. A1209–A1218.
35. Bruce, P. G., Scrosati, B., and Tarascon, J. M. (2008). Nanomaterials for rechargeable lithium batteries, *Angew. Chem. Int. Ed.*, **47**, pp. 2930–2946.

36. López, M. C., Ortiz, G. F., Lavela, P., Alcántara, R., and Tirado, J. L. (2013). Improved energy storage solution based on hybrid oxide materials, *ACS Sustainable Chem. Eng.*, **1**, pp. 46–56.
37. López, M. C., Ortiz, G. F., Lavela, P., Alcántara, R., and Tirado, J. L. (2013). Improved coulombic efficiency in nanocomposite thin film based on electrodeposited-oxidized FeNi-electrodes for lithium-ion batteries, *J. Alloys Compounds*, **557**, pp. 82–90.
38. López, M. C., Ortiz, G. F., Lavela, P., Tirado, J. L. (2013). Towards an all-solid-state battery: Preparation of conversion anodes by electrodeposition-oxidation processes, *J. Power Sources*, **244**, pp. 403–409.
39. Wang, B., Chen, J. S., Wu, H. B. Wang, Z., and Lou, X. W. (2011). Quasiemulsion-templated formation of α-Fe_2O_3 hollow spheres with enhanced lithium storage properties, *J. Am. Chem. Soc.*, **133**, pp. 17146–17148.
40. Zhong, K., Xia, X., Zhang, B., Li, H., Wang, Z., and Chen, L. (2010). MnO powder as anode active materials for lithium ion batteries, *J. Power Sources*, **195**, pp. 3300–3308.
41. Mueller, F., Bresser, D., Paillard, E., Winter, M., and Passerini, S. (2013). Influence of the carbonaceous conductive network on the electrochemical performance of $ZnFe_2O_4$ nanoparticles, *J. Power Sources*, **236**, pp. 87–94.
42. Yao, L., Hou, X., Hu, S., Wang, J., Li, M. C., Su, M., Tade, O., Shao, Z., and Liu, X. (2014). Green synthesis of mesoporous $ZnFe_2O_4$/C composite microspheres as superior anode materials for lithium-ion batteries, *J. Power Sources*, **258**, pp. 305–313.
43. Wang, C., Shao, G., Ma, Z., Liu, S., Song, W., and Song, J. (2014). A facile synthesis of encapsulated $CoFe_2O_4$ into carbon nanofibres and its application as conversion anodes for lithium ion batteries, *J. Power Sources*, **260**, pp. 205–210.
44. Lee, R. C., Lin, Y. P., Weng, Y. T., Pan, H. A., Lee, J. F., and Wu, N. L. (2014). Synthesis of high-performance MnO_x/carbon composite as lithium-ion battery anode by a facile co-precipitation method: Effects of oxygen stoichiometry and carbon morphology, *J. Power Sources*, **253**, pp. 373–380.
45. Sun, X., Xu, Y., Ding, P., Chen, G., Zheng, X., Zhang, R., and Li, L. (2014). The composite sphere of manganese oxide and carbon nanotubes as a prospective anode material for lithium-ion batteries, *J. Power Sources*, **255**, pp. 163–169.

46. Lee, D. H., Seo, S. D., Lee, G. H., Hong, H. S., and Kim, D. W. (2014). One-pot synthesis of Fe_3O_4/Fe/MWCNT nanocomposites via electrical wire pulse for Li ion battery electrodes, *J. Alloys Compounds*, **606**, pp. 204–207.

47. Fang, J., Yuan, Y. F., Wang, L. K., Ni, H. L., Zhu, H. L., Yang, J. L., Guid, J. S., Chen, Y. B., and Guo, S. Y. (2013). Synthesis and electrochemical performances of ZnO/MnO_2 sea urchin-like sleeve array as anode materials for lithium-ion batteries, *Electrochim. Acta*, **112**, pp. 364–370.

48. Sun, Z., Ai, W., Liu, J., Qi, X., Wang, Y., Zhu, J., Zhang, H., and Yu, T. (2014). Facile fabrication of hierarchical $ZnCo_2O_4$/NiO core/shell nanowire arrays with improved lithium ion battery performance, *Nanoscale*, **6**, pp. 6563–6568.

49. Ortiz, G. F., Hanzu, I., Lavela, P., Tirado, J. L., Knauth, P., and Djenizian, T. (2010). A novel architectured negative electrode based on titania nanotube and iron oxide nanowire composites for Li-ion microbatteries, *J. Mater. Chem.*, **20**, pp. 4041–4046.

50. Li, C., Wei, W., Fang, S. M., Wang, H. X., Zhang, Y., Gui, Y. H., and Chen, R. F. (2010). A novel CuO-nanotube/SnO_2 composite as the anode material for lithium ion batteries, *J. Power Sources*, **195**, pp. 2939–2944.

51. Guo, J., Chen, L., Wang, G., Zhang, X., and Li, F. (2014). SnO_2-Fe_2O_3@C composite as fully reversible anode material for lithium-ion batteries, *J. Power Sources*, **246**, pp. 862–867.

52. Yu, Y., Chen, C. H., Shui, J. L., and Xie, S. (2005). Nickel-foam-supported reticular CoO–Li_2O composite anode materials for lithium ion batteries, *Angew. Chem. Int. Ed.*, **44**, pp. 7085–7089.

53. Sun, Y., Du, C., Feng, X. Y., Yu, Y., Lieberwirth, I., and Chen, C. H. (2012). Electrostatic spray deposition of nanoporous CoO/Co composite thin films as anode materials for lithium-ion batteries, *Appl. Surf. Sci.*, **259**, pp. 769–773.

54. Nam, K. T., Kim, D. W., Yoo, P. J., Chiang, C. Y., Meethong, N., Hammond, P. T., Chiang, Y. M., and Belcher, C. (2006). Virus-enabled synthesis and assembly of nanowires for lithium ion battery electrodes, *Science*, **312**, pp. 885–888.

55. Zhao, J., Zhang, S., Liu, W., Du, Z., and Fang, H. (2014). Fe_3O_4/PPy composite nanospheres as anode for lithium-ion batteries with superior cycling performance, *Electrochim. Acta*, **121**, pp. 428–433.

56. Yin, Z., Ding, Y., Zheng, Q., and Guan, L. (2012). CuO/polypyrrole core–shell nanocomposites as anode materials for lithium-ion batteries, *Electrochem. Commun.*, **20**, pp. 40–43.

57. Jung, H. R., Kim, E. J., Park, Y. J., and Shin, H. C. (2011). Nickel-tin foam with nanostructured walls for rechargeable lithium battery, *J. Power Sources*, **196**, pp. 5122–5127.
58. Li, J. T., Swiatowska, J., Seyeux, A., Huang, L., Maurice, V., Sun, S. G., and Marcus, P. (2010). XPS and ToF-SIMS study of Sn–Co alloy thin films as anode for lithium ion battery, *J. Power Sources*, **195**, pp. 8251–8257.
59. Nacimiento, F., Alcántara, R., González, J. R., and Tirado, J. L. (2012). Electrodeposited polyacrylonitrile and cobalt-tin composite thin film on titanium substrate, *J. Electrochem. Soc.*, **159**, pp. A1028–A1033.
60. Ortiz, G. F., López, M. C., Alcántara, R., and Tirado, J. L. (2014). Electrodeposition of copper-tin nanowires on Ti foils for rechargeable lithium micro-batteries with high energy density, *J. Alloys Compounds*, **585**, pp. 331–336.
61. Ui, K., Kikuchi, S., Jimba, Y., and Kumagai, N. (2011). Preparation of Co-Sn alloy film as negative electrode for lithium secondary batteries by pulse electrodeposition method, *J. Power Sources*, **196**, pp. 3916–3920.
62. Nitta, N., and Yushin, G. (2014). High-capacity anode materials for lithium-ion batteries: Choice of elements and structures for active particles, *Part. Part. Syst. Charact.*, **31**, pp. 317–336.
63. Evanoff, K., Khan, J., Balandin, A. A., Magasinski, A., Ready, W. J., Fuller, T. F., and Yushin, G. (2012). Towards ultrathick battery electrodes: Aligned carbon nanotube-enabled architecture, *Adv. Mater.*, **24**, pp. 533–537.
64. Frye, A., Galyon, G. T., and Palmer, L. (2007). Crystallographic texture and whiskers in electrodeposited tin films, *IEEE Trans. Electron. Packaging Manufacturing*, **30**, pp. 2–10.
65. Tabuchi, T., Hochgatterer, N., Ogumi, Z., and Winter, M. (2009). Ternary Sn-Sb-Co alloy film as new negative electrode for lithium-ion cells, *J. Power Sources*, **188**, pp. 552–557.
66. Bourderau, S., Brousse, T., and Schleich, D. M. (1999). Amorphous silicon as a possible anode material for Li-ion batteries, *J. Power Sources*, **81**, pp. 233–236.
67. Fragnaud, P., Nagarajan, R., Schleich, D. M., and Vujic, D. (1995). Thin-film cathodes for secondary lithium batteries, *J. Power Sources*, **54**, pp. 362–366.
68. Wang, J., Du, N., Zhang, H., Yu, J., and Yang, D. (2011). Cu-Sn core–shell nanowire arrays as three-dimensional electrodes for lithium-ion batteries, *J. Phys. Chem. C*, **115**, pp. 23620–23624.

69. Xue, L. J., Xu, Y. F., Huang, L., Ke, F. S., He, Y., Wang, Y. X., Wei, G. Z., Li, J. T., and Sun, S. G. (2011). Lithium storage performance and interfacial processes of three dimensional porous Sn-Co alloy electrodes for lithium-ion batteries, *Electrochim. Acta*, **56**, pp. 5979–5987.
70. Shin, H. C., and Liu, M. L. (2004). Copper foam structures with highly porous nanostructured walls, *Chem. Mater.*, **16**, pp. 5460–5464.
71. Zhang, H., Ye, Y., Shen, R., Ru, C., and Hu, Y. (2013). Effect of bubble behavior on the morphology of foamed porous copper prepared via electrodeposition, *J. Electrochem. Soc.*, **160**, pp. D441–D445.
72. Olurin, O. B., Wilkinson, D. S., Weatherly, G. C., Paserin, V., and Shu, J. (2003). Strength and ductility of as-plated and sintered CVD nickel foams, *Compos. Sci. Technol.*, **63**, pp. 2317–2329.
73. Choi, W. S., Jung, H. R., Kwon, S. H., Lee, J. W., Liu, M., and Shin, H. C. (2012). Nanostructured metallic foam electrodeposits on a nonconductive substrate, *J. Mater. Chem.*, **22**, pp. 1028–1032.
74. Chabi, S., Peng, C., Hu, D., and Zhu, Y. (2014). Ideal three-dimensional electrode structures for electrochemical energy storage, *Adv. Mater.*, **26**, pp. 2440–2445.
75. Fan, X. Y., Ke, F. S., Wei, G. Z., Huang, L., and Sun, S. G. (2009). Sn-Co alloy anode using porous Cu as current collector for lithium ion battery, *J. Alloys Compounds*, **476**, pp. 70–73.
76. Hertzberg, B., Alexeev, A., and Yushin, G. (2010). Deformations in Si-Li anodes upon electrochemical alloying in nano-confined space, *J. Am. Chem. Soc.*, **132**, pp. 8548–8549.
77. González, J. R., Nacimiento, F., Alcántara, R., Ortiz, G. F., and Tirado, J. L. (2013). Electrodeposited $CoSn_2$ on nickel open-cell foam: Advancing towards high power lithium ion and sodium ion batteries, *CrystEngComm*, **15**, pp. 9196–9202.
78. Ortiz, G. F., Lavela, P., Knauth, P., Djenizian, T., Alcántara, R., and Tirado, J. L. (2011). Tin-based composite materials fabricated by anodic oxidation for the negative electrode of Li-ion batteries, *J. Electrochem. Soc.*, **158**, pp. A1094–A1099.
79. Susantyoko, R. A., Wang, X., Fan, Y., Xiao, Q., Fitzgerald, E., Pey, K. L., and Zhang, Q. (2014). Stable cyclic performance of nickel oxide–carbon composite anode for lithium-ion batteries, *Thin Solid Films*, **558**, pp. 356–364.
80. Cheng, J., Pan, Y., Zhu, J., Li, Z., Pan, J., and Ma, Z. (2014). Hybrid network CuS monolith cathode materials synthesized via facile in situ melt-diffusion for Li-ion batteries, *J. Power Sources*, **257**, pp. 192–197.

81. Xi, K., Kidambi, P. R., Chen, R., Gao, C., Peng, X., Ducati, C., Hofmann, S., Kumar, R. V. (2014). Binder free three-dimensional sulfur/few-layer graphene foam cathode with enhanced high-rate capability for rechargeable lithium sulfur batteries, *Nanoscale*, **6**, pp. 5746–5753.

82. Hu, X. F., Han, X. P., Hu, Y. X., Cheng, F. Y., and Chen, J. (2014). Epsilon-MnO_2 nanostructures directly grown on Ni foam: A cathode catalyst for rechargeable Li-O_2 batteries, *Nanoscale*, **6**, pp. 3522–3525.

83. Sun, K., Wei, T. S., Ahn, B. Y., Seo, J. Y., Dillon, S. J., and Lewis, J. A. (2013). 3D printing of interdigitated Li-ion microbattery architectures, *Adv. Mater.*, **25**, pp. 4539–4543.

84. Huang, X., Qi, X., Boey, F., and Zhang, H. (2012). Graphene-based composites, *Chem. Soc. Rev.*, **41**, pp. 666–668.

85. Ambrosi, A., Chua, C. K., Bonanni, A., and Pumera, M. (2014). Electrochemistry of graphene and related materials, *Chem. Rev.*, **114**, pp. 7150–7188.

86. Yoo, E., Kim, J., Hosono, E., Zhou, H. S., Kudo, T., and Honma, I. (2008). Large reversible Li storage of graphene nanosheet families for use in rechargeable lithium ion batteries, *Nano Lett.*, **8**, pp. 2277–2282.

87. Guo, P., Song, H., and Chen, X. (2009). Electrochemical performance of graphene nanosheets as anode material for lithium-ion batteries, *Electrochem. Commun.*, **11**, pp. 1320–1324.

88. Jang, B. Z., Liu, C., Neff, D., Yu, Z., Wang, M. C., Xiong, W., and Zhamu, A. (2011). Graphene surface-enabled lithium ion-exchanging cells: Next-generation high-power energy storage devices, *Nano Lett.*, **11**, pp. 3785–3791.

89. Cao, X., Zheng, B., Rui, X., Shi, W., Yan, Q., and Zhang, H. (2014). Metal oxide-coated three-dimensional graphene prepared by the use of metal-organic frameworks as precursors, *Angew. Chem. Int. Ed.*, **53**, pp. 1404–1409.

90. Hummers, W. S. Jr., and Offeman, R. E. (1958). Preparation of graphitic oxide, *J. Am. Chem. Soc.*, **80**, pp. 1339–1339.

91. Chen, C. M., Zhang, Q., Huang, J. Q., Zhang, W., Zhao, X. C., Huang, C. H., Wei, F., Yang, Y. G., Wang, M. Z., and Su, D. S. (2012). Chemically derived graphene-metal oxide hybrids as electrodes for electrochemical energy storage: pregraphenization or post-graphenization? *J. Mater. Chem.*, **22**, pp. 13947–13955.

92. Luo, J., Cote, L. J., Tung, V. C., Tan, A. T. L., Goins, P. E., Wu, J., and Huang, J. (2010). Graphene oxide nanocolloids, *J. Am. Chem. Soc.*, **140**, pp. 17667–17669.

93. Chartarrayawadee, W., Moulton, S. E., Li, D., Too, C. O., and Wallace, G. G. (2012). Novel composite graphene/platinum electrocatalytic electrodes prepared by electrophoretic deposition from colloidal solutions, *Electrochim. Acta*, **60**, pp. 213–223.
94. Yun, J. H., Wong, R. J., Ng, Y. H., Dub, A., and Amal, R. (2012). Combined electrophoretic deposition-anodization method to fabricate reduced graphene oxide-TiO$_2$ nanotube films, *RSC Adv.*, **2**, pp. 8164–8171.
95. Menéndez, R., Álvarez, P., Botas, C., Nacimiento, F., Alcántara, R., Tirado, J. L., and Ortiz, G. F. (2014). Self-organized amorphous titania nanotubes with deposited graphene film like a new heterostructured electrode for lithium ion batteries, *J. Power Sources*, **248**, pp. 886–893.
96. Yang, S., Feng, X., Ivanovici, S., and Müllen, K. (2010). Fabrication of graphene-encapsulated oxide nanoparticles: Towards high-performance anode materials for lithium storage, *Angew. Chem. Int. Ed.*, **49**, pp. 8408–8411.
97. Zhou, W., Zhu, J., Cheng, C., Liu, J., Yang, H., Cong, C., Guan, C., Jia, X., Fan, H. J., Yan, Q., Li, C. M., and Yu, T. (2011). A general strategy toward graphene@metal oxide core-shell nanostructures for high-performance lithium storage, *Energy Environ. Sci.*, **4**, pp. 4954–4961.
98. Zhang, W., Zeng, Y., Xiao, N., Hng, H. H., and Yan, Q. (2012). One-step electrochemical preparation of graphene-based heterostructures for Li storage, *J. Mater. Chem.*, **22**, pp. 8455–8461.
99. Kim, G. P., Nam, I., Park, S., Park, J., and Yi, J. (2013). Preparation via an electrochemical method of graphene films coated on both sides with NiO nanoparticles for use as high-performance lithium ion anodes, *Nanotechnology*, **24**, pp. 475402/1–475402/8.
100. Wang, D., Choi, D., Li, J., Yang, Z., Nie, Z., Kou, R., Hu, D., Wang, C., Saraf, L. V., Zhang, J., Aksay, I. A., and Liu, J. (2009). Self-assembled TiO$_2$-Graphene hybrid nanostructures for enhanced Li-ion insertion, *ACS Nano*, **3**, pp. 907–914.
101. Wang, D., Kou, R., Choi, D., Yang, Z., Nie, Z., Li, J., Saraf, L. V., Hu, D., Zhang, J., Graff, G. L., Liu, J., Pope, M. A., and Aksay, I. A. (2010). Ternary self-assembly of ordered metal oxide-graphene nanocomposites for electrochemical energy storage, *ACS Nano*, **4**, pp. 1587–1595.
102. Su, Y., Li, S., Wu, D., Zhang, F., Liang, H., Gao, P., Cheng, C., and Feng, X. (2012). Two-dimensional carbon-coated graphene/metal oxide hybrids for enhanced lithium storage, *ACS Nano*, **6**, pp. 8349–8356.

103. Zhu, J., Yang, D., Rui, X., Sim, D., Yu, H., Hoster, H. E., Ajayan, P. M., and Yan, Q. (2013). Facile preparation of ordered porous graphene-metal oxide@C binder-free electrodes with high Li storage performance, *Small*, **9**, pp. 3390–3397.

104. Yue, W., Jiang, S., Huang, W., Gao, Z., Li, J., Ren, Y., Zhao, X., and Yang, X. (2013). Sandwich-structural graphene-based metal oxides as anode materials for lithium-ion batteries, *J. Mater. Chem. A*, **1**, pp. 6928–6933.

105. Park, J., An, K., Hwang, Y., Park, J. G., Noh, H. J., Kim, J. Y., Park, J. H., Hwang, N. M., and Hyeon, T. (2004). Ultra-large-scale syntheses of monodisperse nanocrystals, *Nat. Mater.*, **3**, pp. 891–895.

106. Baek, S., Yu, S. H., Park, S. K., Pucci, A., Marichy, C., Lee, D. C., Sung, Y. E., Piao, Y., and Pinna, N. (2011). A one-pot microwave-assisted non-aqueous sol-gel approach to metal oxide/graphene nanocomposites for Li-ion batteries, *RSC Adv.*, **1**, pp. 1687–1690.

107. Armstrong, A. R., Armstrong, G., Canales, J., and Bruce, P. G. (2004). TiO_2-B Nanowires, *Angew. Chem. Int. Ed.*, **43**, pp. 2286–2288.

108. Lee, J., and Teja, A. S. (2005). Characteristics of lithium iron phosphate ($LiFePO_4$) particles synthesized in subcritical and supercritical water, *J. Supercrit. Fluids*, **35**, pp. 83–90.

109. Zhu, J., Zhu, T., Zhou, X., Zhang, Y., Lou, X. W., Chen, X., Zhang, H., Hng, H. H., and Yan, Q. (2011). Facile synthesis of metal oxide/reduced graphene oxide hybrids with high lithium storage capacity and stable cyclability, *Nanoscale*, **3**, pp. 1084–1089.

110. Han, S. Y., Kim, I. Y., Jo, K. Y., and Hwang, S. J. (2012). Solvothermal-assisted hybridization between reduced graphene oxide and lithium metal oxides: a facile route to graphene-based composite materials, *J. Phys. Chem. C*, **116**, pp. 7269–7279.

111. Park, S. K., Jin, A., Yu, S. H., Ha, J., Jang, B., Bong, S., Woo, S., Sung, Y. E., and Piao, Y. (2014). In situ hydrothermal synthesis of Mn_3O_4 nanoparticles on nitrogen-doped graphene as high-performance anode materials for lithium ion batteries, *Electrochim. Acta*, **120**, pp. 452–459.

112. Chidembo, A., Aboutalebi, S. H., Konstantinov, K., Salari, M., Winton, B., Yamini, S. A., Nevirkovets, Nevirkovets, I. P., and Liu, H. K. (2012). Globular reduced graphene oxide-metal oxide structures for energy storage applications, *Energy Environ. Sci.*, **5**, pp. 5236–5240.

113. Huang, S., Kavan, L., Exnar, I., and Grätzel, M. (1995). Rocking chair lithium battery based on nanocrystalline TiO_2 (anatase), *J. Electrochem. Soc.*, **142**, pp. L142–L144.

114. Dong, S., Wang, H., Gu, L., Zhou, X., Liu, Z., Han, P., Wang, Y., Chen, X., Cui, G., and Chen, L. (2011). Rutile TiO_2 nanorod arrays directly grown on Ti foil substrates towards lithium-ion micro-batteries, *Thin Solid Films*, **519**, pp. 5978–5982.
115. Yang, S., Feng, X., and Müllen, K. (2011). Sandwich-like, graphene-based titania nanosheets with high surface area for fast lithium storage, *Adv. Mater.*, **23**, pp. 3575–3579.
116. Cai, D., Lian, P., Zhu, X., Liang, S., Yang, W., and Wang, H. (2012). High specific capacity of TiO_2-graphene nanocomposite as an anode material for lithium-ion batteries in an enlarged potential window, *Electrochim. Acta*, **74**, pp. 65–72.
117. Zhang, F., Cao, H., Yue, D., Zhang, J., and Qu, M. (2012). Enhanced anode performances of polyaniline-TiO_2-reduced graphene oxide nanocomposites for lithium ion batteries, *Inorg. Chem.*, **51**, pp. 9544–9551.
118. Lee, J. M., Kim, I. Y., Han, S. Y., Kim, T. W., and Hwang, S. J. (2012). Graphene nanosheets as a platform for the 2D ordering of metal oxide nanoparticles: Mesoporous 2D aggregate of anatase TiO_2 nanoparticles with improved electrode performance, *Chem. Eur. J.*, **18**, pp. 13800–13809.
119. Wang, J., Shen, L., Nie, P., Xu, G., Ding, B., Fang, S., Dou, H., and Zhang, X. (2014). Synthesis of hydrogenated TiO_2-reduced-graphene oxide nanocomposites and their application in high rate lithium ion batteries, *J. Mater. Chem. A*, **2**, pp. 9150–9155.
120. Ohzuku, T., Ueda, A., and Yamamoto, N. (1995). Zero-strain insertion material of $Li[Li_{1/3}Ti_{5/3}]O_4$ for rechargeable lithium cells, *J. Electrochem. Soc.*, **142**, pp. 1431–1435.
121. Tao, L., Zai, J., Wang, K., Zhang, H., Xu, M., Shen, J., Su, Y., and Qian, X. (2012). Co_3O_4 nanorods/graphene nanosheets nanocomposites for lithium ion batteries with improved reversible capacity and cycle stability, *J. Power Sources*, **202**, pp. 230–235.
122. Kim, K. H., Jung, D. W., Pham, V. H., Chung, J. S., Kong, B. S., Lee, J. K., Kim, K., and Oh, E. S. (2012). Performance enhancement of Li-ion batteries by the addition of metal oxides (CuO, Co_3O_4)/solvothermally reduced graphene oxide composites, *Electrochim. Acta*, **69**, pp. 358–363.
123. Nethravathi, C., Rajamathi, C. R., Rajamathi, M., Wang, X., Gautam, U. K., Golberg, D., and Bando, Y. (2014). Cobalt hydroxide/oxide hexagonal ring-graphene hybrids through chemical etching of metal hydroxide platelets by graphene oxide: energy storage applications, *ACS Nano*, **8**, pp. 2755–2765.

124. Wang, B., Wu, X. L., Shu, C. Y., Guo, Y. G., and Wang, C. R. (2010). Synthesis of CuO/graphene nanocomposite as a high-performance anode material for lithium-ion batteries, *J. Mater. Chem.*, **20**, pp. 10661–10664.

125. Zhou, G., Wang, D. W., Li, F., Zhang, L., Li, N., Wu, Z. S., Wen, L., Lu, G. Q., and Cheng, H. M. (2010). Graphene-wrapped Fe_3O_4 anode material with improved reversible capacity and cyclic stability for lithium ion batteries, *Chem. Mater.*, **22**, pp. 5306–5313.

126. Hu, A., Chen, X., Tang, Y., Tang, Q., Yang, L., and Zhang, S. (2013). Self-assembly of Fe_3O_4 nanorods on graphene for lithium ion batteries with high rate capacity and cycle stability, *Electrochem. Commun.*, **28**, pp. 139–142.

127. Huang, X., Chen, J., Yu, H., Cai, R., Peng, S., Yan, Q., and Hng, H. H. (2013). Carbon buffered-transition metal oxide nanoparticle-graphene hybrid nanosheets as high-performance anode materials for lithium ion batteries, *J. Mater. Chem. A*, **1**, pp. 6901–6907.

128. Liu, J., Jiang, J., Qian, D., Tan, G., Peng, S., Yuan, H., Luo, D., Wang, Q., and Liu, Y. (2013). Facile assembly of a 3D rGO/MWCNTs/Fe_2O_3 ternary composite as the anode material for high-performance lithium ion batteries, *RSC Adv.*, **3**, pp. 15457–15466.

129. Yue, W., Lin, Z., Jiang, S., and Yang, X. (2012). Preparation of graphene-encapsulated mesoporous metal oxides and their application as anode materials for lithium-ion batteries, *J. Mater. Chem.*, **22**, pp. 16318–16323.

130. Mai, Y. J., Shi, S. J., Zhang, D., Lu, Y., Gu, C. D., and Tu, J. P. (2012). NiO-graphene hybrid as an anode material for lithium ion batteries, *J. Power Sources*, **204**, pp. 155–161.

131. Li, B. J., Cao, H. Q., Shao, J., Zheng, H., Lu, Y. X., Yin, J. F., and Qu, M. Z. (2011). Improved performances of $Ni(OH)_2$@reduced-graphene-oxide in Ni-MH and Li-ion batteries, *Chem. Commun.*, **47**, pp. 3159–3161.

132. Tao, S., Yue, W., Zhong, M., Chen, Z., and Ren, Y. (2014). Fabrication of graphene-encapsulated porous carbon-metal oxide composites as anode materials for lithium-ion batteries, *ACS Appl. Mater. Interfaces*, **6**, pp. 6332–6339.

133. Wang, H., Cui, L. F., Yang, Y. Casalongue, H. S., Robinson, J. T., Liang, Y., Cui, Y., and Dai, H. (2010). Mn_3O_4-graphene hybrid as a high-capacity anode material for lithium ion batteries, *J. Am. Chem. Soc.*, **132**, pp. 13978–13980.

134. Latorre-Sanchez, M., Atienzar, P., Abellán, G., Puche, M., Fornés, V., Ribera, A., and García, H. (2012). The synthesis of a hybrid graphene-nickel/manganese mixed oxide and its performance in lithium-ion batteries, *Carbon*, **50**, pp. 518–525.

135. Bhaskar, A., Deepa, M., Rao, T. N., and Varadaraju, U. V. (2012). Enhanced nanoscale conduction capability of a MoO_2/Graphene composite for high performance anodes in lithium ion batteries, *J. Power Sources*, **216**, pp. 169–178.

136. Kucinskis, G., Bajars, G., and Kleperis. J. (2013). Graphene in lithium ion battery cathode materials: A review, *J. Power Sources*, **240**, pp. 66–79.

137. Ding, Y., Jiang, Y., Xu, F., Yin, J., Ren, H., Zhuo, Q., Long, Z., and Zhang, P. (2010). Preparation of nano-structured $LiFePO_4$/graphene composites by co-precipitation method, *Electrochem. Commun.*, **12**, pp. 10–13.

138. Zhu, X., Hu, J., Wu, W., Zeng, W., Dai, H., Du, Y., Liu, Z., Li, L., Ji, H., and Zhu, Y. (2014). $LiFePO_4$/reduced graphene oxide hybrid cathode for lithium ion battery with outstanding rate performance, *J. Mater. Chem. A*, **2**, pp. 7812–7818.

139. Zhao, H., Pan, L., Xing, S., Luo, J., and Xu, J. (2013). Vanadium oxides-reduced graphene oxide composite for lithium-ion batteries and supercapacitors with improved electrochemical performance, *J. Power Sources*, **222**, pp. 21–31.

140. Prabakar, S. J. R., Hwang, Y. H., Lee, B., Sohn, K. S., and Pyo, M. (2013). Graphene-sandwiched $LiNi_{0.5}Mn_{1.5}O_4$ cathode composites for enhanced high voltage performance in Li ion batteries, *J. Electrochem. Soc.*, **160**, pp. A832–A837.

141. Fang, X., Ge, M., Rong, J., and Zhou, C. (2013). Graphene oxide-coated $LiNi_{0.5}Mn_{1.5}O_4$ as high voltage cathode for lithium ion batteries with high energy density and long cycle life, *J. Mater. Chem. A*, **1**, pp. 4083–4088.

142. Noerochim, L., Wang, J. Z., Wexler, D., Chao, Z., and Liu, H. K. (2013). Rapid synthesis of free-standing MoO_3/Graphene films by the microwave hydrothermal method as cathode for bendable lithium batteries, *J. Power Sources*, **228**, pp. 198–205.

143. Song, Z., Ma, T., Tang, R., Cheng, Q., Wang, X., Krishnaraju, D., Panat, R., Chan, C. K., Yu, H., and Jiang, H. (2014). Origami lithium-ion batteries, *Nat. Commun.*, **5**, doi:10.1038/ncomms4140.

144. Chang, J., Jin, M., Yao, F., Kim, T. H., Le, V. T., Yue, H., Gunes, F., Li, B., Ghosh, A., Xie, S., and Lee, Y. H. (2013). Asymmetric supercapacitors based on graphene/MnO_2 nanospheres and graphene/MoO_3 nanosheets with high energy density, *Adv. Funct. Mater.*, **23**, pp. 5074–5083.

145. Antiohos, D., Pingmuang, K., Romano, M. S., Beirne, S., Romeo, T., Aitchison, P., Minett, A., Wallace, G., Phanichphant, S., and Chen, J. (2013). Manganosite-microwave exfoliated graphene oxide composites for asymmetric supercapacitor device applications, *Electrochim. Acta*, **101**, pp. 99–108.

146. Padhi, A. K., and Nanjundaswamy, K. S. (1997). Phospho-olivines as positive-electrode materials for rechargeable lithium batteries, *J. Electrochem. Soc.*, **144**, pp. 1188–1194.

147. Amatucci, G. G., Schmutz, C. N., Blyr, A., Sigala, C., Gozdz, A. S., Larcher, D., and Tarascon, J. M. (1999). Materials' effects on the elevated and room temperature performance of $C/LiMn_2O_4$ Li-ion batteries, *J. Power Sources*, **69**, pp. 11–25.

148. Yamada, A., Tanaka, M., Tanaka, K., and Skeai, K. (1999). Jahn–Teller instability in spinel Li–Mn–O, *J. Power Sources*, **81**, pp. 73–78.

149. Aoshima, T., Okahara, K., Kiyohara, C., and Shizuka, K. (2001). Mechanisms of manganese spinels dissolution and capacity fade at high temperature, *J. Power Sources*, **97**, pp. 377–380.

150. Chung, S. Y., Bloking, J. T., and Chiang, Y. M. (2002). Electronically conductive phospho-olivines as lithium storage electrodes, *Nat. Mater.*, **1**, pp. 123–128.

151. Burba, C. A., and Frech, R. (2007). Local structure in the Li-ion battery cathode material $Li_x(Mn_yFe_{1-y})PO_4$ for $0 < x \leq 1$ and $y = 0.0$, 0.5 and 1.0, *J. Power Sources*, **172**, pp. 870–876.

152. Yang, J., Wang, J., Wang, D., Li, X., Geng, D., Liang, D., Gauthier, M., Li, R., and Sun, X. (2012). 3D porous $LiFePO_4$/graphene hybrid cathodes with enhanced performance for Li-ion batteries, *J. Power Sources*, **208**, pp. 340–344.

153. Liu, Y., Li, X., Guo, H., Wang, Z., Peng, W., Yang, Y., and Liang, R. (2008). Effect of carbon nanotube on the electrochemical performance of C-$LiFePO_4$/graphite battery, *J. Power Sources*, **184**, pp. 522–526.

154. Wang, Y., Wang, Y., Hosono, E., Wang, K., and Zhou, H. (2008). The design of a $LiFePO_4$/carbon nanocomposite with a core-shell structure and its synthesis by an in situ polymerization restriction method, *Angew. Chem. Int. Ed.*, **47**, pp. 7461–7465.

155. Lim, S., and Cho, J. (2008). PVP-Assisted ZrO_2 coating on $LiMn_2O_4$ spinel cathode nanoparticles prepared by MnO_2 nanowire templates, *Electrochem. Commun.*, **10**, pp. 1478–1481.

156. Li, X., Yang, R., Cheng, B., Hao, Q., Xu, H., Yang, J., and Qian, Y. (2012). Enhanced electrochemical properties of nano-Li_3PO_4 coated on the $LiMn_2O_4$ cathode material for lithium ion battery at 55 degrees C, *Mater. Lett.*, **66,** pp. 168–171.

157. Park, K. S., Son, J. T., Chung, H .T., Kim, S. J., Lee, C. H., Kang, K. T., and Kim, H. G. (2004). Surface modification by silver coating for improving electrochemical properties of $LiFePO_4$, *Solid State Commun.*, **129,** pp. 311–314.

158. Boyano, I., Blázquez, J. A., de Meatza, I., Bengoechea, M., Miguel, O., Grande, H., Huang, Y., and Goodenough, J. B. (2010). Preparation of C-$LiFePO_4$/polypyrrole lithium rechargeable cathode by consecutive potential steps electrodeposition, *J. Power Sources*, **195,** pp. 5351–5359.

159. Her, L. J., Hong, J. L., and Chang, C. C. (2006). Preparation and electrochemical characterizations of poly(3,4-dioxyethylenethiophene)/$LiCoO_2$ composite cathode in lithium-ion battery, *J. Power Sources*, **157,** pp. 457–463.

160. Lepage, D., Michot, C., Liang, G., Gauthier, M., and Schougaard, S. B. (2011). A soft chemistry approach to coating of $LiFePO_4$ with a conducting polymer, *Angew. Chem. Int. Ed.*, **50,** pp. 6884–6887.

161. Macak, J. M., Gong, B. G., Hueppe, M., and Schmuki, P. (2007). Filling of TiO_2nt nanotubes by self-doping and electrodeposition, *Adv. Mater.*, **19,** pp. 3027–3031.

162. Zhang, S. Q., Xie, S., and Chen, C. H. (2005). Fabrication and electrical properties of Li_3PO_4-based composite electrolyte films, *Mater. Sci. Eng. B.*, **121,** pp. 160–1655.

163. Sun, K., and Dillon, S. J. (2011). A mechanism for the improved rate capability of cathodes by lithium phosphate surficial films, *Electrochem. Commun.*, **13,** pp. 200–202.

164. Wu, X., Wang, S., Lin, X., Zhong, G., Gong, Z., and Yang, Y. (2014). Promoting long-term cycling performance of high-voltage Li_2CoPO_4F by the stabilization of electrode/electrolyte interface, *J. Mater. Chem. A*, **2,** pp. 1006–1013.

165. López, M. C., Ortiz, G. F., González, J. R., Alcántara, R., and Tirado, J. L. (2014). Improving the performance of titania nanotube battery materials by surface modification with lithium phosphate, *ACS Appl. Mater. Interfaces*, **6,** pp. 5669–5678.

166. Kuwata, N., Iwagami, N., Tanji, Y., Matsuda, Y., and Kawamura, J. (2010). Characterization of thin-film lithium batteries with stable thin-film Li_3PO_4 solid electrolytes fabricated by ArF excimer laser deposition, *J. Electrochem. Soc.*, **157**, pp. A521–A527.

167. Prosini, P. P., Cento, C., Pozio, A. (2014). Lithium-ion batteries based on titanium oxide nanotubes and $LiFePO_4$, *J. Solid State Electrochem.*, **18**, pp. 795–804.

Chapter 11

Electrochemical Fabrication of Carbon Nanomaterial and Conducting Polymer Composites for Chemical Sensing

Zhanna A. Boeva,[a,b] Rose-Marie Latonen,[a] Tom Lindfors,[a] and Zekra Mousavi[a]

[a]*Faculty of Science and Engineering,*
Johan Gadolin Process Chemistry Centre,
Laboratory of Analytical Chemistry, Åbo Akademi University,
Biskopsgatan 8, 20500 Turku, Finland
[b]*Chemistry Department, Polymer Division,*
M.V. Lomonosov Moscow State University,
Leninskie gory 1, build. 3, 119991 Moscow, Russian Federation

Tom.Lindfors@abo.fi

11.1 Introduction

In recent years, carbon nanotubes (CNTs) and graphene have been extensively studied due to their outstanding electron mobility, mechanical strength and large specific surface area [1]. These properties make them attractive in a wide variety of applications ranging from sensors and actuators [2], catalysts [3], batteries

---

The authors have equal contribution to the book chapter.

Electrochemical Nanofabrication: Principles and Applications (Second Edition)
Edited by Di Wei
Copyright © 2016 Pan Stanford Publishing Pte. Ltd.
ISBN 978-981-4613-86-6 (Hardcover), 978-981-4613-87-3 (eBook)
www.panstanford.com

[4], and hydrogen storage materials to solar cells [5]. Apart from CNTs and graphene, the electrically conducting polymers (ECPs) are efficient materials for various applications such as nanoelectronics and chemical sensors. The ECPs are unique materials since they are both ionic and electronic conductors. Some ECPs also exhibit high redox capacitance, making them interesting for supercapacitor applications and as ion-to-electron transducers in all-solid-state ion-selective electrodes (ISEs) [6]. To further improve the material properties like stability, electrical conductivity, capacitance, and mechanical strength, the ECPs have been fabricated as composites with a wide range of materials such as metal nanoparticles [7], conventional polymers [8], and carbon materials like CNTs, graphenes, and fullerenes [9].

The combination of ECPs and the carbon nanomaterials is greatly expected to contribute to the development of new composites with complementary properties and utilize the synergistic effect arising from the non-electrostatic interactions of their conjugated structures [10]. Different methods of nanofabrication have been recently proposed to prepare composites based on ECPs and carbon materials. In most of these methods, the charged and polar oxygen-containing surface groups (epoxy, hydroxyl, carbonyl and carboxyl) of oxidized CNTs and graphene oxide (GO) function as charge compensating sites in the chemical and electrochemical polymerization of aniline [11], 3,4-ethylenedioxythiophene (EDOT) [12] or pyrrole [13]. The good dispersibility of GO and oxidized CNTs in aqueous solutions is a consequence of their highly oxidized structure providing the carbon surface with a negative charge [14].

For electrochemical sensors, the ECP–CNT/GO composites can be deposited either as thin films on solid substrates or incorporated in conventional host polymeric materials. To further increase the electrical conductivity of the ECP-GO composites, the electrically non-conducting GO in the composites matrix can be converted chemically or electrochemically to reduced GO (RGO), which is electrically conducting [15]. During the reduction, most of the oxygen-containing groups are removed from the GO surface, which partially restores the sp^2 hybridization of the carbon atoms and increases the electrical conductivity of the material.

Due to the synergistic effect between the carbon nanomaterials and ECPs, the formed RGO (but also graphene and CNTs) increases the electrical conductivity of the composite materials. This facilitates more efficient oxidation/reduction of the ECPs, which enhances the electrocatalytic activity and signal amplification of the composites making them attractive for electroanalytical applications. In addition to the improved electrical conductivity, the incorporation of oxidized CNTs or GO in the polymer matrix improves the performance of the ECPs by shortening the ionic diffusion pathways for the negatively charged counterions (GO/CNT), thus improving the electrochemical properties of the ECPs.

In the field of electrochemical sensors, there is a continuous need for new electrode materials. In comparison to chemical oxidative polymerization, the electropolymerization of the carbon-ECP composite materials has been explored to a lesser extent. However, the possibility of precisely directing the electrodeposition of the composite materials on small well-defined electrically conducting surfaces is an advantage in solid-state electrochemical sensing applications. We will therefore discuss below the application of different types of electrochemically prepared ECP-CNT and ECP-graphene composites (including graphene derivatives) in electrochemical sensing. The discussion in the sub-chapters is divided into composite materials containing poly(3,4-ethylenedioxythiophene) (PEDOT), polyaniline (PANI) and polypyrrole (PPy). For the relatively new ECP-graphene composites, the discussion includes also other ECPs.

11.2 Composites of Carbon Nanotubes and Conducting Polymers

The incorporation of CNTs into ECPs, which combines the remarkable and complementary properties of both individual components, has attracted great interest over recent years. Chemical or electrochemical polymerization methods can be used for the fabrication of ECP-CNT composites. However, the electrochemical polymerization is more advantageous as the composite film with controlled thickness, formed uniformly on the substrate surface, can be easily characterized and used in many potential applications such as electrochemical capacitors [16], solar cells [17], and electrochemical sensors [18, 19]. In the electrochemical polymerization, two approaches can be used for preparing the ECP-CNT composites. In the first approach, the monomer is dissolved in a suspension of CNTs and thereafter the polymer is deposited on the substrate surface together with CNTs. In the other approach, a substrate surface covered with CNTs is used for electrochemical deposition of the conducting polymer.

11.2.1 Poly(3,4-Ethylenedioxythiophene)

Most of the practical applications of ECPs depend usually on their stability in ambient conditions. Thus, considerable attention has been paid to thiophene-based ECPs such as PEDOT, which possesses advantageous properties like proper thermal and chemical stability, low oxidation potential and high electrical conductivity in its p-doped state [20]. On the other hand, the combination of the significant characteristics of ECPs with the complementary properties of CNTs was shown to improve the mechanical and electrical properties of the original polymers [21]. It was shown that the morphology of PEDOT doped with CNTs, i.e., PEDOT-CNT, is more porous than PEDOT doped with the commonly used doping material poly(sodium-p-styrenesulfonate) (PSS), i.e., PEDOT-PSS. The unique porous morphology of PEDOT-CNT composite resulted in a higher capacitance and charge injection capacity in comparison with the PEDOT-PSS film [22].

Mousavi et al. have used PEDOT-CNT composite as ion-to-electron transducer in the fabrication of potassium ISEs [19]. In this work, PEDOT was electrochemically synthesized using negatively charged multi-walled CNTs (MWCNTs) as counterions. Results from cyclic voltammetric (CV) and electrochemical impedance spectroscopic (EIS) measurements shown in Fig. 11.1, reveal that the PEDOT-MWCNT film exhibits higher redox capacitance than a film based on PEDOT doped with chloride (Cl^-) ions, i.e., PEDOT-Cl. This sufficiently high redox capacitance is one of the conditions necessary for stable potential in all-solid-state ISEs having an ECP as the solid contact [23].

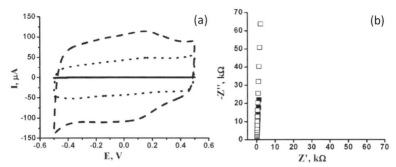

Figure 11.1 (a) Cyclic voltammograms for bare GC (solid line), GC/PEDOT-Cl (dotted line), and GC/PEDOT-MWCNT (dashed line); and (b) impedance spectra of GC/PEDOT-Cl (□), and GC/PEDOT-CNT electrodes (■) (reproduced with permission from the ref. 19, Figs. 7 and 8a, © 2009, Elsevier).

In many cases, chemically or electrochemically synthesized PEDOT-CNT composites have been used for chemical sensing applications without the need for an additional sensing component. The determination of dopamine (DA), which is a neurotransmitter with an important role in the central nervous and hormonal systems, was carried out using sensors based on the PEDOT-CNT composite [24, 25]. The determination of DA by electrochemical methods is limited by interferences from other substances present in physiological samples such as ascorbic acid (AA), which has the redox potential very close to that of DA. Xu et al. have used the PEDOT-MWCNT nanocomposite electrochemically deposited on a carbon paste electrode (CPE) for the fabrication of sensors used in the electrocatalytic detection of DA [24]. The PEDOT-MWCNT modified sensor was easy to prepare and showed long-term stability and appropriate sensitivity and selectivity for the detection of DA. It has been used for measuring DA in the concentration range 0.1 to 20 µM with limit of detection (LOD) 20 nM without any interference from high concentrations of AA present in the samples (Table 11.1). DA was also electrochemically detected using gold microelectrodes coated with the PEDOT-CNT composite prepared by the galvanodynamical method [25]. The polymer electrodeposition was carried out using EDOT aqueous solution containing either PSS or single-walled carbon nanotubes (SWCNTs) suspended in PSS. DA was detected using the square wave voltammetry (SWV) technique at microelectrodes modified with PEDOT-SWCNT composite and with plain PEDOT (without CNTs). Signals from 0.5 and 1 μM DA could be significantly distinguished at the PEDOT-SWCNT based electrode whereas signals of 1 and 5 μM were identified at the plain PEDOT-based electrode. It was shown that the current measured at the PEDOT-SWCNT electrode is about two times higher than the one measured at the plain PEDOT electrode. This was attributed to the high porosity of PEDOT-SWCNT composite in comparison to the PEDOT film without CNTs. According to Samba et al., it was not possible to separate the signal of DA from the mixed signal of AA and DA by using electrodes modified with plain PEDOT. However, they showed that the AA signal at the PEDOT-SWCNT based electrode was shifted to lower potential value and therefore, the peaks of DA and AA could be successfully separated [25].

Table 11.1 Type of composite material, type of analyte, linear response range, and LOD for electropolymerized ECP-CNT composites used in chemical sensing applications

Ref.	Composite	Analyte	Linear range [1] (M)	LOD[1] (M)
47	PANI-MWCNT-CuNP[2]	AA[3]	$(5–600) \times 10^{-6}$	1×10^{-6}
48	PDMA[4]-NH$_2$-MWCNT	AA	$(1–15) \times 10^{-6}$	1×10^{-7}
72	PPyox[5]-SWCNT	AA	$(2–100) \times 10^{-5}$	4.6×10^{-6}
61	PPy-MWCNT-HRP[6]	2-AminoP[7]	$(8–60.8) \times 10^{-6}$	1.5×10^{-6}
51	PANI-MWCNT	AP[8]	10^{-6}–10^{-4}	2.5×10^{-7}
50	PANI-MWCNT-Ch[9]	Bisphenol A	$(1–400) \times 10^{-6}$	1×10^{-10}
57	PoPD[10]-MWCNT	Ca-dob[11]	$(1–10) \times 10^{-7}$ 4.0-400 μM	3.5×10^{-8}
35	PANI-MWCNT-AChE[12]	Carbamate	$(9.9–49.6) \times 10^{-9}$	4.6×10^{-9}
53	PANI-gr[13]-NH$_2$-MWCNT	Celecoxib	N/A	1×10^{-11}
61	PPy-MWCNT-HRP	2-ChloroP[14]	$(1.6–8) \times 10^{-6}$	2.6×10^{-7}
60	PPy-MWCNT-ChEt[15]-ChOx[16]	Cholesterol	$(4–65) \times 10^{-4}$	4×10^{-5}
54	PDPA[17]-gr-NH$_2$-MWCNT	CO	10–200 ppm	0.01 ppm
39	PANI-MWCNT-(CA[18]-Cl[19]-SOx[20])	Creatinine	$(1–75) \times 10^{-4}$	10^{-7}
45	PANI-boronic acid-SWCNT-ssDNA[21]	DA[22]	N/A	4×10^{-11}
46	PANI-boronic acid-SWCNT-ssDNA	DA	$(1–10) \times 10^{-9}$	1×10^{-9}

Ref.	Composite	Analyte	Linear range [1] (M)	LOD[1] (M)
24	PEDOT-MWCNT	DA	$(1-200) \times 10^{-7}$	2.0×10^{-8}
71	PPyox-SWCNT	DA	N/A	3.3×10^{-9}
72	PPyox-SWCNT	DA	$(1-50) \times 10^{-6}$	3.8×10^{-7}
73	PPyox-MWCNT	DA	$(4.0-140) \times 10^{-8}$	1.7×10^{-9}
61	PPy-MWCNT-HRP	2,6-DMOP[24]	$(1.6-19.2) \times 10^{-6}$	2.9×10^{-7}
68	PPy-MWCNT-ssDNA	DNA	$3 \times 10^{-8} - 1.0 \times 10^{-5}$	1.0×10^{-8}
70	PPy-MWCNT-aDNA[25]	DNA	$1.0 \times 10^{-11} - 1.0 \times 10^{-7}$	5.0×10^{-12}
74	PPyox-MWCNT	EN[26]	$(1-80) \times 10^{-7}$	4.0×10^{-8}
59	PoAP[27]-CNT-GOD	Glucose	$2 \times 10^{-5} - 10^{-4}$	10^{-5}
32	PEDOT-CNT-GOD	Glucose	N/A	6.5×10^{-8}
32	PEDOT-CNx[28]-GOD	Glucose	N/A	1.6×10^{-8}
41	PANI-MWCNT-HRP	H_2O_2	$(2-190) \times 10^{-7}$	6.8×10^{-8}
42	PANI-MWCNT-PB[29]	H_2O_2	$8 \times 10^{-9} - 5 \times 10^{-6}$	5×10^{-9}
43	PANI-SWCNT	H_2O_2	$5 \times 10^{-6} - 1 \times 10^{-3}$	1.2×10^{-6}
44	PANI-NH_2-MWCNT	H_2O_2	$(1-20) \times 10^{-8}$	1×10^{-9}
55	PDPA-gr-MWCNT	H_2O_2	$(1-15) \times 10^{-10}$	1×10^{-11}
61	PPy-MWCNT-HRP	HQ[30]	$(1.6-24) \times 10^{-5}$	6.4×10^{-6}
66	PPy-MWCNT-Tiron	Levodopa	$1 \times 10^{-6} - 1 \times 10^{-4}$	1×10^{-7}
61	PPy-MWCNT-HRP	4-MethoxP[31]	$(1.6-81.6) \times 10^{-6}$	1.1×10^{-6}
49	PANI-CSA[32]-MWCNT-aDNA	*Neisseria gonorrhoeae*	$1 \times 10^{-17} - 1 \times 10^{-6}$	1.2×10^{-17}
58	PoPD-MWCNT	NADH[33]	$(2.0-4000) \times 10^{-6}$	0.5×10^{-6}
52	PANI-MWCNT	Nitrite	$5 \times 10^{-6} - 1.5 \times 10^{-2}$	1×10^{-6}
27	PEDOT-MWCNT	Nitrite	$(5-100) \times 10^{-5}$	9.6×10^{-7}
28	PEDOT-MWCNT-FePc[34]	Nitrite	N/A	7.1×10^{-8}
67	PPy-MWNT-PM[35]	Nitrite	$5.0 \times 10^{-6} - 3.4 \times 10^{-2}$	1.0×10^{-6}
56	PDPA-SWCNT-β-CD	4-nitroP[36]	0.1–13.9 µg L^{-1}	0.02 µg L^{-1}
65	MIP[37]-MWCNT-NanoSnO$_2$	Oleanolic acid	$5.0 \times 10^{-8} - 2.0 \times 10^{-5}$ g L^{-1}	8.6×10^{-9} g L^{-1}
40	PANI-MWCNT-OxOx[38]	Oxalate	$(8.4-272) \times 10^{-6}$	3×10^{-6}

(*Continued*)

Table 11.1 (Continued)

Ref.	Composite	Analyte	Linear range [1](M)	LOD[1] (M)
61	PPy-MWCNT-HRP	p-BQ[39]	$(2-16) \times 10^{-8}$	2.7×10^{-8}
61	PPy-MWCNT-HRP	Phenol	$(1.6-14.4) \times 10^{-5}$	3.5×10^{-6}
75	PPyox-MWCNT	PMX[40]	$(1.00-10.0) \times 10^{-8}$	3.3×10^{-9}
37	PANI-AgNP-MWCNT-Lacc[41]	Polyphenol	$(0.1-500) \times 10^{-6}$	1×10^{-7}
36	PANI-AuNP-Ch-MWCNT-SOx[42]	Sulfite	$(7.5-40) \times 10^{-7}$	5×10^{-7}
72	PPyox-SWCNT	Uric acid	$2.0 \times 10^{-6} - 1.0 \times 10^{-4}$	7.4×10^{-7}

[1]Unless otherwise stated
[2]Copper nanoparticle
[3]Ascorbic acid
[4]Poly(2,5-dimethoxy aniline)
[5]Overoxidized PPy
[6]Horseradish peroxidase
[7]2-Aminophenol
[8]Acetaminophen
[9]Chitosan
[10]Poly(o-phenylenediamine)
[11]Calcium dobesilate
[12]Acetylcholine esterase
[13]Grafted
[14]2-Chlorophenol
[15]Cholesterol esterase
[16]Cholesterol oxidase
[17]Poly(diphenylamine)
[18]Creatinine amidohydrolase
[19]Creatine amidinohydrolase
[20]Sarcosine oxidase
[21]Single stranded DNA
[22]Dopamine
[23]β-Cyclodextrin
[24]2,6-Dimethoxyphenol
[25]Amino labeled DNA
[26]Epinephrine
[27]Poly(o-aminophenol)
[28]Nitrogen-doped CNTs
[29]Prussian blue
[30]Hydroquinone
[31]4-Methoxyphenol
[32]Camphor sulfonic acid
[33]Nicotinamide adenine dinucleotide
[34]Iron phthalocyanine
[35]Phosphomolybdic acid
[36]4-Nitrophenol
[37]Oleanic acid imprinted polypyrrole
[38]Oxalate oxidase
[39]p-Benzoquinone
[40]Pemetrexed
[41]Laccase
[42]Sulfite oxidase

Measuring nitrite in food and environmental samples is an important issue as this compound is a human health hazard and may even cause gastric cancer. The determination of nitrite using electrochemical methods offers simple, inexpensive, and faster analysis method compared to some other more expensive and time-consuming techniques such as chromatography or spectrophotometry. Electrochemical determination of nitrite is carried out by either reduction or oxidation methods. However, the oxidation of nitrite at the surface of bare electrodes requires high potentials [26]. A method for lowering the overpotential is using electrodes modified with suitable materials such as ECP-CNT composites. For example, PEDOT electrochemically deposited on MWCNT-modified screen-printed carbon electrodes (SPCE) was used in the amperometric detection of nitrite [27]. This study, carried out by Lin et al., showed that the oxidation potential of nitrite was shifted with 160 mV to negative potential at PEDOT-modified SPCE in comparison with the bare SPCE. This was attributed to the interaction between the negatively charged nitrite ions and the positively charged PEDOT film. In comparison to the PEDOT-modified SPCE, the PEDOT-MWCNT composite film lowered the nitrite oxidation potential still with an additional 100 mV. Amperometric measurements revealed that the PEDOT-MWCNT-SPCE had a linear range of 0.05–1 mM and a LOD of 0.96 µM, whereas the LOD for the PEDOT-modified SPCE was 1.72 µM. According to these results, the presence of CNTs in PEDOT-MWCNT composite enhanced the catalytic properties and the performance of the electrode. In order to further lower the potential for nitrite oxidation and decrease the LOD of the above-mentioned electrodes, iron phthalocyanine (FePc) was introduced into the PEDOT-MWCNT composite [28]. FePc belongs to the transition metal phthalocyanine complexes that are usually used for the detection of environmental and biological compounds such as nitrite, nitrogen dioxide, and adrenalin. To prepare the modified electrode, the MWCNT suspension was first drop-casted on the surface of SPCE and left to dry. Then the FePc solution was drop-casted on the electrode surface. After repeating this procedure for several times, the PEDOT film was electrochemically deposited onto the FePc-MWCNT-modified SPCE by CV. Results from the amperometric detection showed that the nitrite oxidation potential was about 100 mV lower than that of the SPCE modified with the PEDOT-MWCNT composite reported earlier by the same author [27]. The electrocatalytic properties of FePc were considered to be the reason for the lower oxidation potential. In addition, the LOD for the FePc-MWCNT-modified sensor had decreased to 71 nM.

Compared to multilayered arrays and random networks of CNTs where the tube-tube junction limits charge transport, the aligned CNT arrays were found to be more advantageous [29]. Badhulika et al. reported the fabrication of chemical field-effect transistor sensors based on PEDOT-PSS electrochemically deposited on aligned SWCNTs [30]. The SWCNTs were aligned on 3 μm spaced microfabricated Au electrodes by using the alternating current dielectrophoresis technique. The aligned SWCNTs were annealed for 1 h at 300°C in a 5% hydrogen and 95% nitrogen gas mixture in order to remove the extra solvent and to enhance the contact between the SWCNTs and the substrate. The SWCNTs were coated with PEDOT-PSS by using the potentiostatic electropolymerization method. The developed sensors showed enhanced sensitivity over bare SWCNT sensors and were successfully used for the detection of volatile compounds such as methanol, ethanol, and methyl ethyl ketone over a wide dynamic range.

PEDOT-CNT composites were also used for the fabrication of enzyme-based biosensors [31, 32]. Liu et al. reported the immobilization of ascorbate oxidase (AO) enzyme in PEDOT-MWCNT composite prepared by electrochemical polymerization [31]. For preparing the sensor, EDOT was electrochemically polymerized in buffered (pH = 6.5) aqueous solution containing EDOT, MWCNTs, AO, $LiClO_4$, and the surfactant sodium N-lauroylsarcosinate. The resulted enzyme-based sensor was used for the detection of AA by the chronoamperometry method. The determination of AA using PEDOT-CNT modified electrode is based on the oxidation of AA to dehydroascorbic acid and water, which is catalyzed by AO. Results from the chronoamperometric curves recorded in buffered solutions containing different concentrations of AA, revealed that higher concentration of AA results in higher current and short response time. CV measurements carried out on the modified sensor showed a clear oxidation peak at 0.4 V in a buffered solution containing 1 mmol L^{-1} AA. This peak was not observed when using the same sensor in similar buffered solution without AA, confirming that the oxidation of AA is catalyzed by AO immobilized in the PEDOT-MWCNT composite.

In another attempt for fabrication of an enzyme-based biosensor, glucose oxidase (GOD) was immobilized in PEDOT-PSS polymer deposited on carbon cloth (CCl) substrate modified with either CNTs or nitrogen modified CNTs (CNx) [32]. In this work, the doping method was used to prepare CNx via the incorporation of nitrogen and introducing nitrogenated sites into the CNT lattice. Results from this study showed that CCl modified with CNx (CNx-CCl) exhibits higher electroactive area than bare CCl and CCl modified with CNT (CNT-CCl). In addition, CV and EIS measurements showed that CNx-CCl has higher capacitance than CCl and CNT-CCl substrates. In order to decrease the porosity and sorption of the CCl-based material, PEDOT-PSS was electrochemically deposited on the CNT-CCl or CNx-CCl substrates using the CV method. In addition to the decrease in porosity, the conductivity of the CCl-based electrodes increased when PEDOT-PSS was deposited on them. For the immobilization of GOD on top of PEDOT-modified electrodes, a mixture containing enzyme solution and glutaraldehyde (GA) was dropped onto the surface of the electrodes. Results from amperometric determination of glucose revealed that the PEDOT-CNx-CCl electrode with immobilized GOD, i.e., GOD-PEDOT-CNx-CCl, showed higher sensitivity and lower LOD than the other studied electrodes.

11.2.2 Polyaniline

Among the various ECPs, PANI is one of the most attractive polymers due to its high electrical conductivity, good environmental stability and easy preparation both by chemical and electrochemical methods in aqueous solutions. PANI exhibits significant electroactivity but only in acidic media, which limits its application. However, the electroactivity can be retained at neutral pH by using anionic polymeric dopants in the electrosynthesis of PANI, e.g., PSS or polyacrylic acid [33]. Similarly, composite structures of PANI and CNTs can be electroactive at neutral pH due to the bulkiness of the CNT functioning as "polymeric" counterions for PANI. Such composites have also shown to be suitable matrices for enzyme immobilization where the CNTs act as charge-transporting nanoconnectors between the redox centers of the enzyme and the electrode. Therefore, several PANI-CNT biosensors based on enzymatic catalysis have been developed for different analytes [34–41]. Non-enzymatic PANI-CNT sensors have also been reported for several analytes [42–53].

Baravik et al. have prepared a glucose biosensor by coelectropolymerization of aniline-modified SWCNTs and thioaniline-functionalized GOD [34]. The modification steps of SWCNTs and GOD are shown in Fig. 11.2. In this approach a 3D conducting network with high surface area was formed due to CNTs in which aniline bridging units can act as electron-mediating components. GOD was immobilized in the biosensor during the CV electropolymerization and leaching of the enzyme could be avoided in comparison to immobilization of the enzyme after electrochemical synthesis.

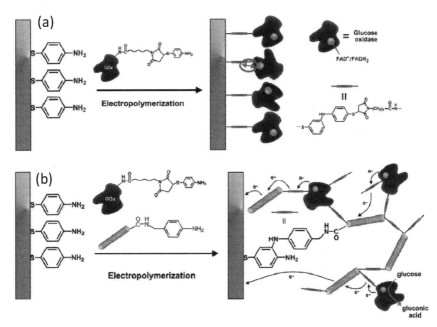

Figure 11.2 (a) Preparation of a bis-aniline-cross-linked GOD monolayer on Au electrode by the electropolymerization of the thioaniline-modified GOD with the thioaniline monolayer-functionalized Au electrode, (b) Preparation of the 3D bis-aniline-cross-linked CNT-GOD composite by the electropolymerization of the thioaniline-modified GOD and aniline-modified CNTs on thioaniline monolayer-functionalized Au electrode (reproduced with permission from the ref. 34, © 2009, American Chemical Society).

Drop-casting of the enzyme solution on the porous PANI-CNT composite film followed by drying has been a common way to prepare biosensors. Direct drop-casting of acetylcholine esterase onto a electropolymerized PANI-c-MWCNT composite (c-MWCNT: Carboxylated MWCNT) was used for determination of the pesticide carbamate. Different detection limits were obtained depending on whether chronoamperometry or SWV was used as the detection method (Table 11.1) [35]. A sulfite biosensor was prepared in a similar way as the carbamate sensor, but Au nanoparticles (AuNPs) dissolved in chitosan were further adsorbed onto the PANI-c-MWCNT composite surface to improve its electrical conductivity and surface area prior to covalent immobilization of the enzyme sulfite oxidase at reduced temperature [36]. The laccase enzyme was also covalently immobilized onto an AgNP-PANI-c-MWCNT electrode and showed improved bioelectrocatalytic activity for the guaiacol oxidation [37].

The N-ethyl-N'-(3-dimethylaminopropyl) carbodiimide (EDC) and N-hydroxysuccinimide (NHS) chemistry has been also utilized for covalent immobilization of enzymes in PANI-CNT composites for biosensor applications. A cholesterol sensor based on cholesterol oxidase (ChOx) covalently bound to an electrophoretically fabricated PANI-c-MWCNT film was reported. This was done via amide bond formation between the NH_2 groups of PANI and the COOH groups of ChOx by using EDC as the coupling agent and NHS as the activator [38]. Similarly, a mixture of three enzymes, creatinine amidohydrolase, creatine amidinohydrolase and sacrosine oxidase, for indirect determination of creatinine [39], and oxalate oxidase for determination of oxalate [40] were immobilized on the PANI-c-MWCNT composite with the EDC-NHS chemistry. The amide bond formation occurred between the NH_2 groups of the enzymes and the COOH groups of c-MWCNTs.

An amperometric hydrogen peroxide (H_2O_2) sensor, with a rather low LOD of 6.8×10^{-8} M, based on bioelectrocatalytic reduction by horseradish peroxidase (HRP) was reported by Luo et al. [41]. HRP was immobilized at a constant potential of 0.7 V and was electrostatically attached on the PANI-MWCNT composite electrode.

A non-enzymatic H_2O_2 sensor based on Prussian Blue (PB), also known as an artificial peroxidase, was accomplished by step-by-step electrodeposition of polyaniline on a glassy carbon electrode (GCE) modified with MWCNTs followed by potentiostatic deposition of PB [42]. By using PB instead of peroxidase, the linear calibration range could be improved and interference due to oxygen prevented. A non-enzymatic H_2O_2 sensor based on PANI-SWCNT electropolymerized in the ionic liquid (IL)

1-butyl-3-methylimidazolium hexafluorophosphate lowered the overpotential of reduction of H_2O_2 [43]. Amperometric detection of H_2O_2 could then be performed at −0.3 V to avoid interference from uric acid (UA), AA and acetaminophen (AP; also known as paracetamol). Similarly, the overpotential of H_2O_2 reduction was lowered by a composite electrode prepared by electropolymerization of aniline with NH_2-functionalized MWCNTs (Fig. 11.3) [44]. The 3D electron-conducting network of the composite film enhanced the H_2O_2 reduction current and lowered the LOD to 1×10^{-9} M.

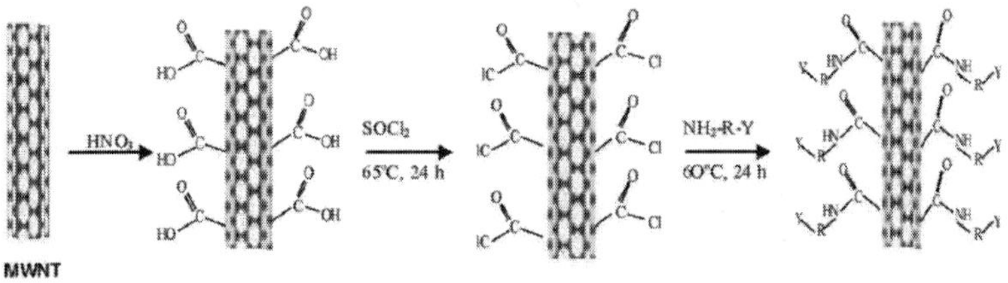

Figure 11.3a Functionalization of MWCNTs; R = poly(ethylene glycol), Y = NH_2 (reproduced with permission from the ref. 44, © 2006, Elsevier).

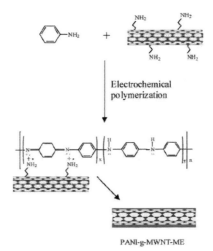

Figure 11.3b Fabrication of PANI grafted MWCNTs (reproduced with permission from the ref. 44, © 2006, Elsevier).

Several PANI-CNT-based composites have been developed to detect neurotransmitters such as DA. Oxidative approaches of DA detection suffer from accuracy of detection because the oxidation products can react with AA and regenerate DA. Two non-oxidative approaches have been developed both based on binding of DA to the boronic acid groups of self-doped poly(anilineboronic acid) and the detection of the change in electrochemical properties of the PANI backbone [45, 46]. For this purpose, PANI-boronic acid was electropolymerized on a 2-aminoethanethiol monolayer modified Au-electrode in the presence of single stranded-DNA (ss-DNA)-SWCNTs dispersion. In one of the approaches, a permselective membrane of Nafion was deposited on the composite film after the electropolymerization [45]. The ss-DNA-SWCNTs act as a molecular template during the electropolymerization and as a stabilizer in the composite film. The large surface area, unique reduction capability and high conductivity of the ss-DNA-SWCNT increased the density and improved the electrochemical activity of the PANI-boronic acid.

For the determination of AA, two non-enzymatic sensors based on PANI-CNT composites were developed that differ from each other in the electrode composition. The catalytic effect of copper nanoparticles (CuNPs) was utilized in one of the approaches [47] and the other was based on an electropolymerized poly(2,5-dimethoxyaniline)-NH_2-MWCNT film [48]. Both sensors showed selective oxidation of AA in the presence of DA and UA at 0.4 and 0.28 V, respectively. The interference from 10 µM glucose, oxalic acid, fructose, lactose, NaCl, sucrose, and tartaric acid was negligible in 0.1 M AA [47].

Sensors for accurate detection of the bacteria *N. gonorrhoeae* causing the sexually transmitted disease gonorrhea have been developed by Singh et al. [49]. The sensor is based on the detection of DNA hybridization using methylene blue as redox indicator. It was prepared by electropolymerization of aniline in presence of camphor sulfonic acid and c-MWCNTs followed by covalent immobilization of amino labeled ss-DNA on the film surface.

Amperometric detection of bisphenol A (BPA), which is used in the production of epoxy resins and polycarbonate plastics, was done with the pencil graphite electrode (PGE) modified with PANI nanorods and MWCNTs [50]. The BPA determination was comparable with the gas chromatographic method coupled with mass spectroscopy, but PANI-c-MWCNT-chitosan-PGE sensor had shorter analysis time and cheaper analytical costs. Another phenolic compound, AP, has been determined in drug samples by a rapid electrochemical method using SWV on a PANI-MWCNT electrode [51]. Na^+, K^+, Fe^{3+}, Cu^{2+}, Al^{3+}, Cl^-, NO_3^-, $H_2PO_4^-$, HPO_4^{2-}, CO_3^{2-}, SO_4^{2-}, 5-chloro-2-benz-oxazolinon, AA, DA, tyrosine, cysteine, hydroquinone (HQ), glucose, and vitamin B2 did not show any interference in the AP detection.

Amperometric nitrite detection at 0 V was conducted with a PANI-MWCNT-Au electrode [52]. The electrode showed a large electrocatalytic activity towards reduction of nitrite compared to the bare Au, MWCNT-Au and PANI-Au electrodes on which the nitrite reduction response was almost undetectable. Na^+, K^+, Ca^{2+}, Mg^{2+}, Al^{3+}, Cl^-, NO_3^-, ClO_3^-, $H_2PO_4^-$, HPO_4^{2-}, CO_3^{2-}, SO_4^{2-} (50-fold excess to nitrite) and Zn^{2+}, Cd^{2+}, Ba^{2+}, and Br^- (10-fold excess), UA and AP did not interfere with the nitrite detection, whereas Fe^{3+}, Cu^{2+}, I^-, BrO_3^-, IO_3^-, and AA were interfering.

Celecoxib (CEL), 4-[5-(4-methylphenyl)-3-(trifluoromethyl)-1H-pyrazol-1-yl]benzenesulfonamide, used for the relief from symptoms of ankylosing spondylitis, a chronic inflammatory disease of the axial skeleton, has been determined electrochemically by a PANI grafted NH_2-MWCNT electrode [53]. Aniline was electrochemically polymerized in presence of NH_2-MWCNTs on an indium tin oxide (ITO) electrode. The composite film had a large surface area and showed a stripping DPV signal from CEL reduction after an accumulation time of only 30 s at −0.5 V. With stripping SWV, ten times higher peak currents were detected.

Many functional derivatives of PANI are also used in combination with CNTs for sensor fabrication. For example, Santhosh et al. have made an amperometric sensor for carbon monoxide based on polydiphenylamine (PDPA) grafted to NH_2-MWCNT [54]. CNTs were NH_2-functionalized and electrochemically co-deposited by CV together with diphenylamine. The composite material showed pronounced electrocatalytic activity for oxidation of CO in comparison to the NH_2-MWCNT and PDPA modified electrodes. For the composite electrode, the peak current for CO oxidation was enhanced and shifted with ca. 0.16 V to lower potential. No changes in the current response were observed in mixtures of CH_4, C_3H_8, NO_x, SO_2, and C_2H_5OH with CO. The same composite material was used for the amperometric H_2O_2 detection [55]. The sensor showed a linear response to H_2O_2 between 0.5–19 mM. The addition of AP, AA, and UA as interferents did not significantly change the amperometric response of the PDPA grafted HN_2-MWCNT electrode. Moreover, the PDPA-MWCNT-β-CD (β-CD: β-cyclodextrin) composite electrode showed a LOD of 0.02 µg L^{-1} for the detection of 4-nitrophenol, which is an environmental pollutant and important compound for chemical and pharmaceutical industry [56].

A sensor for the determination of calcium dobesilate, a drug used for the treatment of blood vessels, was fabricated by CV from a solution containing o-phenylenediamine, MWCNT and sodium dodecylbenzene-sulfonate [57]. The composite material of poly(o-phenylenediamine) (PoPD) and MWCNTs had larger effective surface area compared to the neat PoPD, and the peak oxidation current for calcium dobesilate was two times higher than the sum of the currents on the plain MWCNT and the PoPD electrodes. PoPD deposited onto GCE, pre-modified with MWCNT, is also used for determination of β-nicotinamide adenine dinucleotide (NADH) [58]. The main problem in NADH detection on bare GCE is the considerable overpotential (>1 V) causing fouling of the electrode. PoPD-MWCNT decreased the oxidation potential of NADH allowing the amperometric detection of β-nicotinamide at 0.07 V.

An amperometric biosensor for the detection of glucose was fabricated by Pan et al. by immobilization of GOD in the electropolymerized poly(o-aminophenol) (PoAP) and CNT composite film matrix [59]. Due to the higher surface area and good electrical conductivity of CNTs, the electrode had a LOD of 0.01 mM, and higher sensitivity than the biosensor based on only PoAP-GOD.

11.2.3 Polypyrrole

Due to the high electrical conductivity, biocompatibility, environmental stability, good mechanical properties, and the easiness of preparation even in aqueous media, PPy is one of the ECPs that has received a great deal of attention. Significant amount of research has been particularly dedicated to the characterization of PPy-CNT composites and their application in various chemical sensors and biosensors [60–75].

Glucose biosensors based on PPy-CNT composites have been prepared and studied using different fabrication methods and characterization techniques [62–64]. Shirsat et al. have reported the fabrication of amperometric glucose biosensors based on layer-by-layer assembled SWCNT and PPy multilayer film [62]. For preparing the multilayer structure, a layer of SWCNT was first deposited by vacuum filtration on a Pt coated polyvinylidene fluoride membrane. Thereafter, a layer of PPy was electrochemically deposited using the galvanostatic method. Two- (PPy-SWCNT), three- (PPy-SWCNT-PPy), and four-layer (SWCNT-PPy-SWCNT-PPy) structures were prepared. After the immobilization of the GOD enzyme on the surface of the assembled films, the resulted electrodes were used for the amperometric determination of glucose. According to the reaction scheme below, the oxidation of glucose to gluconic acid and H_2O_2 is catalyzed by GOD immobilized in the assembled film.

$$\text{Glucose} + O_2 \xrightarrow{\text{GOD}} \text{Gluconic acid} + H_2O_2$$

The anodic current resulting from the oxidation of H_2O_2 produced in this reaction is used for determining the amount of glucose in the sample. According to the CV measurements, no significant difference can be seen in the CVs of the two-, three-, and four-layer structures. However, only electrodes based on the two-layer structure showed good analytical performance as illustrated in Fig. 11.4. This finding indicates that the two-layer structure, with uniform and porous surface morphology, is a better matrix for the immobilization of GOD than the three- and four-layer structures.

As was mentioned earlier, negatively charged CNTs can act as counterions in the electrochemical deposition of ECPs [19]. This straightforward method has been used for preparation of amperometric enzyme electrodes via entrapment of the enzyme in the resulting ECP-CNT composite [63, 64]. Wang et al. developed amperometric glucose biosensors based on the PPy-MWCNT-GOD composite [63]. The composite was prepared by a simple one-step electrochemical method in which pyrrole was electropolymerized at a constant potential of 0.7 V in the presence of c-MWCNTs and GOD. Results from the CV measurements showed that the incorporated c-MWCNTs act as counterions that maintain the electrical neutrality of the film. The influence of different parameters, such as the amount of the used MWCNTs, and the polymerization time, on the electrode response was studied in this work to optimize the electrode performance. It was shown that the current response of the electrode increases as the concentration of MWCNTs increases up to 0.5 mg ml^{-1}. Thereafter, the current decreases for concentration higher than 0.5 mg ml^{-1} and becomes almost constant above the concentration of 1.0 mg ml^{-1}. The increasing of the polymerization time from 5 to 10 min caused a rise in the current response. However, polymerization times higher than 10 min had no significant effect on the response. The PPy-MWCNT-GOD film was overoxidized by repetitive potential cycles between −0.2 and +1.3 V. As a result of increasing the overoxidation cycles, the glucose response increased steadily, and then it slightly decreased after the 50th cycle. The optimized PPy-MWCNT-GOD based electrode with the best analytical performance exhibited sensitivity of 2.33 nA mM^{-1} and a LOD of 0.2 mM. Furthermore, the addition of 0.1 mM UA and AA had no effect on the electrode response in 2 mM glucose.

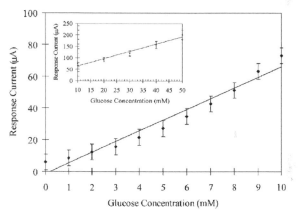

Figure 11.4 The relationship between response current and glucose concentration for SWCNT-PPy-GOD (two-layer structure on Pt coated PVDF membrane) electrode in 0.1 M PBS (pH 7.4). The inset indicates the response from 10 mM to 50 mM (reproduced with permission from the ref. 62, © 2008 WILEY-VCH Verlag GmbH & Co. KGaA, Weinheim).

Using a similar one-step preparation protocol, Tsai et al. also reported the incorporation of c-MWCNTs and GOD in the electrochemical deposition of PPy on GCE. The formed PPy-MWCNT-GOD composite film was used for glucose biosensing

[64]. Various concentrations of GOD were added during the electropolymerization of PPy in order to study the effect of the GOD amount in the PPy-MWCNT-GOD composite on the performance of the biosensors. The optimal glucose sensor based on PPy-MWCNT-GOD composite was prepared using 1.5 mg mL^{-1} GOD. The resulting amperometric sensor had a short response time of ca. 8 s and a sensitivity of 95 nA mM^{-1}.

DNA detection is important because of its usefulness in fields such as forensic investigations, and medical diagnostics of pathogenic and genetic diseases. Using traditional DNA detection techniques is usually time consuming and requires expensive equipments. For this reason, DNA biosensors with unique properties such as portability, high sensitivity, and low cost are extremely important tools that have recently received great attention. Different approaches have been used for the fabrication of DNA biosensors based on PPy-CNT composites [68–70]. Cai et al. developed a biosensor that was used in indicator-free DNA hybridization detection using AC impedance measurement [68]. The prepared biosensor is based on a DNA-doped PPy film electrochemically deposited on a GCE modified with c-MWCNTs. The c-MWCNTs, dispersed in organic solvent, were first drop-casted on the GCE surface and the extra solvent was then evaporated using an infrared lamp. The electrodeposition of PPy doped with the DNA probes was carried out in a mixture of pyrrole and the oligonucleotide probe using CV. The comparison of CVs recorded at bare GCE and MWCNT-GCE revealed an enhancement in the electrode surface area as a result of modifying the surface with MWCNTs. Comparing the electropolymerization of PPy loaded with oligonucleotide probes showed that the growth current on the MWCNT-GCE was about 30 times higher than that of bare GCE. This result was attributed to the high surface area of CNTs and to the lower nucleation energy necessary for starting the polymerization of pyrrole. The working principle of the biosensor presented in this study is based on measuring the difference in impedance values before and after the hybridization with the different concentrations of the complementary target sequence. Results from the impedance measurements showed a linear relationship between the decreased values of logarithmic impedance and the complementary oligonucleotide in the range

1.0×10^{-5}–3.0×10^{-8} M. The detection limit was 1.0×10^{-8} M with a 5 times higher sensitivity than that measured with a similar impedance platform without CNTs.

In another attempt for the fabrication of impedance-based DNA biosensors, ss-DNA probes were linked onto the PPy-MWCNT composite by using a cross-linking agent [70]. PPy-MWCNT composite was first deposited on the GCE using CV. Thereafter, amino functionalized ss-DNA (NH_2-ssDNA) probes were attached to the PPy-MWCNT film by using EDC for cross-linking the amine in the DNA probe and the carboxylic acid group in the composite film. The hybridization reaction taking place on the DNA biosensor lowered the impedance of the composite film. The lower electron transfer resistance of the double-stranded DNA in comparison with the ss-DNA is believed to be the reason for this decrease in the impedance. The improved selectivity and sensitivity of the fabricated biosensor make it suitable for detecting complementary DNA sequences down to 5.0×10^{-12} M without the need for a hybridization marker.

As it was shown by Wang et al., overoxidation of the PPy-MWCNT-GOD film to some extent improved the glucose response [63]. Sensors based on overoxidized PPy-CNT composites (PPyox-CNT) were used for the detection of species such as DA [71–73], nitrite, UA, and AA [72], epinephrine [74], and the anti-cancer drug pemetrexed [75]. The oxidation of the biological compounds AA, DA and UA usually gives overlapping oxidation peaks. Therefore, the separate determination of these species can be challenging. Li et al. reported the fabrication of the PPyox-SWCNT composite. CV was used for the electrochemical deposition of PPy-SWCNT on GCE in a polymerization solution containing pyrrole and SWCNT dispersed in sodium dodecyl sulfate (SDS) [72]. The overoxidation of the film was carried out in phosphate buffer solution (PBS) by applying a potential of 1.8 V for 250 s. The PPyox-SWCNT-GCE was used in the electrocatalytic oxidation of nitrite as an amperometric sensor that showed higher faradic responses than bare GCE or the PPyox-GCE. Results from CV measurement revealed that AA, DA, and UA in a mixture give overlapped voltammetric peaks at bare GCE. However, these overlapped peaks were successfully resolved into three separate peaks at the PPyox-SWCNT-GCE as shown in Fig. 11.5.

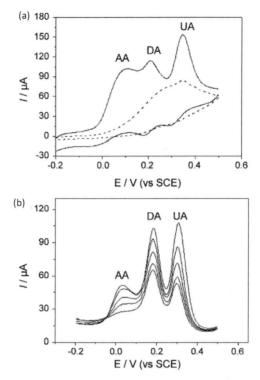

Figure 11.5 (a) CVs in a solution of 0.1M PBS (pH 4.0) containing 0.1 mM AA + 0.1 mM DA + 0.1 mM UA at PPyox-SWNTs-GCE (solid curve) and bare GCE (dashed curve). (b) DPVs at PPyox-SWNTs-GCE in solutions containing 10, 20, 30, 40, 50 μM AA; 5, 10, 20, 30, 40 μM DA; 5, 10, 20, 40, 60 μM UA, respectively (reproduced with permission from the ref. 72, © 2007, Elsevier).

Another amperometric DA sensor based on PPyox-MWCNT composite was developed by Tu et al. [73]. In this work, the deposition of PPyox-MWCNT film on gold substrate was carried out potentiostatically in a solution containing pyrrole and functionalized MWCNTs dispersed in water. The overoxidation of the PPy in the composite was carried out in a NaOH solution using CV. The PPyox-MWCNT modified electrode shows good sensitivity, selectivity and stability with a LOD of 1.7 nM.

As for the DA detection, the determination of the epinephrine (EN; adrenalin) hormone in the presence of other coexisting species such as AA and UA can be a challenging task. However, a sensor based on the composite PPyox-MWCNT was successfully used for the determination of EN in the presence of AA and UA [74]. For preparing the sensor, the PPy-MWCNT composite film was deposited on the GCE by CV from a solution of pyrrole and MWCNTs dispersed in sodium dodecyl benzenesulfonate. The electrochemical oxidation was carried out in PBS (pH = 6.0) using CV between –0.2 V and 1.3 V for 30 cycles. The authors reported an excellent sensitivity and selectivity for the composite electrode during the determination of EN in the presence of interferences such as UA and AA.

11.3 Composites of Graphene Derivatives and Conducting Polymers

Composites of ECPs with graphene derivatives for chemical sensors applications have attracted significant attention in recent years due to the strong potential of the carbon materials to improve the sensing characteristics of already known conducting materials such as PEDOT, PANI, PPy and many other less common ECPs. The main idea of the combination of the ECPs with graphene is to increase the specific surface area of the electrode, which creates more reactive sites for the electrochemical oxidation and reduction of the specific analyte. This approach has been realized successfully with the use of GO due to its excellent dispersibility in water. Being dispersible in aqueous monomer solutions of the ECPs, GO can be easily electrodeposited as thin composite films onto conducting substrates together with the ECP. As GO has a negative surface charge even at relatively low pH, it is capable to act as a bulky counterion for the positively charged ECP, thus enabling the formation of the ECP-GO composite films. Composite materials of PEDOT and GO were recently prepared electrochemically in this way from an aqueous solution of GO and EDOT without additional counterions [12]. One important feature of the composites of graphene derivatives and the ECPs in electrochemical sensing applications is their good electrochemical stability withstanding numerous oxidation and reduction cycles without significant signal degradation [76, 77]. It has been shown that the incorporation of GO in the PEDOT matrix improves both the potential cycling stability and mechanical properties of PEDOT. For example, the

PEDOT-GO film lost only 6.2% and 5.4% of its anodic charge in the three-electrode and two-electrode cell, respectively, upon 10000 oxidation and reduction cycles [76].

Because GO is a poor conductor, the electrical conductivity of ECP-GO can be increased by the reduction of GO to RGO in the composite film. Despite of the large variety of available chemical reduction methods for GO, the electrochemical reduction is a quick, clean, practical, and safe method for the reduction of GO to RGO [78]. For the electro-chemical reduction of GO in ECP-GO composites, they are exposed to sufficiently negative potentials resulting in irreversible reduction of GO to RGO [79], whereas the ECP can be brought back to the electrically conducting state by applying a positive potential to the film.

Electrochemically prepared composite materials of graphene derivatives and PANI have been reported to a lesser extent [80–84] than the PEDOT and PPy based composites. This is due to the pronounced pH sensitivity of PANI [85] that usually limits its applicability in chemical sensing applications operating at (close to) neutral pH. The electrically conducting emeraldine salt (ES) form of PANI is stable at acidic pH (≤ 3), but converts to the non-conducting emeraldine base (EB) form at higher pH. In some cases the ES to EB transition can be shifted to higher pH values with bulky hydrophobic counterions [33, 86]. For example, the exfoliated graphite and PANI composite film prepared by chemical oxidative polymerization had enhanced electrocatalytic activity that resulted in considerable signal amplification and lowered the offset potential for the ascorbate oxidation with ca. 0.1 V at pH 7.3 [87]. In this particular work, both the PANI-graphite and PANI-graphene composites showed higher electroactivity at neutral pH than neat PANI, which is demonstrating the advantage of incorporating exfoliated graphene/graphite in the ECP matrix.

11.3.1 Poly(3,4-Ethylenedioxythiophene)

Currently, the determination of biologically derived molecules is becoming more important in analytical chemistry and clinical analysis and many papers have been therefore devoted to biomarker determination. For example, the detection of DA, AA, and UA have attracted significant attention in recent years. As was mentioned earlier, DA has an important role as signaling molecule in the central nervous systems, and its dysfunction has been connected to Parkinson's disease and schizophrenia. However, AA and UA interfere with the determination of DA. Consequently, the weak electrochemical response of DA must be separated from AA and UA, and amplified, e.g., with graphene materials.

Wang et al. have shown that significantly enhanced electroactivity was obtained for the PEDOT-GO composite after electrochemical reduction of GO to RGO in the ECP matrix [88]. In this work, PEDOT-GO was deposited by electrochemical polymerization using CV on a GCE surface from a solution containing GO and EDOT. After that, the electrochemical reduction of GO in PEDOT-GO was carried out in PBS at pH 7.4 by applying the potential of −0.9 V to the composite film for 900 s. The GCE, PEDOT-GO, and PEDOT-RGO were used for voltammetric detection of DA. The authors have shown that poor electroactivity with a badly defined reduction process of DA was observed at the GCE. However, much higher electroactivity (higher currents) with well-defined oxidation and reduction peak potentials was observed for PEDOT-GO. This was ascribed to the higher electroactivity of PEDOT-GO and its larger effective surface area that provides more active sites to improve the electron transfer between the GCE substrate and DA. Compared to PEDOT-GO, the PEDOT-RGO further lowered the DA oxidation potential resulting in higher currents in the CV. It shows that PEDOT-RGO has improved electrocatalytic activity for DA oxidation compared to PEDOT-GO. This is due to the higher conductivity of RGO compared to GO. The amperometric detection of DA was therefore done at 0.2 V with the PEDOT-RGO films. As shown in Table 11.2, the composite film had a linear response range between 0.1–175 µM with the LOD of 3.9×10^{-8} M. Neither AA nor UA interfered with the DA detection when a solution containing 10 µM DA was spiked with AA and UA with the same concentration.

Weaver et al. also reported the superior detection of DA with a PEDOT-GO film electrode compared to the bare GCE [89]. The composite was electropolymerized with a constant current on a GC substrate from an aqueous solution containing GO and EDOT. It is nicely shown that in comparison to the bare GCE electrode, the oxidation of AA (1 mM), DA (100 µM) and UA (100 µM) in PBS solution can be separated with the PEDOT-GO electrode without significant peak overlapping resulting in higher peak currents for the DA oxidation (Fig. 11.6). It is proposed that the significant lowering of the oxidation potential (compared to bare GCE and neat PEDOT) and the decreased width of the AA oxidation peak arise from faster electron transfer to the composite film or improved electrocatalytic activity towards the AA oxidation. Weaver et al. suggest that the interaction of PEDOT-GO and DA is driven by electrostatic interactions resulting in an increased sensitivity for the DA oxidation that had a linear response range of 1–40 µM and a LOD of 8.3×10^{-8} M.

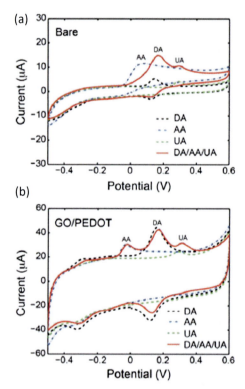

Figure 11.6 Electrochemical oxidation of DA in the presence of interfering species. CVs of the (a) bare GCE and (b) PEDOT-GO modified electrode in solutions containing 100 mM DA, 1 mM AA, 100 mM UA alone or in combination. Three separate oxidation peaks are discernible on the PEDOT-GO electrode, but not the bare GCE, in solutions containing all three analytes (reproduced with permission from the ref. 89, © 2014, The Royal Society of Chemistry).

AA has an important role in the biosynthesis of e.g. collagen and neurotransmitters, iron metabolism and immune system. Also for the AA detection, the electrochemical biosensors offers simplicity, high selectivity and sensitivity compared to conventional methods of analysis. This has been utilized in the electrochemical biosensor based on PEDOT-RGO composite with AO immobilized in it [90] resulting in a nanocomposite with high surface area and conductivity, good biocompatibility and fast redox properties. The electrode based on PEDOT-RGO composite film was fabricated by electropolymerization from an aqueous solution consisting of GO, EDOT, LiClO$_4$, and AO. In this case, the negatively charged GO sheets, the ClO$_4^-$ anions and the AO were simultaneously incorporated in the polymer matrix as charge compensating counter ions during the polymerization process. The advantage of the fabrication method utilizing CV is that it promotes both the PEDOT-GO film growth at more positive potentials and also the electrochemical reduction of GO to RGO in the negative potential regime. Therefore, the PEDOT-RGO biosensor film can elegantly be formed in one-step. AA detection, which was carried out at 0.4 V, had a linear response range of 5–480 µM and a LOD of 2×10^{-6} M. The sensor had a short response time and much higher sensitivity for the AA oxidation than the PEDOT-AO film.

Both AA and rutin, which is a flavonoid, have been simultaneously detected with a PEDOT-MeOH-GO composite film electrode that was electropolymerized by CV from an aqueous solution of EDOT-MeOH, GO and LiClO$_4$ [91]. EDOT-MeOH is a functionalized monomer with higher solubility in water compared to EDOT. The presence of both GO and perchlorate anions indicates that they were both incorporated as negatively charged counterions in the polymer film during electropolymerization. CV characterization showed that AA and rutin cannot be separated at the bare GCE while the separation is possible with PEDOT-MeOH-GO that gave the highest oxidation currents for both AA and rutin of all types of PEDOT electrodes studied. The LOD for rutin was as low as 6×10^{-9} M, whereas it was 2×10^{-6} M for AA.

Nie et al. developed a PEDOT-RGO nitrite sensor having a linear response range, LOD, and response time of 0.3–600 µM, 1×10^{-7} µM, and 8 s, respectively [92]. Like with DA and AA, the electrochemical methods have an advantage over the traditional spectrophotometric, chromatographic and chemiluminescent methods because of their simplicity, high sensitivity and selectivity, and inexpensive fast detection. In the paper by Nie et al., PEDOT-GO was deposited in one-step on the GCE substrate by scanning the potential between −1.5 and 1.1 V for 10 cycles in an aqueous solution of GO, EDOT, and LiClO$_4$. Similarly to ref [90], simultaneous film formation of PEDOT-GO and reduction of GO to RGO takes place in this deposition procedure. The nitrite detection (oxidation of NO_2^- to NO_3^-) was carried out by CV in PBS at pH 6.0. In comparison to a bare GCE, a considerable decrease in the overpotential (190 mV) and 1.75-fold enhancement of the peak current was observed for the PEDOT-RGO electrode. Moreover, it was shown that the nitrite sensor had good reproducibility and stability. The PEDOT-RGO sensor response was not either affected by a 200-fold excess of K^+, Na^+, Ag^+, Mg^{2+}, Ca^{2+}, Al^{3+}, Zn^{2+}, Cl^-, SO_4^{2-}, and PO_4^{3-}, and a 20-fold excess of BrO_3^-, IO_3^-, H_2O_2, glucose, methanol, and UA, indicating that the sensor had also good selectivity.

In another approach, the nitrite sensors were fabricated by first drop casting a solution of RGO dispersed in ethanol on the GCE [93]. PEDOT was then electropolymerized with CV on the dry RGO layer (RGO and PEDOT are probably grown into each other). Cobalt nanoparticles (CoNP) were then electrodeposited by CV on top of the PEDOT-RGO layer. In the PEDOT-RGO-CoNP films, the synergistic effect of PEDOT and RGO is combined with the electrocatalytic activity of the CoNPs, which is beneficial in lowering the nitrite oxidation potential. Cobalt-based compounds are known for their high electrocatalytic activity for glucose, L-glutathione, H_2O_2, phosphate, and methanol (see ref. 10–18 in [93]). The excellent electrocatalytic effect of PEDOT-RGO-CoNP compared to bare GCE, RGO-CoNP and PEDOT-RGO is illustrated in Fig. 11.7. The figure shows the CVs measured in 0.5 mM nitrite in PBS at pH 6.5. The incorporation of CoNPs in the PEDOT-RGO matrix increased the oxidation peak current and lowered the nitrite oxidation potential compared to PEDOT-RGO and bare GCE with ca. 200 and 500 mV, respectively. Moreover, it was confirmed with EIS that the PEDOT-RGO-CoNP film had a lower charge transfer resistance than both the neat PEDOT and PEDOT-CoNP films further confirming the electrocatalytic role of the CoNPs, and their synergistic effect with the PEDOT-RGO matrix.

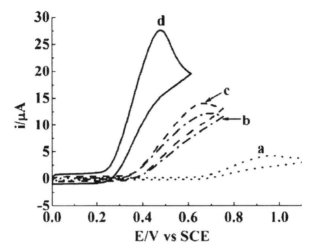

Figure 11.7 CVs of the (a) bare GCE, (b) CoNP-RGO/GCE, (c) PEDOT-RGO/GCE and (d) PEDOT-RGO-CoNP/GCE in 0.1 M PBS (pH 6.5) containing 0.5 mM nitrite. Scan rate: 0.025 V s^{-1} (reproduced with permission from the ref. 93, © 2012, Springer-Verlag).

Catechol (CT) and HQ and are isomers of dihydroxybenzene that usually coexist and can interfere with each other due to the similar properties and structure. Si et al. have prepared a PEDOT-GO sensor with good electrocatalytic activity, stability and acceptable sensitivity for the detection of HQ and CT [94]. The composite film was formed by electropolymerization of EDOT in the presence of GO by using CV. The electrochemical activity towards HQ and CT was studied by CV and DPV (Fig. 11.8). CV measurements reveal that the oxidation and reduction peaks of HQ and CT can be separated from each other on the PEDOT-GO film while only one broad oxidation peak and two badly resolved reduction peaks were observed on GCE. The redox reactions are reported to be reversible on PEDOT-GO, but quasi-reversible on GCE indicating a higher electrocatalytic activity for PEDOT-GO. The synergistic effect of GO and PEDOT and the good adhesion of the composite to the GCE substrate was proposed to reduce the intercalation distance and accelerate the charge transfer. This facilitates the separation of the close oxidation peaks of HQ and CT.

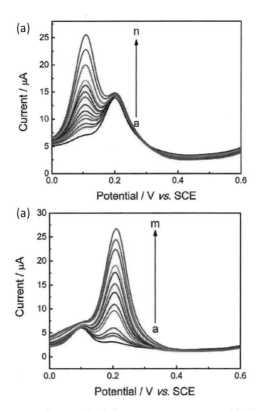

Figure 11.8 DPVs of HQ with different concentrations (a) ((a–n) 10, 15, 20, 30, 40, 50, 60, 80, 100, 120, 140, 200, 260 and 320 µM) in the presence of 50 µM CT; (b) DPVs of CT with different concentrations ((a–m) 10, 15, 20, 30, 50, 60, 80, 100, 120, 140, 200, 260, 320 and 400 µM) in the presence of 50 µM HQ (reproduced with permission from the ref. 94, © 2012, Elsevier).

In another study, Si et al. reported recently that PEDOT-GO could be used for the electrochemical detection of AP [95]. The detection is based on oxidation of AP to *N*-acetyl-*p*-quinone-imine. The PEDOT-GO film was synthesized at a constant potential of 1.2 V for 24 s from an aqueous solution containing EDOT and GO as the only counterion. It was found that the GO sheets were covered with PEDOT nanodots. The relatively rough PEDOT-GO layer was different from the uniform PEDOT-GO layer reported by the same authors in ref. [94] and it was expected to be beneficial for the AP detection.

In comparison with the GCE and neat GO, the oxidation potential of AP at PEDOT-GO based electrode was lower, and the electrode gave a linear response in the range 10–60 µM of AP at pH 4.8 (LOD: 5.7 × 10^{-7} M) that allows the selective detection of AP in urine. The composite film sensor had good selectivity to AP in the presence of a 100-fold higher concentration of Mg^{2+}, Zn^{2+}, Pb^{2+}, Cd^{2+}, Ni^{2+}, Cu^{2+}, K^+, Ce^{3+}, $Cr_2O_4^-$, MoO_3^{2-}, Cl^-, and SO_4^{2-}. Moreover, it is shown with the DPV technique that AP can be detected in the presence of AA and DA with a good resolution (Fig. 11.9).

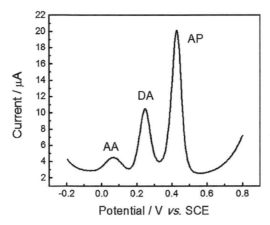

Figure 11.9 DPVs on PEDOT-GO/GCE with AP (0.1 mM), DA (0.1 mM) and AA (0.1 mM) in 0.1 M PBS (pH = 4.8) (reproduced with permission from the ref. 95, © 2014, Elsevier).

A composite film based on PEDOT and sulfonated RGO was prepared with CV for the detection of L-cysteine that plays an essential role in the regulation of biological activity of certain proteins, peptides and enzymes [96]. The RGO was prepared by pre-reduction of GO with $NaBH_4$ at 80°C (1 h) to remove most of the oxygen-containing surface functional groups. Then the partially reduced sheets were sulfonated (s-RGO) - to render them water soluble - with the aryl diazonium salt of sulfanilic acid, followed by post-reduction with hydrazine to remove the remaining functional groups on GO. The PEDOT-s-RGO film was thereafter synthesized from an aqueous solution of EDOT and sulfonated RGO by CV. Finally, AuNPs were electrodeposited with CV on the surface of the PEDOT-sulfonated RGO composite. The composite film of PEDOT-s-RGO-AuNP utilizes the biocompatibility, high conductivity and good catalytic activity of AuNPs in combination with the good reversibility and electrical properties of PEDOT-sulfonated RGO. Although AuNPs do not form a continuous electrically conducting pathway in PEDOT-s-RGO, they form numerous active sites that facilitate the charge transfer inside the composite film matrix and provide a good contact with the GCE substrate [97]. The PEDOT-s-RGO-AuNP sensor had a linear response to L-cysteine between 0.1–382 µM and a LOD of 2×10^{-8} M. Furthermore, the PEDOT-s-RGO-AuNPs film electrode had good selectivity with no interferences in the detection of L-cysteine in the presence of two-fold excess of tyrosine, tryptophan, glutathione, methionine, cysteine and UA, and 30-fold excess of 16 different amino acids, citric acid and glucose, and 100-fold excess of Na^+, K^+, Cl^-, NO_3^-, $H_2PO_4^-$, HPO_4^{2-}, and SO_4^{2-}.

In contrast to graphene, which is a zero–band gap semiconductor, the introduction of nitrogen to the graphene structure opens up the band gap transforming graphene into an n- or p-type semiconductor, thus providing pathways for efficient electron transfer processes that can be utilized in, for example, biosensors and supercapacitors. The introduction of electron-rich nitrogen heteroatoms increases the free charge carrier density (i.e., the conductivity) of the graphene network [98]. The defects induced by nitrogen doping at the edge planes and the surface of the N-doped RGO sheets give rise to enhanced electron transfer resulting in a considerable signal amplification compared to undoped graphene [99]. Prathish et al. have shown that nitrogen doped graphene (RGO) modified GCE has excellent electrocatalytic regeneration of the enzyme cofactors NADH and flavin adenine dinucleotide (FAD) [100]. In their work, the N-doped RGO was prepared by thermal reduction of graphite oxide. The formed N-doped RGO was then dispersed in 1% (w/v) chitosan dissolved in 1% (v/v) acetic acid to form a 0.1% dispersion. The composite of PEDOT and N-doped RGO had the best charge transfer properties when a layer of N-doped RGO was formed by drop casting on top of an electropolymerized PEDOT film. However, pure N-doped RGO films formed on a GCE showed the highest sensitivity of all studied electrode types (including composites with PEDOT) demonstrating that N-doped graphene can advantageously be applied without modifications as a ready-to-use electrode material.

Table 11.2 Type of composite material, type of analyte, linear response range, and LOD for electropolymerized ECP-GO/RGO composites used in chemical sensing applications

Ref.	Composite	Analyte	Linear range[1] (M)	LOD[1] (M)
90	PEDOT-RGO	AA[2]	$(5–480) \times 10^{-6}$	2.0×10^{-6}
118	PAPT-RGO[3]	AA	$(0.7–9) \times 10^{-4}$	1.4×10^{-5}
91	PEDOT-f-GO[4]	AA	$8 \times 10^{-6}–10^{-3}$	2×10^{-6}
104	PPyox[5]-RGO	Adenine	$(0.06–100) \times 10^{-6}$	2.0×10^{-8}
115	PPDCA[6]-RGO	Adenine	$(5–450) \times 10^{-8}$	2.0×10^{-8}
110	poly(DPB)-RGO-AuNP[7]	Aflatoxin B1	$(3.2–320) \times 10^{-15}$	1×10^{-15}
95	PEDOT-GO	AP[8]	$(1–6) \times 10^{-5}$	5.7×10^{-7}
113	PpABSA[9]-RGO	Cadmium	$1–70\ \mu g\ L^{-1}$	$0.05\ \mu g\ L^{-1}$
94	PEDOT-GO	Catechol	$(2–400) \times 10^{-6}$	1.6×10^{-6}
88	PEDOT-RGO	DA[10]	$(0.1–175) \times 10^{-6}$	3.9×10^{-8}
103	PPy-RGO	DA	$(0.1–150) \times 10^{-6}$	2.3×10^{-8}
89	PEDOT-GO	DA	$(1–40) \times 10^{-6}$	8.3×10^{-8}
120	PTh-RGO-ADG-BSA[11]	Ethanol	$(5–100) \times 10^{-8}$	3×10^{-7}
104	PPyox-RGO	Guanine	$(0.04–100) \times 10^{-6}$	1×10^{-8}
115	PPDCA-RGO	Guanine	$(1–60) \times 10^{-7}$	2×10^{-8}
80	PANI-GO-HRP[12]	H_2O_2	$(1–160) \times 10^{-6}$	8×10^{-8}
83	PANI-f-RGO[13]	H_2O_2	$(2.5–35) \times 10^{-5}$	1×10^{-6}
101	PPy-RGO	Hg^{2+}	$(0.1–110) \times 10^{-9}$	3×10^{-11}
94	PEDOT-GO	HQ[14]	$(2.5–200) \times 10^{-6}$	1.6×10^{-6}
116	PCz[15]-RGO	Imidacloprid	$(3–10) \times 10^{-6}$	2.2×10^{-7}
114	PEDOTM[16]-GO	Indole-3-HAc	$(0.05–40) \times 10^{-6}$	8.7×10^{-8}
96	PEDOT-SRGO[17]	L-cysteine	$(0.1–382) \times 10^{-6}$	2×10^{-8}
111	Poly(DPB)-RGO-AuNP	Microcystin-LR	$(1–80) \times 10^{-16}$	3.7×10^{-17}
121	RGO-poly(JUG-co-JUGA)[18]-DNA	microRNA	$10^{-15}–10^{-8}$	5×10^{-15}
112	PPy-co-PPA-RGO-PtNP[19]	Myoglobin	$0.01–1\ \mu g\ mL^{-1}$	$4\ ng\ mL^{-1}$
120	PTh-RGO-ADG-BSA	NADH	$(1–390) \times 10^{-5}$	1×10^{-7}
107	PPy-RGO-AgNP[20]-Ch[21]-MWCNT	Neomycin	$(0.9–700) \times 10^{-8}$	7.6×10^{-9}
92	PEDOT-RGO	Nitrite	$(0.3–600) \times 10^{-6}$	1×10^{-7}

(*Continued*)

Table 11.2 (Continued)

Ref.	Composite	Analyte	Linear range[1] (M)	LOD[1] (M)
93	PEDOT-RGO-Co	Nitrite	$(0.5-240) \times 10^{-6}$	1.5×10^{-7}
109	PPy-GO-PANI-Ch	Oxalate	$(1-400) \times 10^{-6}$	1×10^{-6}
118	PAPT-RGO	Piroxicam	$(5-80) \times 10^{-5}$	5.3×10^{-6}
117	RGO-poly(1,5-diaminonaphtalene)	Propranolol	$(0.1-750) \times 10^{-6}$	2.0×10^{-8}
106	PPy-GO-BiCoPC-PoPD[22]	Q2CA[23]	$(0.01-100) \times 10^{-6}$	2.1×10^{-9}
119	PNAANI[24]-RGO	Rhodamine B 1-aminopyrene	$(1-26) \times 10^{-8}$	8.2×10^{-9}
91	PEDOT-f-GO	Rutin	$(0.2-100) \times 10^{-7}$	6×10^{-9}
118	PAPT-RGO	Tyrosine	$(5-60) \times 10^{-5}$	6.0×10^{-6}
118	PAPT-RGO	Uric acid	$(3-50) \times 10^{-5}$	1.9×10^{-6}

[1]Unless otherwise stated
[2]Ascorbic acid
[3]Poly(3'-(2-aminopyrimidyl)-2,2':5',2'-terthiophene)), reduced graphene oxide
[4]f = MeOH, graphene oxide
[5]Overoxidized polypyrrole
[6]Poly(2,6-pyridine-dicarboxylic acid)
[7]Poly(DPB) = poly(2,5-di-(2-thienyl)-1-pyrrole-1-(p-benzoic acid); AuNP = gold nanoparticles
[8]Acetaminophen
[9]Poly(p-aminobenzenesulfonic) acid
[10]Dopamine
[11]PTh = polythionine; ADG = alcohol dehydrogenase; BSA = bovine serum albumin
[12]GO = Graphite oxide; HRP = Horseradish peroxidase
[13]f-RGO = Carboxylated reduced graphite oxide
[14]Hydroquinone
[15]Polycarbazole
[16]Poly(hydroxymethylated-3,4-ethylenedioxythiophene)
[17]Sulfonated RGO
[18]Poly(dihydronaphthalen-2(3)-yl) propanoic acid-co-5-hydroxy-1,4-naphthoquinone)
[19]PPy-co-PPA = poly(pyrrole-propylic acid); PtNP = platinum nanoparticles
[20]Silver nanoparticles
[21]Chitosan
[22]BiCoPC = binuclear phthalo-cyanine cobalt (II) sulfonate; PoPD = poly(o-phenylenediamine)
[23]Quinoxaline-2-carboxylic acid
[24]Poly(N-acetylaniline)

11.3.2 Polyaniline

As we previously mentioned, the composite materials of PANI and graphene derivatives have been studied to a lesser extent due to the pH-sensitivity of PANI. Feng et al. have applied the electrochemically prepared PANI-graphite oxide composite film for H_2O_2 detection [80]. The composite film was prepared by first drop casting a suspension of graphite oxide and aniline onto a GCE or by spin coating it on an ITO substrate followed by drying in a vacuum oven. The electrodeposition of the graphite oxide-PANI was then carried out by CV between −1.3 and 1.0 V in 1.0 M H_2SO_4. The wide potential range allows the simultaneous formation of PANI on the electrode surface at anodic potentials and the reduction of graphite oxide incorporated in the polymer matrix at sufficiently cathodic potentials. Finally, the composite film was dipped into HRP solution to immobilize the enzyme in the film matrix. The cross-sectional image of the composite film showed that it had a layered structure, which is possibly caused by the self-assembly effect of the graphite oxide sheets during the simultaneous electropolymerization and reduction process. In PBS at pH 7.0, the PANI-graphite oxide-HRP composite film had oxidation and reduction peaks for H_2O_2 at −0.212 and −0.280 V, respectively. According to Feng et al., these peaks are detectable with CV due to direct electron transfer between the catalytic centers of HRP (Fe(II)/Fe(III)) and the electrode substrate. For the reduction of H_2O_2 at −0.3 V in PBS (pH 7.0), the PANI-graphite oxide-HRP electrode had a LOD of 8×10^{-8} M. This is much better than for both the neat PANI and graphite oxide films indicating that PANI-graphite oxide-HRP had much higher electrocatalytic activity. The sensor had a fast response reaching 98% of the steady state current value within 5 s and a good stability for at least 100 potential cycles. No interference from AA and UA could either be observed.

Zheng et al. have reported the preparation of a nanocomposite consisting of PANI, RGO and palladium nanoparticles (PdNPs) [82]. First, aniline was electropolymerized by CV in a solution containing different amounts of the aniline-reduced graphite oxide complex. The PdNPs were then deposited on the composite film surface electrochemically. It was shown that by using the PANI-RGO-PdNP composite film it is possible to distinguish between the oxidation and reduction peaks of HQ and CT voltammetrically.

Composites of PANI-PSS and carboxylated reduced graphite oxide have been used for the detection of H_2O_2 [83]. The PANI-PSS was electropolymerized by CV on a drop cast reduced carboxylated graphite oxide layer from an aqueous solution of aniline, PSS and H_2SO_4. The sensor had a linear response for H_2O_2 between 25–350 µM with a LOD of 1.0×10^{-6} M and showed no interferences for DA, AA and UA.

11.3.3 Polypyrrole

In sensor applications, most of the electrochemically prepared composite materials based on PPy and graphene derivatives are used for detection of biologically derived species, such as DA, toxins, antibiotic and their metabolites, while only a very few papers are devoted to the detection of inorganic ions. For example, recently, it has been reported that the composite material of PPy and electrochemically reduced GO can be used for selective detection of Hg^{2+} [101]. It is known that PPy alone can bind the Hg^{2+} ions selectively unlike Cu^{2+}, Pb^{2+}, Cd^{2+}, and Zn^{2+}, due to its capability to form four coordination complexes with Hg^{2+} ions [102]. The combination of PPy with RGO can potentially improve the binding properties of the composite because of the higher specific surface area of the modified electrode and additional binding capability of Hg^{2+} ions to the RGO. Therefore, Wang et al. have proposed that the arrangement of RGO and PPy into 3D structures (Fig. 11.10) will significantly improve the electrochemical detection of Hg^{2+} ions compared with already known plain structures of PPy-RGO nanocomposites [101]. For example, the peak current for the plain PPy-RGO is found to be 50 times lower than for the 3D PPy-RGO measured with square wave anodic stripping voltammetry (SWASV) in a 0.1 µM Hg^{2+} solution.

Most commonly, the composite materials of PPy and graphene derivatives are used for the detection of biologically derived molecules. Currently, many attempts have been done for simultaneous electrochemical detection of DA and AA. When DA is voltammetrically detected in biological samples, UA or uridine is detected along with high AA concentrations. To overcome this interference, Si et al. have suggested the use of a PPy-RGO composite [103]. To prepare the composite, GO was first mixed with pyrrole followed by electropolymerization by CV between −0.2

and 0.8 V. After the electropolymerization, the incorporated GO is irreversibly reduced to RGO by cycling the potential between −0.1 and −1.0 V. The PPy-RGO electrode can be then successfully used for simultaneous DPV determination of DA, UA, and AA. Interestingly, both bare GCE and RGO show rather poor selectivity towards AA and DA: their oxidation peaks in DPV measurements cannot be separated although RGO has higher oxidation peak currents due to its higher specific surface. The GCE modified with PPy shows very good peak separation but poor sensitivity. However, the PPy-RGO composite can detect all three species in human serum due to its sufficient selectivity and sensitivity.

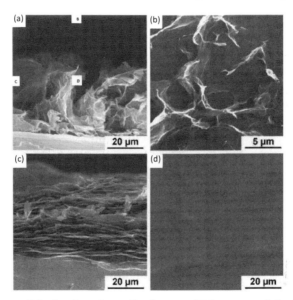

Figure 11.10 Side (a, c) and top (b, d) view SEM images of the 3D PPy-RGO (a, b), and the plain PPy-RGO (c, d) electrodes (reproduced with permission from the ref. 101, © 2014, Springer-Verlag).

Other biomarkers can be detected directly with electrodes modified with composites of PPy and graphene derivatives. For example, electrodeposited overoxidized PPy incorporating electrochemically reduced GO was successfully utilized for simultaneous detection of adenine and guanine [104].

The composite materials of ECPs and RGO enhance the electrochemical response originating from the oxidation of the nucleobases and possess LODs as low as 20 nM and 10 nM for adenine and guanine, respectively. In another approach, the guanine detection was carried out with composite films prepared by electrochemical deposition of pyrrole with stacked graphene platelet nanofibers (SGNF) onto screen-printed carbon electrodes [105]. The composite electrode can be used for determination of guanine by CV, and shows better sensitivity to guanine compared to the screen-printed electrode modified with PPy alone. Additionally, the oxidation potential of guanine on the PPy-SGNF electrode was 230 mV lower compared to the neat PPy electrode, which indicates that SGNF facilitates the guanine oxidation. The PPy-SGNF electrode can also be used for amperometric detection of H_2O_2, another important biomarker, that showed also higher sensitivity compared to the electrode modified only with PPy.

Yang et al. present a more elaborate sensor construction for the detection of biomarkers [106] with a composite material of PPy, GO and binuclear phthalocyanine cobalt (II) sulfonate, which is electrochemically deposited on a GCE. On top of the composite electrode, o-phenylenediamine is electrochemically polymerized in presence of the quinoxaline-2-carboxylic acid (Q2CA) as a template, which is later washed out with a mixture of acetic acid and methanol. The resulting molecularly imprinted (MIP) electrode was used for recognition of Q2CA, a metabolite of carbadox, a banned antibiotic with a potentially carcinogenic effect introduced in animals. The template recognition is carried out with the use of SWV and by having the non-imprinted electrode as a reference. The authors show that the MIP electrode had higher selectivity to Q2CA compared to other similar species like methyl-3-quinoxaline-2-carboxylic acid, olaquindox, carbadox, and mequindox. Additionally, they showed that the non-imprinted electrode was not sensitive to Q2CA thus confirming the effect of molecular imprinting. Another interesting MIP electrode was fabricated by Lian et al. [107] using neomycin as a template. To do this, silver nanoparticles (AgNPs) stabilized with chitosan were cast onto the surface of a Au electrode that was then coated with a mixture of reduced GO and MWCNTs. After that, PPy was electrochemically polymerized on top of this layer in the presence of the neomycin template resulting in a molecularly imprinted surface and, finally, the template was removed from PPy layer by washing. The authors have shown that the MIP electrode is highly selective to neomycin compared to the non-imprinted electrode, and neomycin can be selectively detected in the presence of gentamicin, streptomycin, kanamycin, and erythromycin.

Big classes of electrodes modified with PPy and graphene derivatives contain also biological components incorporated in their structure for the detection of biomarkers. Mainly these biosensors have an enzyme or antibody as the biological component covalently immobilized onto the electrode surface. For example, for the detection of glucose, the GOD is immobilized on the surface of the electrode providing a selective binding of glucose with the protein followed by the oxidation of the sugar and the release of H_2O_2 as the oxidation product. For electrochemical sensor applications, GO is an excellent immobilization substrate because it has sufficient amount of functional groups on its surface that can form covalent or electrostatic bonds with the NH-groups of any protein.

A glucose biosensor working in millimolar concentration range, potentially suitable for blood sugar measurements, was recently fabricated electrochemically [108]. In this work, PPy, RGO and GOD were deposited onto the electrode substrate simultaneously. This procedure makes it possible to form the composite material through a one-step electrochemical synthesis. It is shown that the sensitivity of the electrodes to H_2O_2 is directly related to the RGO content in the electropolymerization solution. Also other critical parameters such as the thickness of the biocomposite film, enzyme content, and the applied H_2O_2 detection potential for the amperometric determination of glucose were optimized. The PPy-GO-GOD electrode had good selectivity to glucose in the presence of UA and AA.

Recently, an oxalate biosensor based on the composite of graphite oxide, PPy, PANI and chitosan was reported by Devi et al. [109]. PPy was first deposited electrochemically together with graphite oxide on a Au electrode. Aniline was then electrochemically polymerized in the presence of chitosan on the PPy and graphite oxide composite. Eventually, oxalate oxidase was covalently immobilized on the composite film surface via the amino groups of chitosan by using glutaraldehyde chemistry. The composite biosensor was used for amperometric detection of oxalate both in buffer solution and in human serum. It was found to be selective to oxalate in the presence of glucose, UA, AA, cysteine, acetic acid, ethanol, and citric acid and showed a linear response range from 1 to 400 µM. The biosensor has potential application areas within diagnostics of acute and chronic kidney diseases.

Composite films consisting of PPy, graphene derivatives and proteins has been used for the detection of toxins [110, 111], and protein biomarkers such as myoglobin [112]. The fabrication of an aflatoxin B1 biosensor consists of a five step procedure: first RGO is deposited electrochemically onto the surface of a Au electrode, then the conducting polymer poly(2,5-di-(2-thienyl)-1-pyrrole-1-(p-benzoic acid)) (polyDPB) is electrochemically deposited on the RGO surface [112]. On the layered structure of polyDPB-RGO, AuNPs were deposited potentiostatically and the aflatoxin B1 antibody was then covalently attached to the carboxylic groups of the ECP using carbodiimide chemistry. The composite film is thereafter coated with a dispersion of chitosan in IL. It is shown that aflatoxin B1 can be detected impedimetrically in femtomolar concentrations due to the high specificity of the binding of the toxin with its antibody. The aflatoxin B1 sensor does not show any sensitivity to the other toxins of the aflatoxin family and had very good signal reproducibility.

A sensor for the detection of the toxin microcystin-LR is basically constructed in a similar way: RGO is first immobilized on the GCE followed by electrodeposition of polyDPB and AuNPs. Finally, a new AuNP layer was electrodeposited over the polyDPB layer [111]. The composite film was covered with chitosan dispersed in IL and the microcystin-LR antibody was covalently attached to the carboxyl groups of polyDPB. The microcystin-LR was detected by DPV in the concentration range of 0.1–8 fM. In [110] and [111], it was necessary to use a specific IL, RGO, and AuNP to obtain the lowest signal to noise ratio as well as a low LOD. The RGO and the AuNPs participate in signal amplification while the IL creates a microenvironment preserving the activity of the immobilized antibody.

The myoglobin electrochemical biosensor is another biosensor type having an antibody in its structure as well as PPy and graphene derivatives. Myoglobin is a small protein that is a cardiac marker released by the human body over a short period of time after the onset of a myocardial infarction. Its detection is therefore crucial in diagnosing heart attacks in their very early stage. For the fabrication of the sensors for myoglobin, ITO glass plates are modified with 3-aminopropyl-triethoxysilane to create a self-assembly monolayer [112]. This monolayer is then used for electrostatic immobilization of GO on the electrode surface followed by the reduction of GO by CV. PPy is then electropolymerized with pyrrolepropylic acid accompanied with the electrodeposition of platinum nanoparticles (PtNPs) onto the pre-modified ITO. Finally, the myoglobin protein antibody is covalently attached to the PPy-co-PPA by using the same carbodiimide chemistry as for microcystin LR and aflatoxin B1. The myoglobin biosensor had a linear response range of 10 ng mL^{-1}– 1 μg mL^{-1} and a LOD of 4 ng mL^{-1}. As in [111], the PtNPs embedded in the ECP matrix enhance the electron transfer between the antigen–antibody complex and the electrode substrate.

11.3.4 Other ECPs

Besides PANI, PEDOT and PPy, some other composites of ECPs and graphene derivatives have been fabricated electrochemically.

For the detection of inorganic ions, such as Cd^{2+}, Wang et al. have proposed electropolymerization of p-aminobenzene sulfonic acid (p-ABSA) [113]. First, GO is electrochemically reduced to RGO and simultaneously deposited at a constant potential onto a GCE surface. Electrodeposition of poly-ABSA was then carried out voltammetrically in the presence of RGO and the composite film was used for SWASV detection of Cd^{2+}. Therefore, tin (replacing bismuth or mercury) was electrodeposited on top of the poly(p-ABSA) layer. The composite electrode showed good stripping performance for the analysis of Cd^{2+}, wider potential window compared to bare GCE and GCE covered with only Sn, and gave a linear response from 1.0 to 70.0 μg L^{-1} with a detection limit of 0.05 μg L^{-1}.

For the detection of organic species, such as pollutants, phytohormones, metabolites, biomarkers, and nucleic acids different approaches have been reported. For the detection of indole-3-acetic acid, a phytohormone, the GCE was modified with the GO and poly(hydroxymethylated-3,4-ethylenedioxythiophene) (PEDOTM) composite [114]. The composite is prepared by cycling the electrode potential between −0.9 and 1.1 V in a mixture of GO and PEDOTM monomer. The composite showed higher currents for the oxidation of indole-3-acetic acid compared to the neat

PEDOTM electrode. In a similar way, the electrodes modified with chemically reduced GO and poly(2,6-pyridinedicarboxylic acid) (polyPDCA) were prepared for voltammetric determination of adenine and guanine. First, GO was chemically reduced in a separate process and then the RGO suspension in N,N-dimethylformamide was cast onto the GCE. After drying, the polyPDCA film was electropolymerized onto the RGO modified surface of the electrode [115]. It was shown that simultaneous measurement of guanine and adenine was possible by LSV. The electrode showed also good performance for detection of guanine and adenine within native DNA molecules. The same preparation procedure was exploited for modification of the GCE with RGO and polycarbazole [116] and RGO with poly(1,5-diaminonapthalene) [117]. By using the CV and DPV techniques, these modified electrodes were used for the detection of imidacloprid, an insecticide, and propanolol, respectively.

For the detection of the anti-arthritis drug Piroxicam, another ECP, poly-(3'-(2-aminopyrimidyl)-2,2':5',2"-terthiophene)) (PAPT) was electrochemically formed followed by casting and drying of GO on top of it [118]. GO was then reduced electrochemically to RGO. The composite electrode was used for voltammetric determination of Piroxicam both alone and in the presence of its major interferents: AA, tyrosine and UA both in model and real samples. The authors have shown that all four species can be detected simultaneously due to the good peak separation on PAPT and good signal amplification by RGO.

Zhu et al. proposed more complicated biosensor construction for selective binding of rhodamine B and 1-aminopyrene [119]. To do this, first GO is electrochemically reduced on the surface of the GCE and then N-acetylaniline is polymerized on top of it. After this, β-CD was covalently attached to the composite surface by electrooxidation. This sensor was used for voltammetric determination of 1-aminopyrene, an electrochemically active polycyclic aromatic hydrocarbon deleterious for living organisms. The authors show that β-CD binds rhodamine B selectively, but in the presence of 1-aminopyrene, the rhodamine B molecules are displaced by 1-aminopyrene. This displacement results in a decreased oxidation peak current of rhodamine B and the appearance of an oxidation peak current for 1-aminopyrene. The current is linearly dependent on the 1-aminopyrene concentration between 10–260 nM.

A few electrochemically fabricated biosensors having ECPs other than PANI, PPy and PEDOT were reported recently. One of them has alcohol dehydrogenase as a biorecognition element [120]. The enzyme is covalently attached with the use of GA to the electrochemically deposited polythionine-RGO. The surface of the polythionine-RGO, which remained unoccupied with enzyme after its immobilization, is blocked with bovine serum albumin to exclude parasitic signals from interferents. The ethanol detection with the composite electrode is carried out by adding ethanol to the analyte solution containing the coenzyme NADH. The sensor response was relatively accurate in the concentration range of 0.05 to 1.0 mM. Another biosensor is fabricated by drop-casting chemically reduced GO onto the GCE followed by electrochemical copolymerization of 3-(5-Hydroxy-1,4-dioxo-1,4-dihydronaphthalen-2(3)-yl) propanoic acid with 5-hydroxy-1,4-naphthoquinone. The composite film was then used as a substrate for immobilization of oligodeoxyribonucleotide (oligomeric ss-DNA), which acts as a recognition element (probe) for the detection of microRNA (target) [121]. The fabricated sensor is selective due to the ability of the target and probe to form a double helix provided that their sequences are complementary to each other. The DNA sensor functions according to the "on-off-on" principle, and the signal detection is done by SWV; the current increases (the sensor response is turned "on") upon the formation of a double helix between the target and the probe.

11.4 Conclusions

Electrochemically nanofabricated composite materials of ECPs and carbon materials, such as CNTs and graphene derivatives, broaden the possibilities of the electrochemical detection of a wide range of analytes. The combination of the electrocatalytic properties of the ECPs with the large surface area of the carbon materials is advantageously used in simultaneous voltammetric detection of biomarkers, e.g., DA, AA, and UA. Compared to the ECP-carbon materials composite the conventional GCE or metallic electrodes show insufficient peak separation making the simultaneous detection of the biomarkers difficult. The high surface area of the carbon materials contributes to signal amplification and more efficient electron transfer compared to the ECPs alone that lowers the LOD of the composite based sensors. Most importantly, the optimization of the nanofabrication conditions allows fine-tuning of the morphology, composition, and hence, the electrochemical properties of the ECP-carbon materials based sensors.

References

1. Geim, A. K., and Novoselov, K. S. (2007). The rise of graphene, *Nat. Mater.*, **6**, pp. 183–191.
2. Oueiny, C., Berlioz, S., and Perrin, F.-X. (2014). Carbon nanotube-polyaniline composites, *Prog. Polym. Sci.*, **39**, pp. 707–748.
3. Park, S., Shao, Y., Wan, H., Rieke, P. C., Viswanathan, V. V., Towne, S. A., Saraf, L. V., Liu, J., Lin, Y., and Wang, Y. (2011). Design of graphene sheets-supported Pt catalyst layer in PEM fuel cells, *Electrochem. Commun.*, **13**, pp. 258–261.
4. Wang, D., Choi, D., Li, J., Yang, Z., Nie, Z., Kou, R., Hu, D., Wang, C., Saraf, L. V., Zhang, J., Aksay, I. A., and Liu, J. (2009). Self-assembled TiO_2-graphene hybrid nanostructures for enhanced Li-ion insertion, *ACS Nano*, **3**, pp. 907–914.
5. Morales-Torres, S., Pastrana-Martínez, L. M., Figueiredo, J. L., Faria, J. L., and Silva, A. M. T. (2012). Design of graphene-based TiO_2 photocatalysts—A review, *Environ. Sci. Pollut. Res.*, **19**, pp. 3676–3687.
6. Lindfors, T., Szücs, J., Sundfors, F., and Gyurcsányi, R. E. (2010). Polyaniline nanoparticle-based solid-contact silicone rubber ion-selective electrodes for ultratrace measurements, *Anal. Chem.*, **83**, pp. 9425–9432.
7. Tsakova, T. (2008). How to affect number, size, and location of metal particles deposited in conducting polymer layers, *J. Solid State Electrochem.*, **12**, pp. 1421–1434.
8. Lu, C., Tang, Y., Huan, S., Li, S., Li, L., and Wang, Y. (2010). Enhanced charge transportation in semiconducting polymer/insulating polymer composites: The role of an interpenetrating bulk interface, *Adv. Funct. Mater.*, **20**, pp. 1714–1720.
9. Chen, G. Z., Shaffer, M. S., Coleby, D., Dixon, G., Zhou, W., Fray, D. J., and Windle, A. H. (2000). Carbon nanotube and polypyrrole composites: Coating and doping, *Adv. Mater.*, **12**, pp. 522–526.

10. Baibarac, M., and Gómez-Romero, P. (2006). Nanocomposites based on conducting polymers and carbon nanotubes: From fancy materials to functional applications, *J. Nanosci. Nanotechnol.*, **6**, pp. 289–302.
11. Zhang, K., Zhang, L. L., Zhao, X. S., and Wu, J. (2010). Graphene/polyaniline nanofiber composites as supercapacitor electrodes, *Chem. Mater.*, **22**, pp. 1392–1401.
12. Österholm, A., Lindfors, T., Kauppila, J., Damlin, P., and Kvarnström, C. (2012). Electrochemical incorporation of graphene oxide into conducting polymer films, *Electrochim. Acta*, **83**, pp. 463–470.
13. Yang, Y., Wang, C., Yue, B., Gambhir, S., Too, C. O., and Wallace, G. G. (2012). Electrochemically synthesized polypyrrole/graphene composite film for lithium batteries, *Adv. Energy Mater.*, **2**, pp. 266–272.
14. Dreyer, D. R., Park, S., Bielawski, C. W., and Ruoff, R. (2010). Chemistry of graphene oxide, *Chem. Soc. Rev.*, **39**, pp. 228–240.
15. Chua, C. K., and Pumera, M. (2014). Chemical reduction of graphene oxide: A synthetic chemistry viewpoint, *Chem. Soc. Rev.*, **43**, pp. 291–312.
16. Peng, C., Jin, J., and Chen, G. Z. (2007). A comparative study on electrochemical co-deposition and capacitance of composite films of conducting polymers and carbon nanotubes, *Electrochim. Acta*, **53**, pp. 525–537.
17. Keru, G., Ndungu, P. G., and Nyamori, V. O. (2014). A review on carbon nanotube/polymer composites for organic solar cells, *Int. J. Energy Res.*, **38**, pp. 1635–1653.
18. Agüí, L., Yáñez-Sedeño, P., and Pingarrón, J. M. (2008). Role of carbon nanotubes in electroanalytical chemistry, *Anal. Chim. Acta*, **622**, pp. 11–47.
19. Mousavi, Z., Bobacka, J., Lewenstam, A., and Ivaska, A. (2009). Poly(3,4-ethylenedioxythiophene) (PEDOT) doped with carbon nanotubes as ion-to-electron transducer in polymer membrane-based potassium ion-selective electrodes, *J. Electroanal. Chem.*, **633**, pp. 246–252.
20. Groenendaal, L., Jonas, F., Freitag, D., Pielartzik, H., and Reynolds, J. R. (2000). Poly(3,4-ethylenedioxythiophene) and its derivatives: Past, present, and future, *Adv. Mater.*, **12**, pp. 481–494.
21. Lota, K., Khomenko, V., and Frackowiak, E. (2004). Capacitance properties of poly(3,4-ethylenedioxythiophene)/carbon nanotubes composites, *J. Phys. Chem. Solids*, **65**, pp. 295–301.
22. Gerwig, R., Fuchsberger, K., Schroeppel, B., Link, G. S., Heusel, G., Kraushaar, U., Schuhmann, W., Stett, A., and Stelzle, M. (2012). PEDOT-

CNT composite microelectrodes for recording and electrostimulation applications: Fabrication, morphology, and electrical properties, *Front. Neuroeng.*, doi:10.3389/fneng.2012.00008.

23. Nikolskii, B. P., and Materova, E. A. (1985). Solid contact in membrane ion-selective electrodes, *Ion-Sel. Electrode Rev.*, **7**, pp. 3–39.

24. Xu, G., Li, B., Cui, X. T., Ling, L., and Luo, X. (2013). Electrodeposited conducting polymer PEDOT doped with pure carbon nanotubes for the detection of dopamine in the presence of ascorbic acid, *Sens. Actuators, B*, **188**, pp. 405–410.

25. Samba, R., Fuchsberger, K., Matiychyn, I., Epple, S., Kiesel, L., Stett, A., Schuhmann, W., and Stelzle, M. (2014). Application of PEDOT-CNT microelectrodes for neurotransmitter sensing, *Electroanalysis*, **26**, pp. 548–555.

26. Newbery, J. E., and Lopez de Haddad, M. P. (1985). Amperometric determination of nitrite by oxidation at a glassy carbon electrode, *Analyst*, **110**, pp. 81–82.

27. Lin, C.-Y., Vasantha, V. S., and Ho, K.-C. (2009). Detection of nitrite using poly(3,4-ethylenedioxythiophene) modified SPCEs, *Sens. Actuators B*, **140**, pp. 51–57.

28. Lin, C.-Y., Balamurugan, A., Lai, Y.-H., and Ho, K.-C. (2010). A novel poly(3,4-ethylenedioxythiophene)/iron phthalocyanine/multi-wall carbon nanotubes nanocomposite with high electrocatalytic activity for nitrite oxidation, *Talanta*, **82**, pp. 1905–1911.

29. Shekhar, S., Stokes, P., and Khondaker, S. I. (2011). Ultrahigh density alignment of carbon nanotube arrays by dielectrophoresis, *ACS Nano*, **5**, pp. 1739–1746.

30. Badhulika, S., Myung, N. V., and Mulchandani, A. (2014). Conducting polymer coated single-walled carbon nanotube gas sensors for the detection of volatile organic compounds, *Talanta*, **123**, pp. 109–114.

31. Liu, M., Wen, Y., Li, D., He, H., Xu, J., Liu, C., Yue, R., Lu, B., and Liu, G. (2011). Electrochemical immobilization of ascorbate oxidase in poly(3,4-ethylenedioxythiophene)/multiwalled carbon nanotubes composite films, *J. Appl. Polym. Sci.*, **122**, pp. 1142–1151.

32. Barsan, M. M., Carvalho, R. C., Zhong, Y., Sun, X., and Brett, C. M. A. (2012). Carbon nanotube modified carbon cloth electrodes: Characterisation and application as biosensors, *Electrochim. Acta*, **85**, pp. 203–209.

33. Hu, Z., Xu, J., Tian, Y., Peng, R., Xian, Y., Ran, Q., and Jin, L. (2009). Layer-by-layer assembly of polyaniline nanofibers/poly(acrylic acid)

multilayer film and electrochemical sensing, *Electrochim. Acta*, **54**, pp. 4056–4061.

34. Baravik, I., Tel-Vered, R., Ovits, O., and Willner, I. (2009). Electrical contacting of redox enzymes by means of oligoaniline-cross-linked enzyme/carbon nanotube composites, *Langmuir*, **25**, pp. 13978–13983.

35. Cesarino, I., Moraes, F. C., and Machado, S. A. S. (2011). A biosensor based on polyaniline-carbon nanotube core-shell for electrochemical detection of pesticides, *Electroanalysis*, **23**, pp. 2586–2593.

36. Rawal, R., Chawla, S., Dahiya, T., and Pundir, C. S. (2011). Development of an amperometric sulfite biosensor based on a gold nanoparticles/chitosan/multiwalled carbon nanotubes/polyaniline-modified gold electrode, *Anal. Bioanal. Chem.*, **401**, pp. 2599–2608.

37. Rawal, R., Chawla, S., and Pundir, C. S. (2011). Polyphenol biosensor based on laccase immobilized onto silver nanoparticles/multiwalled carbon nanotube/polyaniline gold electrode, *Anal. Biochem.*, **419**, pp. 196–204.

38. Dhand, C., Arya, S. K., Datta, M., and Malhotra, B. D. (2008). Polyaniline-carbon nanotube composite film for cholesterol biosensor, *Anal. Biochem.*, **383**, pp. 194–199.

39. Yadav, S., Kumar, A., and Pundir, C. S. (2011). Amperometric creatinine biosensor based on covalently coimmobilized enzymes onto carboxylated multiwalled carbon nanotubes/polyaniline composite film, *Anal. Biochem.*, **419**, pp. 277–283.

40. Yadav, S., Devi, R., Kumari, S., Yadav, S., and Pundir, C. S. (2011). An amperometric oxalate biosensor based on sorghum oxalate oxidase bound carboxylated multiwalled carbon nanotubes-polyaniline composite film, *J. Biotechnol.*, **151**, pp. 212–217.

41. Luo, X., Killard, A. J., Morrin, A., and Smyth, M. R. (2006). Enhancement of a conducting polymer-based biosensor using carbon nanotube-doped polyaniline, *Anal. Chim. Acta*, **575**, pp. 39–44.

42. Zou, Y., Sun, L., and Xu, F. (2007). Prussian Blue electropolymerized on MWCNTs-PANI hybrid composites for H_2O_2 detection, *Talanta*, **72**, pp. 437–442.

43. Wang, Q., Yun, Y., and Zheng, J. (2009). Nonenzymatic hydrogen peroxide sensor based on a polyaniline-single walled carbon nanotubes composite in a room temperature ionic liquid, *Microchim. Acta*, **167**, pp. 153–157.

44. Santosh, P., Manesh, K. M., Gopalan, A., and Lee, K.-P. (2006). Fabrication of a new polyaniline grafted multi-wall carbon nanotube

modified electrode and its application for electrochemical detection of hydrogen peroxide, *Anal. Chim. Acta*, **575**, pp. 32–38.

45. Ma, Y., Ali, S. R., Dodoo, A. S., and He, H. (2006). Enhanced sensitivity for biosensors: Multiple functions of DNA-wrapped single-walled carbon nanotubes in self-doped polyaniline nanocomposites, *J Phys. Chem. B*, **110**, pp. 16359–16365.

46. Ali, S. R., Ma, Y., Parajuli, R. R., Balogun, Y., Lai, W. Y.-C., and He, H. (2007). A nonoxidative sensor based on a self-doped polyaniline/carbon nanotube composite for sensitive and selective detection of the neurotransmitter dopamine, *Anal. Chem.*, **79**, pp. 2583–2587.

47. Chauhan, N., Narang, J., Rawal, R., and Pundir, C. S. (2011). A highly sensitive non-enzymatic ascorbate sensor based on copper nanoparticles bound to multi walled carbon nanotubes and polyaniline composite, *Synth. Met.*, **161**, pp. 2427–2433.

48. Ragupathy, D., Park, J. J., Lee, S. C., Kim, J. C., Gomathi, P., Kim, M. K., Lee, S. M., Ghim, H. D., Rajendran, A., Lee, S. H., and Jeon, K. M. (2011). Electrochemical grafting of poly(2,5-dimethoxy aniline) onto multiwalled carbon nanotubes nanocomposite modified electrode and electrocatalytic oxidation of ascorbic acid, *Macromol. Res.*, **19**, pp. 764–769.

49. Singh, R., Dhand, C., Sumana, G., Verma, R., Sood, S., Gupta, R. K., and Malhotra, B. D. (2010). Polyaniline/carbon nanotubes platform for sexually transmitted disease detection, *J. Mol. Recognit.*, **23**, pp. 472–479.

50. Poorahong, S., Thammakhet, C., Thavarungkul, P., Limbut, W., Numnuam, A., and Kanatharana, P. (2012). Amperometric sensor for detection of bisphenol A using a pencil graphite electrode modified with polyaniline nanorods and multiwalled carbon nanotubes, *Microchim. Acta*, **176**, pp. 91–99.

51. Li, M., and Jing, L. (2007). Electrochemical behavior of acetaminophen and its detection on the PANI-MWCNTs composite modified electrode, *Electrochim. Acta*, **52**, pp. 3250–3257.

52. Guo, M., Chen, J., Li, J., Tao, B., and Yao, S. (2005). Fabrication of polyaniline/carbon nanotube composite modified electrode and its electrocatalytic property to the reduction of nitrite, *Anal. Chim. Acta*, **532**, pp. 71–77.

53. Manesh, K. M., Santosh, P., Komathi, S., Kim, N. H., Park, J. W., Gopalan, A. I., and Lee, K.-P. (2008). Electrochemical detection of celecoxib at a polyaniline grafted multiwall carbon nanotubes modified electrode, *Anal. Chim. Acta*, **626**, pp. 1–9.

54. Santhosh, P., Manesh, K. M., Gopalan, A., and Lee, K. P. (2007). Novel amperometric carbon monoxide sensor based on multi-wall carbon nanotubes grafted with polydiphenylamine-Fabrication and performance, *Sens. Actuators B*, **125**, pp. 92–99.
55. Santhosh, P., Manesh, K. M., Lee, K. P., and Gopalan, A. I. (2006). Enhanced electrocatalysis for the reduction of hydrogen peroxide at new multiwall carbon nanotube grafted polydiphenylamine modified electrode, *Electroanalysis*, **18**, pp. 894–903.
56. Zarei, K., Teymori, E., and Kor, K. (2014). Very sensitive differential pulse adsorptive stripping voltammetric determination of 4-nitrophenol at poly (diphenylamine)/multi-walled carbon nanotube-β-cyclodextrin-modified glassy carbon electrode, *Int. J. Environ. Anal. Chem.*, **94**, pp. 1407–1421.
57. Zhang, X., Wang, S., Jia, L., Xu, Z., and Zeng, Y. (2008). An electrochemical sensor for determination of calcium dobesilate based on PoPD/MWNTs composite film modified glassy carbon electrode, *J. Biochem. Biophys. Methods*, **70**, pp. 1203–1209.
58. Zeng, J. Gao, X., Wei, W., Zhai, X., Yin, J., Wu, L., Liu, X., Liu, K., and Gong, S. (2007). Fabrication of carbon nanotubes/poly(1,2-diaminobenzene) nanoporous composite via multipulse chronoamperometric electropolymerization process and its electrocatalytic property toward oxidation of NADH, *Sens. Actuators B*, **120**, pp. 595–602.
59. Pan, D., Chen, J., Yao, S., Tao, W., and Nie, L. (2005). An amperometric glucose biosensor based on glucose oxidase immobilized in electropolymerized poly(o-aminophenol) and carbon nanotubes composite film on a gold electrode, *Anal. Sci.*, **21**, pp. 367–371.
60. Singh, K., Solanki, P. R., Basu, T., and Malhotra, B. D. (2012). Polypyrrole/multiwalled carbon nanotubs based biosensor for cholestrol estimation, *Polym. Adv. Technol.*, **23**, pp. 1084–1091.
61. Ozoner, S. K., Erhan, E., and Yilmaz, F. (2011). Enzyme based phenol biosensors. In *Environmental Biosensors*, Somerest, V., ed., Chapter 15 (In Tech, Rijeka), pp. 319–340.
62. Shirsat, M. D., Too, C. O., and Wallace, G. G. (2008). Amperometric glucose biosensor on layer by layer assembled carbon nanotube and polypyrrole multilayer film, *Electroanalysis*, **20**, pp. 150–156.
63. Wang, J., and Musameh, M. (2005). Carbon-nanotubes doped polypyrrole glucose biosensor, *Anal. Chim. Acta*, **539**, pp. 209–213.
64. Tsai, Y.-C., Li, S.-C., and Liao, S.-W. (2006). Electrodeposition of polypyrrole–multiwalled carbon nanotube–glucose oxidase

nanobiocomposite film for the detection of glucose, *Biosens. Bioelectron.*, **22,** pp. 495–500.

65. Zhang, Z., Luo, L., Chen, H., Zhang, M.,Yang, X., Yao, S., Li, J., and Peng, M. (2011). A polypyrrole-imprinted electrochemical sensor based on nano-SnO_2/multiwalled carbon nanotubes film modified carbon electrode for the determination of oleanolic acid, *Electroanalysis*, **23**, pp. 2446–2455.

66. Shahrokhian, S., and Asadian, E. (2009). Electrochemical determination of L-dopa in the presence of ascorbic acid on the surface of the glassy carbon electrode modified by a bilayer of multi-walled carbon nanotube and polypyrrole doped with tiron, *J. Electroanal. Chem.*, **636**, pp. 40–46.

67. Tang, Q., Luo, X., and Wen, R. (2005). Construction of a heteropolyanion-containing polypyrrole/carbon nanotube modified electrode and its electrocatalytic property, *Anal. Lett.*, **38**, pp. 1445–1456.

68. Cai, H., Xu, Y., He, P.-G., and Fang, Y.-Z. (2003). Indicator free DNA hybridization detection by impedance measurement based on the DNA-doped conducting polymer film formed on the carbon nanotube modified electrode, *Electroanalysis*, **15**, pp. 1864–1870.

69. Cheng, G., Zhao, J., Tu, Y., He, P., and Fang, Y. (2005). A sensitive DNA electrochemical biosensor based on magnetite with a glassy carbon electrode modified by multi-walled carbon nanotubes in polypyrrole, *Anal. Chim. Acta*, **533**, pp. 11–16.

70. Xu, Y., Ye, X., Yang, L., He, P., and Fang, Y. (2006). Impedance DNA biosensor using electropolymerized polypyrrole/multiwalled carbon nanotubes modified electrode, *Electroanalysis*, **18**, pp. 1471–1478.

71. Peairs, M. J., Ross, A. E., and Venton, B. J. (2011). Comparison of Nafion- and overoxidized polypyrrole-carbon nanotube electrodes for neurotransmitter detection, *Anal. Methods*, **3**, pp. 2379–2386.

72. Li, Y., Wang, P., Wang, L., and Lin, X. (2007). Overoxidized polypyrrole film directed single-walled carbon nanotubes immobilization on glassy carbon electrode and its sensing applications, *Biosens. Bioelectron.*, **22**, pp. 3120–3125.

73. Tu, X., Xie, Q., Jiang, S., and Yao, S. (2007). Electrochemical quartz crystal impedance study on the overoxidation of polypyrrole–carbon nanotubes composite film for amperometric detection of dopamine, *Biosens. Bioelectron.*, **22**, pp. 2819–2826.

74. Shahrokhian, S., and Saberi, R.-S. (2011). Electrochemical preparation of over-oxidized polypyrrole/multi-walled carbon

nanotube composite on glassy carbon electrode and its application in epinephrine determination, *Electrochim. Acta*, **57**, pp. 132–138.

75. Karadas, N., and Ozkan, S. A. (2014). Electrochemical preparation of sodium dodecylsulfate doped over-oxidized polypyrrole/multi-walled carbon nanotube composite on glassy carbon electrode and its application on sensitive and selective determination of anticancer drug: Pemetrexed, *Talanta*, **119**, pp. 248–254.

76. Lindfors, T. (2015). Potential cycling stability of composite films of graphene derivatives and poly(3,4-ethylenedioxythiophene), *Electroanalysis*, **27**, pp. 727–732.

77. Lindfors, T., and Latonen, R.-M. (2014). Improved charging/discharging behavior of electropolymerized nanostructured composite films of polyaniline and electrochemically reduced graphene oxide, *Carbon*, **69**, pp. 122–131.

78. Ramesha, G. K., and Sampath, S. (2009). Electrochemical reduction of oriented graphene oxide films: An in situ Raman spectroelectrochemical study, *J. Phys. Chem. C*, **113**, pp. 7985–7989.

79. Lindfors, T., Österholm, A., Kauppila, J., and Pesonen, M. (2013). Electrochemical reduction of graphene oxide in electrically conducting poly(3,4-ethylenedioxythiophene) composite films, *Electrochim. Acta*, **110**, pp. 428–436.

80. Feng, X.-M., Li, R.-M., Ma, Y.-W., Chen, R.-F., Shi, N.-E., Fan, Q.-L., and Huang, W. (2011). One-step electrochemical synthesis of graphene/polyaniline composite film and its applications, *Adv. Funct. Mater.*, **21**, pp. 2989–2996.

81. Bo., Y., Yang, H., Hu, Y., Yao, T., and Huang, S. (2011). A novel electrochemical DNA biosensor based on graphene and polyaniline nanowires, *Electrochim. Acta*, **56**, pp. 2676–2681.

82. Zheng, Z., Du, Y., Feng, Q., Wang, Z., and Wang, C. (2012). Facile method to prepare Pd/graphene-polyaniline nanocomposite and used as new electrode material for electrochemical sensing, *J. Mol. Catal.*, **353–354**, pp. 80–86.

83. Prasannakumar, S., Manjunatha, R., Nethravathi, C., Suresh, G. S., and Rajamathi, M. (2012). Non-enzymatic reduction of hydrogen peroxide sensor based on (polyaniline-polystyrene sulphonate)—carboxylated graphene modified graphite electrode, *Portugaliae Electrochim. Acta*, **30**, pp. 371–383.

84. Radhapyari, K., Kotoky, P., Das, M. R., and Khan, R. (2013). Graphene-polyaniline nanocomposite based biosensor for detection of

antimalarial drug artesunate in pharmaceutical formulation and biological fluids, *Talanta*, **111**, pp. 47–53.

85. Lindfors, T., and Ivaska, A. (2002). pH sensitivity of polyaniline and its substituted derivatives, *J. Electroanal. Chem.*, **531**, pp. 43–52.

86. Lindfors, T., Sandberg, H., and Ivaska, A. (2004). Influence of lipophilic additives on the emeraldine base-emeraldine salt transition of polyaniline, *Synth. Met.*, **142**, pp. 231–242.

87. Boeva, Z. A., Milakin, K., Pesonen, M., Ozerin, A. N., Sergeyev, V. G., and Lindfors, T. (2014). Dispersible composites of exfoliated graphite and polyaniline with improved electrochemical behavior for solid-state chemical sensor applications, *RSC Adv.*, **4**, pp. 46340–46350.

88. Wang, W., Xu, G., Cui, X. T., Sheng, G., and Luo, X. (2014). Enhanced catalytic and dopamine sensing properties of electrochemically reduced conducting polymer nanocomposites doped with pure graphene oxide, *Biosen. Bioelectron.*, **58**, pp. 153–156.

89. Weaver, C. L., Li, H., Luo, X., and Cui, X. T. (2014). A graphene oxide/conducting polymer nanocomposite for electrochemical dopamine detection: origin of improved sensitivity and specificity, *J. Mater. Chem. B*, **2**, pp. 5209–5219.

90. Lu, L., Zhang, O., Xu, J., Wen, Y., Duan, X., Yu, H., Wu, L., and Nie, T. (2013). A facile one-step redox route for the synthesis of graphene/poly(3,4-ethylenedioxythiophene) nanocomposite and their applications in biosensing, *Sens. Actuators, B*, **181**, pp. 567–574.

91. Wu, L.-P., Zhang, L., Lu, L.-M., Duan, X.-M., Xu, J.-K., and Nie, T. (2014). Graphene oxide doped poly(hydroxymethylated 3,4-ethylenedioxythiophene): enhanced sensitivity for electrochemical determination of rutin and ascorbic acid, *Chin. J. Polym. Sci.*, **8**, pp. 1019–1031.

92. Nie, T., Zhang, O., Lu, J., Xu, J., Wen, Y., and Qiu, X. (2013). Facile synthesis of poly(3,4-ethylenedioxythiophene)/graphene nanocomposite and its application for determination of nitrite, *Int. J. Electrochem. Sci.*, **8**, pp. 8708–8718.

93. Wang, Q., and Yun, Y. (2012). A nanomaterial composed of cobalt nanoparticles, poly(3,4-ethylenedioxythiophene) and graphene with high electrocatalytic activity for nitrite oxidation, *Microchim. Acta*, **177**, pp. 411–418.

94. Si, W., Lei, W., Zhang, Y., Xia, M., Wang, F., and Hao, Q. (2012). Electrodeposition of graphene oxide doped poly(3,4-ethylenedioxythiophene) film and its electrochemical sensing of catechol and hydroquinone, *Electrochim. Acta*, **85**, pp. 295–301.

95. Si, W., Lei, W., Han, Z., Zhang, Y., Hao, Q., and Xia, M. (2014). Electrochemical sensing of acetaminophen based on poly(3,4-ethylenedioxythiophene)/graphene oxide composites, *Sens. Actuators, B*, **193**, pp. 823–829.

96. Wang, X., Wen, Y., Lu, L., Xu, J., Zhang, L., Yao, Y., and He, H. (2014). A novel L-cysteine electrochemical sensor using sulfonated graphene-poly(3,4-ethylenedioxythiophene) composite film decorated with gold nanoparticles, *Electroanalysis*, **26**, pp. 648–655.

97. Feng, X., Huang, H., Ye, Q., Zhu, J.-J., and Hou, W. (2007). Ag/polypyrrole core-shell nanostructures: Interface polymerization, characterization and modification by gold nanoparticles, *J. Phys. Chem. C*, **2007**, pp. 8463–8468.

98. Liu, H., Liu, Y., and Zhu, D. (2011). Chemical doping of graphene, *J. Mater. Chem.*, **21**, 3335–3345.

99. Wang, Y., Shao, Y. Y., Matson, D. W., Li, J. H., and Lin, Y. H. (2010). Nitrogen-doped graphene and its application in electrochemical biosensing, *ACS Nano*, **4**, pp. 1790–1798.

100. Prathish, K. P., Barsan, M. M., Geng, D., Sun, X., and Brett, C. M. A. (2013). Chemically modified graphene and nitrogen-doped graphene: Electrochemical characterization and sensing applications, *Electrochim. Acta*, **114**, pp. 533–542.

101. Wang, M., Yuan, W., Yu, X., and Shi, G. (2014). Picomolar detection of mercury (II) using a three-dimensional porous graphene/polypyrrole composite electrode, *Anal. Bioanal. Chem.*, **406**, pp. 6953–6956.

102. Zhao, Z.-Q., Chen, X., Yang, Q., Liu, J.-H., and Huang, X.-J. (2012). Selective adsorption toward toxic metal ions results in selective response: Electrochemical studies on a polypyrrole/reduced graphene oxide nanocomposite, *Chem. Commun.*, **48**, pp. 2180–2182.

103. Si, P., Chen, H., Kannan, P., and Kim, D.-H. (2011). Selective and sensitive determination of dopamine by composites of polypyrrole and graphene modified electrodes, *Analyst*, **136**, pp. 5134–5138.

104. Gao, Y. S., Xua, J. K., Lu, L. M., Wu, L. P., Zhang, K. X., Nie, T., Zhu, X. F., and Wu, Y. (2014). Overoxidized polypyrrole/graphene nanocomposite with good electrochemical performance as novel electrode material for the detection of adenine and guanine, *Biosens. Bioelectron.*, **62**, pp. 261–267.

105. Scott, C. L., Zhao, G., and Pumera, M. (2010). Stacked graphene nanofibers doped polypyrrole nanocomposites for electrochemical sensing, *Electrochem. Commun.*, **12**, pp. 1788–1791.

106. Yang, Y., Fang, G., Wang, X., Pan, M., Qian, H., Liu, H., and Wang, S. (2014). Sensitive and selective electrochemical determination of quinoxaline-2-carboxylic acid based on bilayer of novel poly(pyrrole) functional composite using one-step electropolymerization and molecularly imprinted poly(o-phenylenediamine), *Anal. Chim. Acta*, **806**, pp. 136–143.

107. Lian, W., Liu, S., Yu, J., Li, J., Cui, M., Xu, W., and Huang, J. (2013). Electrochemical sensor using neomycin-imprinted film as recognition element based on chitosan-silver nanoparticles/graphene-multiwalled carbon nanotubes composites modified electrode, *Biosens. Bioelectron.*, **44**, pp. 70–76.

108. Lin, E.-R., Chiu, C.-J., and Tsai, Y.-C. (2014). One-step electrochemical synthesis of polypyrrole-graphene-glucose oxidase nanobiocomposite for glucose sensing, *J. Electrochem. Soc.*, **161**, pp. B243–B247.

109. Devi, R., Relhan, S., and Pundir, C. S. (2013). Construction of a chitosan/polyaniline/graphene oxide nanoparticles/polypyrrole/Au electrode for amperometric determination of urinary/plasma oxalate, *Sens. Actuators, B*, **186**, pp. 17–26.

110. Linting, Z., Ruiyi, L., Zaijun, L., Qianfang, X., Yinjun, F., and Junkang, L. (2012). An immunosensor for ultrasensitive detection of aflatoxin B1 with an enhanced electrochemical performance based on graphene/conducting polymer/gold nanoparticles/the ionic liquid composite film on modified gold electrode with electrodeposition, *Sens. Actuators, B*, **174**, pp. 359–365.

111. Ruiyi, L., Qianfang, X., Zaijun, L., Xiulan, S., and Junkang, L. (2013). Electrochemical immunosensor for ultrasensitive detection of microcystin-LR based on graphene-gold nanocomposite/functional conducting polymer/gold nanoparticle/ionic liquid composite film with electrodeposition, *Biosens. Bioelectron.*, **44**, pp. 235–240.

112. Puri, N., Niazi, A., Srivastava, A. K., and Rajesh. (2014). Biointerfacial impedance characterization of reduced graphene oxide supported carboxyl pendant conducting copolymer based electrode, *Electrochim. Acta*, **123**, pp. 211–218.

113. Wang, Z., Wang, H., Zhang, Z., Yang, X., and Liu, G. (2014). Sensitive electrochemical determination of trace cadmium on a stannum film/poly(*p*-aminobenzene sulfonic acid)/electrochemically reduced graphene composite modified electrode, *Electrochim. Acta*, **120**, pp. 140–146.

114. Feng, Z. L., Yao, Y. Y., Xu, J. K., Zhang, L., Wang, Z. F., and Wen, Y. P. (2014). One-step co-electrodeposition of graphene oxide doped

poly(hydroxymethylated-3,4-ethylenedioxythiophene) film and its electrochemical studies of indole-3-acetic acid, *Chinese Chem. Lett.,* **25**, pp. 511–516.

115. Li, C., Qiu, X., and Ling, Y. (2013). Electrocatalytic oxidation and the simultaneous determination of guanine and adenine on (2,6-pyridinedicarboxylic acid)/graphene composite film modified electrode, *J. Electroanal. Chem.,* **704**, pp. 44–49.

116. Lei, W., Wu, Q., Si, W., Gu, Z., Zhang, Y., Deng, J., and Hao, Q. (2013). Electrochemical determination of imidacloprid using poly(carbazole)/chemically reduced graphene oxide modified glassy carbon electrode, *Sens. Actuators B,* **183**, pp. 102–109.

117. Gupta, P., Yadav, S. K., Agrawal, B., and Goyal, R. N. (2014). A novel graphene and conductive polymer modified pyrolytic graphite sensor for determination of propranolol in biological fluids, *Sens. Actuators B,* **204**, pp. 791–798.

118. Noh, H.-B., Revin, S. B., and Shim, Y.-B. (2014). Voltammetric analysis of anti-arthritis drug, ascorbic acid, tyrosine, and uric acid using a graphene decorated-functionalized conductive polymer electrode, *Electrochim. Acta,* **139**, pp. 315–322.

119. Zhu, G., Wu, L., Zhang, X., Liu, W., Zhang, X., and Chen, J. (2013). A new dual-signalling electrochemical sensing strategy based on competitive host-guest interaction of a β-cyclodextrin/poly(N-acetylaniline)/graphene-modified electrode: Sensitive electrochemical determination of organic pollutants, *Chem. Eur. J.,* **19**, pp. 6368–6373.

120. Li, Z., Huang, Y., Chen, L., Qin, X., Huang, Z., Zhou, Y., Meng, Y., Li, J., Huang, S., Liu, Y., Wang, W., Xie, Q., and Yao, S. (2013). Amperometric biosensor for NADH and ethanol based on electroreduced graphene oxide-polythionine nanocomposite film, *Sens. Actuators B,* **181**, pp. 280–287.

121. Tran, H. V, Piro, B., Reisberg, S., Duc, H. T., and Pham, M. C. (2013). Antibodies directed to RNA/DNA hybrids: An electrochemical immunosensor for microRNAs detection using graphene-composite electrodes, *Anal. Chem.,* **85**, pp. 8469–8474.

Index

AA *see* ascorbic acid
AAO *see* anodized aluminum oxide
AAO templates 25–26, 32–33
  conical 25
AB diblock copolymers 67
acetonitrile (ACN) 117
ACN *see* acetonitrile
activated carbons 264, 285
active materials 15, 20, 60, 179–180, 194, 227, 243, 269, 344, 374–375, 397
adenine 453–454, 458
AFM *see* atomic force microscope/atomic force microscopy
AgCl 148–149, 154, 158
alignment 62, 68–78, 86, 308
alkane thiols 39
amino acids 255, 448
amorphous materials 397
amorphous ntTiO$_2$ 201–206, 208–209, 211
anaphoresis 382–383
anatase 32, 91–92, 100, 180, 188, 190–192, 195–200, 205–206, 213, 239, 242, 305–307, 367–368, 387–388, 397
anatase TiO$_2$ 32, 92, 195, 199–200, 206, 239, 242, 306–307, 387–388
anatase TiO$_2$ nanotubes 199–200, 206

aniline 12, 23, 34, 36, 122, 418, 428, 430–432, 451–452, 455
  electropolymerization of 430–431
aniline monomer 264, 266
anode materials 226–227, 229–230, 239, 242, 244, 346, 369, 381, 386, 390–391, 397
  active 381
  high-performance 386, 389, 391
anodization 5, 9, 11, 16, 18, 24–27, 30–31, 181–184, 187–190, 192–193, 195, 204–205, 209, 213, 366–368, 383, 387
anodization experiments 182, 188
anodized aluminum oxide (AAO) 24, 45
AO *see* ascorbate oxidase
arrays 8–10, 16, 18, 21–28, 30–32, 35–36, 38, 42–43, 60, 62, 70, 78, 80–81, 94–96, 100–102, 425–426
  nanoparticle 100–102
  synthesized titanium dioxide 62
as-formed ntTiO$_2$ 190, 204–205, 213
ascorbate oxidase (AO) 426, 443

ascorbic acid (AA) 143, 260, 300, 424, 450
atomic force microscope/atomic force microscopy (AFM) 9, 330, 334, 338, 344, 347, 350
Au 298
Au NPs 298–299, 375, 429, 447–448, 450, 456

batteries 80, 114, 172, 174, 177, 179, 194, 199, 223, 225–226, 232–233, 236, 263, 269, 293, 328, 335–336, 343, 346, 366–367, 369, 380, 392, 397–399, 417
  commercial 369–370
  lithium batteries 174, 176, 181, 227, 230–231
  lithium ion 171–172, 174–176, 178–180, 182, 184, 186, 188, 190, 192–196, 198, 200, 202, 204, 206, 210–213, 223–228, 230, 232–234, 236–240, 242–245, 270, 328, 335–338, 339–347, 349, 366, 368–369, 370, 380–381, 383, 386–387, 389–392
  rechargeable 335–337
  rocking-chair 177, 376
  secondary 335–336
battery applications 199, 201, 203, 205, 207, 209, 211, 365, 380–381, 383, 385, 387, 389, 391, 393, 397
battery architectures 365, 376–377, 379

BCP see block copolymers
BCP templating technology 103
binder materials 271
bioelectroanalytical chemistry 149
biomarkers 453–455, 457, 459
biomass 255
biomaterials 293
biosensors 293, 428, 433, 436, 448, 455–456, 459
  enzyme-based 426
blend 61, 63, 69, 87
block copolymer architectures 63
block copolymer films 59, 82
block copolymer thin films 67, 69, 71, 73
block copolymers 24, 42, 59–60, 72, 75, 87, 97
bovine serum albumin (BSA) 303, 450, 459
BSA see bovine serum albumin
bulk heterojunction solar cell 60, 96

calcium dobesilate 424, 433
carbon 258, 332, 371–372, 387, 393–394
carbon-based materials 270, 294, 388
carbon materials 227–229, 252, 264, 270, 371, 418, 439, 459–460
  hard 228–229
  nongraphitized 227–228
  traditional porous 264
  various 264
carbon nanoparticles 126–127
carbon nanostructures 154, 156, 260, 262

carbon nanotubes (CNTs) 8, 24,
    35, 43, 113, 123, 125, 129,
    156, 199, 243, 258, 264,
    294, 309, 372, 393–394,
    417–423, 425–429,
    431–433, 435–437, 459
carbon spheres, hollow 230, 238
carbonaceous material 270
cathode materials 230–232,
    242–243, 391, 393–395
CCG *see* chemically converted
    graphene
cells, nanowire 100
characterization of titania
    nanotubes layers 188–189,
    191, 193, 195, 197
charge transport 97–99, 252,
    268, 276, 426
chemical defects 202, 208
chemical industrial processes
    146
chemical vapor deposition (CVD)
    258–259, 292, 369, 378, 380
  plasma-enhanced 369
chemically converted graphene
    (CCG) 291–292, 294, 296,
    298, 300, 302, 304–306,
    308, 310, 312, 314, 316
chitosan 424, 429, 448, 450,
    454–456
choline 255
citric acid 371, 448, 456
CNTs *see* carbon nanotubes
composites
  graphene-based 382
  graphene-nanoparticle 293
  graphene-porphyrin/phthalo-
    cyanine 310
  graphene–SiO_2 309
  polymer-based 294
conducting polymer nanotube
    33

conducting polymer nanowires
    22–23, 32, 41
conducting polymers 9, 12,
    20–22, 32–35, 42, 45, 113,
    117, 119, 121–122,
    128–129, 251–253,
    256–257, 263–264, 266,
    268–271, 274–275, 395,
    418–421, 423, 425, 427,
    429, 431, 433, 439
  synthesis of 22, 35, 113, 117,
    123, 128–129, 252
conducting PPy polymer
    nanowires 85
conducting substrate 62, 75,
    81–82, 90
conjugated polymers 252, 256
constant thickness (CT) 86
constant thickness gyroid
    morphology 88
copolymer block 69, 77
copolymer film 67–68, 70, 72,
    75, 77, 79–83, 85, 87, 89,
    91, 93, 95
copolymer gyroid morphology
    86
copolymers 24, 42, 45, 59–62,
    65–69, 72, 75, 77, 80, 87, 97
copper 8, 10, 19, 142
covalent immobilization 429,
    431
crystalline anatase 206
crystalline materials 208
crystalline $ntTiO_2$ 204–206,
    208–209, 213
crystalline structure 199–201,
    208, 231–232
crystallized $ntTiO_2$ 202–203
CT *see* constant thickness
Cu_2O nanoparticles 350
CV *see* cyclic voltammetry
CVD *see* chemical vapor
    deposition

cyclic voltammetry (CV) 266, 309, 420, 425, 432–433, 437, 439, 441, 443–445, 447, 451–452, 454, 457
cylindrical pore size range 79

degree of polymerization 63, 78
devices, nanowire 100, 102
diazonium salts 312–313
diblock copolymers 62, 65, 67, 75, 87
dielectric constant 71–72
dielectric contrast 72–73, 77
DNA biosensors 436–437
DNA probes 436–437
domain spacing 65, 67–68, 70–71, 80
dopamine 424, 450
dopants 3, 39, 125, 267, 427
DSSCs *see* dye-sensitized solar cells
dye-sensitized solar cells (DSSCs) 30
dynamics of photogenerated electrons 101

EBSD *see* electron back scatter diffraction
EC-ALE *see* electrochemical atomic layer epitaxy
EC-DPN *see* electrochemical "dip-pen" nanolithography
ECDs *see* electrochromic devices
ECP *see* electrically conducting polymer
ECP-carbon materials 459–460
ECP-CNT composites 419–420
ECP-graphene composites 419

ECs *see* electrochemical capacitors
EDLCs *see* electrical double layer capacitors
EIS *see* electrochemical impedance spectroscopic
electric field alignment 71–74, 77
electric vehicles 176, 335, 393
electrical double layer capacitors (EDLCs) 270
electrically conducting polymers (ECPs) 418–420, 433–434, 439–440, 453, 456–459
electroactive material 371
electroactivity 275–276, 385, 427
electroanalytical chemistry 139–140, 163, 291
  traditional 163
electrocatalysis 142, 152–153, 155–157, 315
electrocatalysts 39, 156–157, 159
electrocatalytic activity 121, 141, 143, 155
electrochemical anodization 30–31, 195, 213
electrochemical applications 25, 115–116, 129, 252, 336
electrochemical atomic layer epitaxy (EC-ALE) 2–3, 45
electrochemical biosensors 443
electrochemical capacitors (ECs) 269, 271, 274, 420
electrochemical cell 9, 13, 15, 22, 62, 93
electrochemical deposition 7, 15, 27, 36–39, 42, 61, 80, 83–84, 90, 103, 309, 373, 384, 420, 434–435, 437, 454

electrochemical deposition methods 7, 39
electrochemical detection 446, 452, 459
electrochemical "dip-pen" nanolithography (EC-DPN) 13
electrochemical etching 11, 15–16, 76, 80–81
electrochemical exfoliation 252, 262, 274–275
electrochemical functionalization 36, 123, 140–141
electrochemical gyroid replication 90
electrochemical impedance spectroscopic (EIS) 420, 444
electrochemical machining 19–20
electrochemical micromachining 17–19
electrochemical nanolithography 9, 11–13
electrochemical oxidation 439
electrochemical polymerization 252–253, 265, 418–420, 426, 441
electrochemical replication 59–62, 64, 66–68, 70, 72, 74–76, 78, 80, 82, 84, 86, 88, 90, 94–96, 98
  film-based 62
electrochemical replication of gyroid templates 95
electrochemical replication of self-assembled block copolymer nanostructures 59–60, 62, 64, 66, 68, 70, 72, 74, 76, 78, 80, 82, 84, 86, 88
electrochemical replication of Ti 90
electrochemical technique, bipolar 43, 45
electrochemical window 7, 115–116, 163, 256–257, 269, 274
electrochemistry 2, 14–15, 17, 38–40, 43–45, 114–117, 128, 140, 150–151, 155, 171, 255, 260, 380
  bipolar 43–44
  electrolyte-free 150–151
electrochromic devices (ECDs) 269, 275–276
electrode materials 263, 335–336, 338, 344, 366, 370–371, 376, 388, 392, 419
  active 273, 337, 344, 380
  composite 269, 344
  pseudocapacitive 270
electrodeposition 253, 266, 271, 274, 301, 369–370, 377–380, 394, 396, 419, 436, 451, 456–457
  metal 7, 40–41
electrodes
  Au 428, 454–456
  cycled 211, 213
  freestanding thin-film composite 271
  gate 23
  graphite 8, 126–127, 178–180, 261–262, 390, 431
  junction 23
  multilayer films 148
  nanoarchitectured 389
  nanostructured 181
  nanotubular titanium dioxide 396
  ntTiO$_2$ 201, 207
  PEDOT 443
  porous 270, 379
  screen-printed carbon 454

supercapacitor 272, 274
electron back scatter diffraction (EBSD) 377
electroplating 377–378
electropolymerization 12, 20, 23, 34, 36, 42, 84, 117, 119–121, 140, 256–257, 267–269, 271–272, 275, 395–396, 419, 426, 428, 430–431, 436, 443, 445, 451–453, 455, 457
  localized 12
electrosynthesis 118–119, 122–123, 140–141
emeraldine base 440
emeraldine salt 440
enzymes 426–429, 434, 447, 451, 455, 459
epinephrine 437–438
epoxy resins 294–295, 431
EQE *see* external quantum efficiency
ethyl 2-bromoacetate 329
excitons 96–97
external quantum efficiency (EQE) 100

FAD *see* flavin adenine dinucleotide
Faradaic reactions 39, 194–195
FIB *see* focused ion beam
films 5–7, 11–12, 24, 27–28, 34–40, 67–71, 73–74, 77–80, 95–97, 117–119, 121, 127–128, 148, 181–182, 193, 257–258, 260, 265–267, 269, 275–277, 299, 308–309, 374–375, 377, 379, 383–384, 396–397, 439, 441, 444

  as-synthesized aligned nanotube 35
  composite 34, 97
  electrodeposited 40
  nanoconical 25
  nanoporous AAO template 27
  oxide 5, 11, 24, 30, 39, 90, 93, 95, 127, 180–181, 204, 374, 379, 383–384, 390, 451
  templated silica 88
flavin adenine dinucleotide (FAD) 448
fluorescent nanomaterials 127–128, 262
focused ion beam (FIB) 19, 43
free energy of mixing 63–64

GCE *see* glassy carbon electrode
GISAXS *see* grazing incidence small angle X-ray scattering
GISAXS fitting 91
glassy carbon electrode (GCE) 7, 143, 149, 429, 433, 435–437, 439, 441, 444–445, 447–448, 451, 453–454, 456–459
gluconic acid 434
glucose 432–434, 444, 448, 455
glucose biosensors, amperometric 434
gold 36, 38, 40–41, 68–69, 142
graphene 45, 126–127, 154–155
  Congo red-functionalized 310
  hydrophobic pristine 384
  indolizine-modified 329
  inorganic structure–modified 354

ionic liquid-functionalized 311
nitrogen-doped 391
polymer-functionalized 293, 296
pre-synthesized 308
pristine 259, 292
pyridine-functionalized 314
single-sheet 384
two-dimensional 310
undoped 448
graphene-based composite materials 315
graphene-based heterostructures 383
graphene-based materials 259, 296, 315, 328, 335, 337, 339, 341, 343, 345, 347, 354
graphene-based materials research 315
graphene-based nanocomposite 292–293, 295, 297, 299, 301, 303, 305, 307, 309, 311, 313, 328, 354
graphene-based nanosheets 297
graphene-based nanostructures 327
graphene-CdSe QDs 308
  as-prepared 308
graphene electrosynthesis 262
graphene fillers 252, 295
graphene films 299, 301, 383–384, 390
  seed-modified 299
graphene flakes 256, 261–262, 301
graphene heterostructures 385
graphene nanosheets 154, 156, 274–275, 328, 385, 388–389, 393
graphene networks, three-dimensional 381, 386

graphene oxide 259, 271, 292, 294, 328, 346, 383, 390, 418, 450
graphene paper 265–266, 272
graphene pyridine-functionalized 313
graphene quantum dots hybrids 307
graphene sheets 126–127, 129, 227, 257–258, 262, 264, 293–294, 296, 298–302, 305, 308, 312, 314–315, 328, 330, 338–339, 383, 386
  as-prepared 303
  conductive 344
  diazonium-functionalized 312
  functionalized 384
  large area 258
graphite 4, 126–127, 178–180, 212, 226–228
graphite oxide 328, 384, 386, 448, 450–451, 455
  carboxylated reduced 450, 452
  exfoliation of 259
  reduction of 451
graphite oxide films 451
grazing incidence small angle X-ray scattering (GISAXS) 90
guanine 449, 453–454, 458

HAADF *see* high-angle annular dark-field
hard carbon spheres (HCSs) 229
HCSs *see* hard carbon spheres
HEVs *see* hybrid electric vehicles
high-angle annular dark-field (HAADF) 300

high-resolution transmission electron microscopy (HRTEM) 228
high temperature annealing 91, 94, 96
highly oriented pyrolytic graphite (HOPG) 7–8, 12, 42
HOPG *see* highly oriented pyrolytic graphite
host polymeric materials 418
HRTEM *see* high-resolution transmission electron microscopy
hybrid electric vehicles (HEVs) 172, 176, 225, 232–233, 244, 270, 393
hybrid supercapacitor 195
hybrids
  graphene-based 350
  graphene-bimetallic 298
  graphene-GdS 307
  graphene–quantum dots 297
hydrogen storage materials 418

ILFG *see* ionic liquid functionalized graphene
ILs *see* ionic liquids
in-plane nanowires 80–81
indole-3-acetic acid 457
indolizine 329, 331–332, 334
interfaces 40, 60, 64–65, 67, 69, 72–73, 88, 154
  water/oil 153–154
interfacial tension 65
ion-selective electrodes (ISEs) 418
ionic liquid 251–252, 254–258, 260, 262–264, 266–270, 272, 274–278, 299, 305, 311, 429, 456

ionic liquid functionalized graphene (ILFG) 267–268, 275–276
ionic liquids *see* ILs
ionic liquids (ILs) 114–115, 119, 121, 123, 125, 127, 139, 143, 146, 153, 163, 252
iron phthalocyanine 424–425
iron-porphyrin 313–314
ISEs *see* ion-selective electrodes

L-cysteine 447–448
lamellar microdomains 64, 69, 73
lamellar periods 68–69
LEDs *see* light-emitting diodes
$LiClO_4$ 426, 443–444
$LiFePO_4$ 391, 393, 395, 398
LIGA technique 15, 17
light-emitting diodes (LEDs) 28
lithiation 206, 229–231, 239
lithium cells 199–200, 204, 206, 209, 373, 375, 379, 389, 396
lithium insertion 177, 181, 195, 198–199, 203–204, 226–227, 387
lithium intercalation 180, 197–198, 200, 202, 234, 387
lithium ion battery cathode materials 391
lithium ions 177, 202, 207–208, 224–226, 230–231, 233–234, 367, 381, 393, 395, 398
lithium phosphate 396–397

materials
  annealed 201, 208

bulk 17, 67, 87, 96, 196, 259, 270, 336, 366
carbon-based 123, 177, 225, 229, 270, 294, 388
electroactive 113–114, 116, 118, 120, 122, 124, 126, 128, 200, 251–252, 254, 256, 258, 260, 262, 264, 266, 268, 270, 272, 274, 276, 278, 371
electrodeposited 42, 453
micro-sized 236
mesoporous TiO_2 spheres 239–240, 242
metal electrodeposition 40–41
metal nanoparticles 5, 7, 34, 125, 129, 270, 298, 418
metal nanowires 8, 15, 27, 85
metal organic framework (MOF) 313–314, 386, 389
metal oxide nanoparticles 293, 305, 381, 385–386
 graphene-encapsulated 383
metal-oxide-semiconductor field effect transistors (MOSFET) 30
metal oxide–graphene composites 386
metal oxides 305, 373, 383, 386
 graphene-based 385
methanol 426, 444, 454
methanol electrooxidation 156–157
methylene blue 350
micro-supercapacitor 272
microphase morphology 64, 67, 97
microphase separation 61, 65–66, 77–78
 thermodynamics of 78
modified ITO substrate 123–124

MOF see metal organic framework
graphene–metalloporphyrin 313
MOSFET see metal-oxide-semiconductor field effect transistors
multiwalled carbon nanotubes (MWCNTs/ MWNTs) 125, 389, 420, 426, 429–430, 432–433, 435–436, 439, 449, 454
MWCNT-modified screen-printed carbon electrodes 425
MWCNTs/MWNTs see multiwalled carbon nanotubes

N-doped graphene 448
N-doped graphene aerogel 306
N-succinimidyl acrylate 141
nanoarchitectured materials 381, 385
nanoarchitectured titania materials 387
nanoarchitectures, graphene/metal oxide 385
nanocomposites
 graphene-NPs 308
 graphene-PtAu alloy 303
 graphene-Si NPs 308
 graphene–metal NPs 298, 301
 graphene–TiO_2 306
nanofabrication 1–2, 9, 14–15, 20, 24, 39, 41–43, 45, 59, 113, 139, 171, 223, 251, 291, 327, 365, 387, 417–418, 460
nanoflowers 337
nanoparticles

amorphous TiO_2 306–307
graphene-based 297
graphene/platinum 304
graphene-wrapped TiO_2 306
graphene–metal 297, 301
graphene–metal oxide 297
linker-free graphene/CdSe 308
nanorod arrays 26, 28, 78
nanosheets, graphene-based titania 387
nanostructured electrode materials 172, 223–224, 226, 228, 230, 232–244
nanostructured titania 180–181
nanotube arrays 26–27, 30–31, 35, 193, 195
nanotube morphology 27, 187–189, 204, 211–213, 367
nanotubes, titanium oxide 399
nanowire arrays 8, 32, 62, 82, 84, 100
nanowire cell 100
nanowire devices 100, 102
nanowires 4, 8, 15, 21–24, 27, 29, 32–35, 41–42, 45, 61, 80–86, 98, 100, 195, 198, 264–265, 341, 369, 372–373, 387, 397
  free-standing platinum 83
neutralized ITO glass substrates 85
noncovalent compositing 309
nontoxic materials 270
nt-TiO_2 181–183, 189–190, 192, 200–202, 204, 206–208, 210–211, 214, 366–369, 383, 397

o-phenylenediamine 433, 454

ODT see order–disorder transition
open-ended nanotubes 26
order–disorder transition (ODT) 64, 66
ordered nanostructures 24, 62
organic electroactive materials 113–114, 116, 118, 120, 122, 124, 126, 128
organic solvents 77, 117–119, 225
  conventional 115, 117–119
oriented nanowires 21–22
ORR see oxygen reduction reactions
oxygen reduction reactions (ORR) 313

PANI see polyaniline
PANI films 22, 121–122
PANI nanotubules 34
PANI nanowires 22–23, 33
patterned Au substrate 85–86
PBS see phosphate buffer solution
PEDOT 257, 267–268, 272, 274–275, 419–420, 425, 439–440, 444–445, 447–448, 457, 459
  conducting polymer 278
  neat 442, 444
  plain 422
PEDOT films 422, 425
  charged 425
  electropolymerized 448
PEDOT- ILFG nanocomposite films 275
pencil graphite electrode (PGE) 431
PFIL see polyelectrolyte-supported ILs

PFIL-AuNPs 157, 159
PFIL-modified electrode
    assembly 150–151
PGE *see* pencil graphite
    electrode
phosphate buffer solution
    (PBS) 435, 437, 439,
    441, 444–445, 447, 451
physical vapor deposition
    (PVD) 8, 204
phytohormones 457
PLD *see* pulsed laser
    deposition
PLL *see* poly-L-lysine
PMMA homopolymer 70, 84–85
poly-L-lysine (PLL) 296
polyaniline 274, 294–295, 346,
    419, 427, 429, 451
polyaniline (PANI) 12, 21,
    33–36, 38–39, 84, 117,
    121–124, 128, 141
polyelectrolyte 119, 143,
    146–147, 150, 157,
    160, 296
polyethyleneimine 384
polymer composites 123, 125,
    127, 293–296, 309, 417
polymer films 34, 128, 269
polymer matrix 34, 62, 82–83,
    419, 443, 451
polypyrrole 12, 21, 23, 33, 42,
    150
polypyrrole films 118
polystyrene 36, 79–80
porous gyroid 87–88, 90
porous gyroid morphology 88
PPP *see* poly(para-phenylene)
proteins 303, 447, 455–456
protons 336
Prussian blue 148
pseudocapacitors 270
pulsed laser deposition
    (PLD) 397

PVD *see* physical vapor
    deposition
pyridine 308, 313–314, 329
pyridinium salt 329–332
pyrrole 372, 418, 434,
    436–439, 452, 454

QDs *see* quantum dots
QHE *see* quantum Hall effect
quantum dots (QDs) 194, 307
quantum Hall effect (QHE) 292,
    328

random copolymer brushes 69
recombination events 101–102
reduced graphene oxide (RGO)
    259–260, 264, 266–268,
    271–273, 275–277,
    329–334, 344–345, 347,
    349–350, 382–383,
    440–441, 443–444,
    447–448, 450–453,
    455–458
RGO *see* reduced graphene
    oxide
rhodamine 350, 458
room-temperature ionic liquids
    (RTILs) 254
RTILs *see* room-temperature
    ionic liquids
rutile 305, 367, 387
rutin 443

SAMs *see* self-assembled
    monolayers
scanning electron microscopy
    (SEM) 74

scanning probe microscopy (SPM) 9
scanning tunneling microscopy (STM) 9
screen-printed carbon electrodes (SPCE) 425, 454
SDBS *see* sodium dodecyl benzene sulfonate
SDS *see* sodium dodecyl sulfate
secondary ion mass spectroscopy (SIMS) 68
SEI *see* solid electrolyte interface
selective degradation 62
self-assembled monolayers (SAMs) 11, 24, 39, 45
self-organized TiO_2 nanotubes 186, 189, 212–213, 373, 383, 393–394, 396
SEM *see* scanning electron microscopy
semiconducting nanocompounds 4–5
semiconductors 4, 7, 9, 11, 27, 30, 38, 44, 102
sensors
  chemical 418, 433, 439
  chemical field-effect transistor 426
  electrochemical 293, 418–420
  nitrite 444
  non-enzymatic H_2O_2 429
SERS *see* surface-enhanced Raman scattering
silicon 11, 19, 22, 40, 82, 142, 183, 191, 199, 204
silicon substrates 11, 68, 189–190, 213
SIMS *see* secondary ion mass spectroscopy
single crystal substrates 2

single-walled carbon nanotubes (SWNTs/ SWCNTs) 36, 123–124, 140–141, 153–155, 329, 426, 428, 434, 437
Sn nanoparticles 229–230, 238, 349
sodium dodecyl benzene sulfonate (SDBS) 309
sodium dodecyl sulfate (SDS) 305, 437
solar cells 28, 30–31, 62, 97, 101, 114, 150, 171, 193–194, 263, 315, 418, 420
  nanostructured 62, 194
  photoelectrochemical 193
solid electrolyte interface (SEI) 179, 341, 348
SPCE *see* screen-printed carbon electrodes
spherical voids 36–37
SPM *see* scanning probe microscopy
square wave anodic stripping voltammetry (SWASV) 452
STM *see* scanning tunneling microscopy
structural compression 91
supercapacitor materials 270
supercapacitors 22, 193–195, 243, 252, 255, 263, 269–270, 274, 293, 315, 352, 382, 391–392, 448
surface-enhanced Raman scattering (SERS) 297
surface-grafted brush 73
SWASV *see* square wave anodic stripping voltammetry
SWNTs/ SWCNTs *see* single-walled carbon nanotubes

TEM *see* transmission electron micrograph/transmission electron microscopy
templates 21, 24–27, 32–33, 35–39, 41–42, 45, 59, 61, 75–81, 83, 85–87, 89–91, 93, 95, 97, 296, 308, 370, 384
 bicontinuous gyroid copolymer 86
 carbon nanotube 35
 copolymer 76, 83
 latex 36–38, 45
 molecular 41, 77, 79, 97
 PMMA 79, 85
 polymer 38, 83, 90–91, 94
 porous PS 86
TiO_2 *see* titanium dioxide
TiO_2 arrays 31, 101–102
TiO_2 nanotube arrays 30–31, 193, 195
TiO_2 nanotubes 30–31, 180, 182–187, 189, 195, 198–200, 204, 206, 212–213, 366, 373, 383, 393–394, 396
 self-assembled 182–183, 185, 187
TiO_2 nanowires 32, 369
titania 190, 193–194, 205
 nanostructured 180–181
titania nanotube layer 189, 203, 205, 212–213
titania nanotubes 188–189, 191, 193, 195, 197, 199, 203, 206–208, 211, 383, 388, 398
 self-organized 207

titanium 30, 32, 182, 184–185, 191, 196–198, 202, 204, 213, 227
titanium dioxide (TiO_2) 27, 30–31, 45, 95, 98–99, 178, 180, 182, 185, 190, 192–200, 204–205, 209–211, 242
titanium oxides 195
transition metal oxides 230, 235, 370–371, 379, 384–385, 388, 392
transition metals 98, 225, 230–232
transmission electron micrograph/transmission electron microscopy (TEM) 3, 330, 334, 338–339, 344, 347, 350, 368

water-ethylene glycol system 298
wet chemical method 299

X-ray diffraction (XRD) 188
X-ray photoelectron spectroscopy 330, 332
XRD *see* X-ray diffraction

zinc oxide (ZnO) 5, 27, 30, 45
ZnO *see* zinc oxide
ZnO films 28
ZnO nanowire 29